Mechanical Engineering and Technology

Mechanical Engineering and Technology

Editor: Bilroy Lockhart

NYRESEARCH
P R E S S

New York

Published by NY Research Press
118-35 Queens Blvd., Suite 400,
Forest Hills, NY 11375, USA
www.nyresearchpress.com

Mechanical Engineering and Technology
Edited by Bilroy Lockhart

International Standard Book Number: 978-1-63238-534-5 (Hardback)

Cataloging-in-Publication Data

Mechanical engineering and technology / edited by Bilroy Lockhart.
    p. cm.
Includes bibliographical references and index.
ISBN 978-1-63238-534-5
1. Mechanical engineering. 2. Technology. 3. Machinery. 4. Welding--Equipment and supplies.
I. Lockhart, Bilroy.
TJ145 .M43 2017
621--dc23

Printed in the United States of America.

Contents

Permissions

List of Contributors

Index

Preface

Over the recent decade, advancements and applications have progressed exponentially. This has led to the increased interest in this field and projects are being conducted to enhance knowledge. The main objective of this book is to present some of the critical challenges and provide insights into possible solutions. This book will answer the varied questions that arise in the field and also provide an increased scope for furthering studies.

Mechanical engineering deals primarily with the design and manufacture of machines and mechanical systems. This book on mechanical engineering presents a multi-disciplinary approach to the field. Thermodynamics, structural analysis, mechanics, etc. are some of the key areas of mechanical engineering that have been thoroughly discussed in this text. The prospects of mechanical engineering are vast and technology is upgraded constantly. The topics included in this book reflect the technological progress that has been made and the corresponding theoretical breakthroughs that have also occurred. The book elucidates innovative models and concepts around prospective developments with respect to this field. It will be of great help to students, experts, researchers and engineers in the fields of manufacturing, electronics and other related areas.

I hope that this book, with its visionary approach, will be a valuable addition and will promote interest among readers. Each of the authors has provided their extraordinary competence in their specific fields by providing different perspectives as they come from diverse nations and regions. I thank them for their contributions.

Editor

Numerical study on flow length in injection molding process with high-speed injection molding

Pham Son Minh[*], Tran Minh The Uyen

Department of Mechanical Engineering, HCMC University of Technology and Education, Ho Chi Minh City, Vietnam

Email address:

minhps@hcmute.edu.vn (P. S. Minh)

Abstract: In thin-wall injection molding, due to the very fast polymer melt heat transfer to the mold wall, the freeze layer appears quickly during the filling stage. In this study, high-speed injection molding (up to 1400 mm/s injection speed) was studied. A mold of spiral shape, 0.4 mm thick, is used to verify the ability of melt filling under different injection speeds. Simulation by Moldflow software was also performed for verification. The result shows that when injection speeds vary for 100 mm/s, 500 mm/s, 1000 mm/s, and 1400 mm/s, the flow length to thickness ratio was increased with the value of 335, 467.5, 605, and 640, respectively. The simulation results also show that the heat transfer coefficient between hot melt and mold wall has a strong influence on the flow length, especially with the high-speed injection molding. In general, slower injection speed requires a higher heat transfer coefficient, whereas higher injection speed require only a lower heat transfer coefficient.

Keywords: High-Speed Injection Molding, Flow Length to Thickness Ratio, Heat Transfer Coefficient

1. Introduction

Injection molding is a popular technology for manufacturing. However, as products become thinner and smaller, it is difficult to manufacture them using conventional injection molding (CIM), because heat transfers rapidly from the melt to mold wall due to part's thinness. Increasing the mold temperature, melt temperature, or packing pressure, it increases the cycle time. There are two definitions of "thin-wall" parts: a part with a thickness below 1 mm across an area greater than 50 cm2 [1, 2]; and a flow length to thickness (L/t) ratio greater than 100:1 or 150:1 [3, 4]. To fill the cavity in an extremely short time before the formation of the skin layer in thin-wall parts, injection machine manufacturers have developed machines for high-speed injection, and this process was called high-speed injection molding (HSIM) [5, 6]. The object of HSIM is to fill the cavity in an extremely short time. Hence, a higher injection pressure, a stable controller, and rigid steel for the injection machine and mold are necessary.

HSIM rapidly fills the mold cavity, enabling it to more precisely mold thin-wall parts or parts with micro structures [7 – 9]. For example, there are many micro structures on light guide plates, making transcription uniformity very important. The usual method is to increase the mold or melt temperature

to improve the transcription uniformity. However, increasing the temperature possibly increases the cycle time. Yokoi [10, 11] found that HSIM was useful in improving both the transcription ratio and transcription uniformity of a light guide plate. HSIM permits the melt to fill the cavity as rapidly as possible, before the formation of the solidified skin layer. Despite the potential usefulness of HSIM in the manufacture of precision parts, few studies have explored HSIM. Manufacturers generally use trial and error approaches to obtain the properties of HSIM. This is unsystematic and wastes time and money.

With this in mind, we use CAE (Computer-Aided Engineering) to anticipate issues that may arise in production using HSIM. CAE technology has been used for years to assist in part design, playing an important role. It can simulate and anticipate manufacturing conditions to prevent defects before manufacturing. By comparing the CIM and thin-wall injection molding the result shows that injection speed and pressure were key parameters [1, 4, 5, and 7]. Marco [3] anticipated the mold deformation and thickness changes of thin-wall parts, and that high injection pressure during molding affected the precision of thin-wall parts. Though conventional trial and error methods do not easily determine the factors which affect thin-wall part manufacture, they can easily be anticipated through simulation.

Over the past few years, the thin-wall injection molding speed was about 500 mm/s [1 – 10], but with products becoming ever smaller and thinner, 500mm/s proved too slow to satisfactorily mold them. In response, injection machine manufacturers began developing new injection machines with speeds exceeding 500 mm/s. However, at present, HSIM is used only by those who have special requirements, and there have been few reported investigations of HSIM. When the injection speed becomes high-speed, many conventional conditions must be altered. The goals of this study thus include (1) using flow length study to visualize high speed injection characteristic through both numerical and experimental methods; (2) investigating the relationship between the heat transfer coefficient and injection speed. Since the behavior of the material is changed by high injection speed, the conditions of the simulation should also be modified. We first use the spiral model to experimentally obtain the flow length at injection speeds ranging from 100 mm/s to 1400 mm/s, and then heat transfer coefficients that affect the thermal conductivity in simulation are varied from 5000 to 20,000 W/m2 °C by Moldflow software to anticipate the flow length. Finally, the results of the experiments and simulations will be compared and discussed.

2. Experiment and Simulation Work

2.1. Experimental Work

The high-speed injection molding machine in this study was a High Speed High Pressure Injection Molding Machine TX-SC Series by CHUAN LIH FA. On this machine the arrangement of the accumulator, the control valve, and the injection unit eliminates delays in control and losses by piping, and by adopting the V-line system, only the plunger is under control. This machine can reach a maximum injection speed of 1500 mm/s and a maximum injection pressure of 343 MPa. The injection speed control is an important property of the machine that goes from zero to maximum speed in 0.1 s and from maximum speed to zero in 0.215 s.

The material used was Acrylonitrile Butadiene Styrene (ABS), type PA-756, supplied by Chi-Mei. The experimental model was a spiral shape 6 mm long and 0.4 mm thick, as shown in Figure 1, and was used to measure the flow length. After the injection molding process was finished, the part will be rejected, then, the melt flow length will be measured. If the flow length is longer, it means the material has better flow behavior [7 - 9]. The experimental mold temperature, shown in Table 1, was set at 30 °C, while melt temperature was 240 °C. Since these experiments were designed for verifying the melt flow length, so, the packing pressure was ignored, and the melt volume was fixed at a ram position of 20 mm. The injection speed conditions were selected as 100, 500, 1000, and 1400 mm/s, and the injection pressure was 343 MPa.

2.2. Simulation Work

To ensure the accuracy of the simulation, the heat transfer coefficient (HTC) between the mold and melt is a key point, especially for thin-wall parts. Therefore, this parameter was varied in the simulation. According to introduction to HTC in the Moldflow software, during the mold filling in injection molding, as shown in Figure 2 [12], due to the thermal contact resistance between mold and melt, the mold-melt interface temperature (T_{mt}) is usually higher than the set mold temperature (T_{mb}). The difference between T_{mb} and T_{mt} depends on the thermal properties of the mold metal and the melt, as well as on the process conditions. Generally speaking, T_{mt} may not be greatly different from T_{mb}. This means that T_{mb} may be used to represent T_{mt} directly in the CAE molding simulation. However, for cases involving thin-wall parts, high speed filling, or melt with highly temperature-dependent viscosity, the use of T_{mt} tends to underestimate the mold-melt interface temperature, which further leads to an overestimate of the injection pressure. It is often convenient to work with the HTC, which is defined by an expression analogous to Newton's law of cooling:

$$\frac{dQ}{dt} = h \times A(\Delta T) \tag{1}$$

Where:
Q = thermal energy in Joules
h = heat transfer coefficient
A = Surface area of the heat being transferred
ΔT = Difference in temperature between the solid surface and surrounding fluid area.

Figure 1. *Spiral shaped mold insert with a thickness of 0.4 mm*

Table 1. *Boundary conditions in experiment and simulation.*

Boundary conditions	Experiment	Simulation
Material	Chi-Mei, PA-756	Chi-Mei, PA-756
Melt temperature	240 °C	240 °C
Mold temperature	30 °C	30 °C
Filling control	20 mm	Injection velocity by Injection Pressure vs. Ram position
Injection pressure	343 MPa	343 MPa
Injection speed	100, 500, 1000, 1400 mm/s	100, 500, 1000, 1400 mm/s

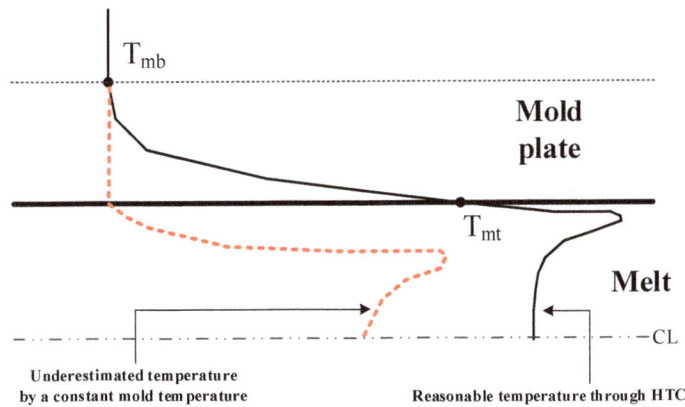

Figure 2. *The schematic of thermal conductivity during filling stage [12].*

In the simulation model shown in Figure 3, the total element count was 1,050,225, with element types of Tetra. The parameters of the simulation, as shown in Table 1, were the same as in the experiments. Normally, the HTC value set to the default 5000 W/m2 °C. In this study, we simulate high-speed injection molding. HTC would thus be set at 5000, 10000, 15000, and 20000 W/m2 °C under different injection speeds to determine which HTC approaches the experimental results.

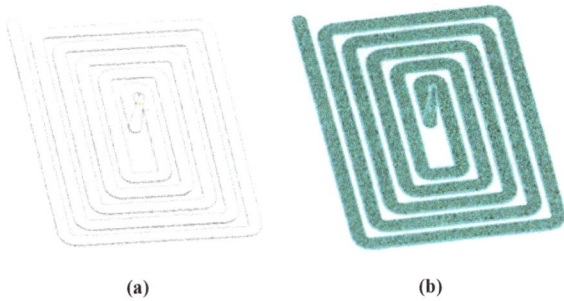

Figure 3. *(a) CAD model. (b) The meshing model in Moldflow simulation.*

3. Results and Discussions

In this paper, the part thickness is only 0.4 mm, meaning that the heat of polymer could more easily be dissipated by the mold wall. By comparing with the same time for melt flowing, with a slower injection speed, the freeze layer will defend seriously the melt [1 - 3]. On the contrary, with a higher injection speed, the polymer will flow faster, as a result, the effect of freeze layer could be reduced [5, 6]. On the other hand, the shear heating also impact on the melt temperature, as well as the melt viscosity. With the higher injection speed, the higher shear rate will appear. So, the melt will be heated by the shear heating effect [10, 11], which will create higher melt temperature, so, the melt will flow easier. So, in general, according to the effect of freeze layer and shear rate heating, the higher injection speed will let the longer flow length.

In this research, the melt of ABS – PA756 will be filled into cavity with the melt temperature of 240 °C, mold temperature of 30 °C, and the injection pressure of 343 MPa. After the molding process was finished, the melt flow length will be measured by the length of molding part. For observing the influence of injection speed on the ability of melt filling, four injection speed will be applied. Each type of injection speed, the molding process will be run with 10 cycles for reaching to the stable stage. Then, in the next 10 processes, the molding part will be collected and its length values will be used for comparing and discussing. The length of parts under different injection speed were shown in Figure 4. The melt flow length comparison with different injection speeds are shown in Figure 5. According to these results, with the common injection speed (lower than 500 mm/s [1, 8, 9]), when the injection speed increases from 100 mm/s to 500 mm/s, the melt flow length was increased 39.7%. However, with high-speed injection molding, when the injection speed increases to 1400 mm/s, the flow length could reach to 91% longer than the case of 100 mm/s. With the case of 1400 mm/s injection speed, the flow length to thickness (L/t) reaches to 640. This ratio shows a bright improvement with the cases of common injection speed [1 - 8]. In the other injection speed, the experiment show that the L/t ratios are 335, 468, 605, and 640 with the injection speed of 100 mm/s, 500 mm/s, 1000 mm/s, 1400 mm/s, respectively.

For researching the influence of heat transfer coefficient between melt and cavity surface, a simulation model was created with the same size of experiment model. In addition, the boundary condition was set the same with experiment. The model size and mesh model are shown in Figure 3. By simulation, four injection speed will be simulated. With each injection speed value, the heat transfer coefficient between melt and cavity surface will change by four values: 5000, 10000, 15000, and 20000 W/m2 °C. Then, the length of melt flow will be measured and compared with experiment results.

Figure 6 shows the comparison of melt flow length between simulation and experiment under different heat transfer coefficient. Based on this figure, with the normal injection speed (100 mm/s and 500 mm/s), the real heat transfer coefficient is much higher than 20000 W/m2 °C, so, this is the reason that the simulation case has a longer flow length than the experiment cases. On the contrary, with the high injection speed cases (1000 mm/s and 1500 mm/s), the real heat transfer coefficient is varied from 15000 W/m2 °C to lower than 10000 W/m2 °C. On the other hand, these results also show that, with the simulation step, the higher injection speed should have a smaller HTC in the set up.

In general, for improvement of the simulation accuracy, the heat transfer coefficient should have a proper modification in cases of high-speed injection molding. If the HTC is too high, the melt flow length in simulation will be shorter than the real value. However, if the HTC is too low, the melt flow length in simulation will be longer than the real value.

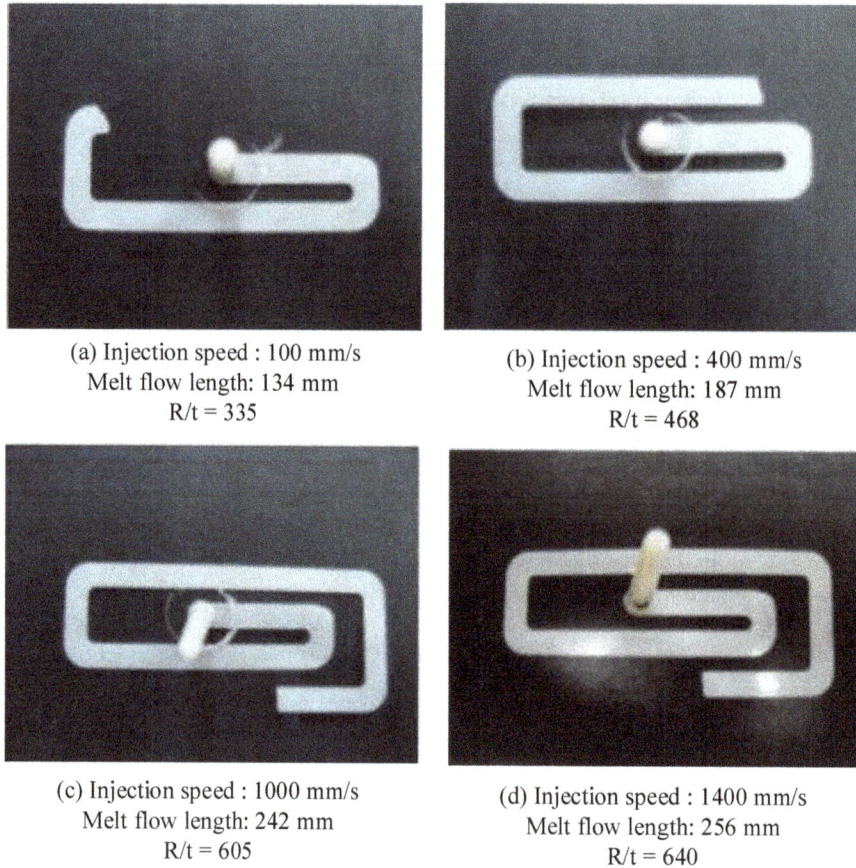

(a) Injection speed : 100 mm/s
Melt flow length: 134 mm
R/t = 335

(b) Injection speed : 400 mm/s
Melt flow length: 187 mm
R/t = 468

(c) Injection speed : 1000 mm/s
Melt flow length: 242 mm
R/t = 605

(d) Injection speed : 1400 mm/s
Melt flow length: 256 mm
R/t = 640

Figure 4. *Melt flow length by experiment with injection speeds varies from 100 mm/s to 1400 mm/s.*

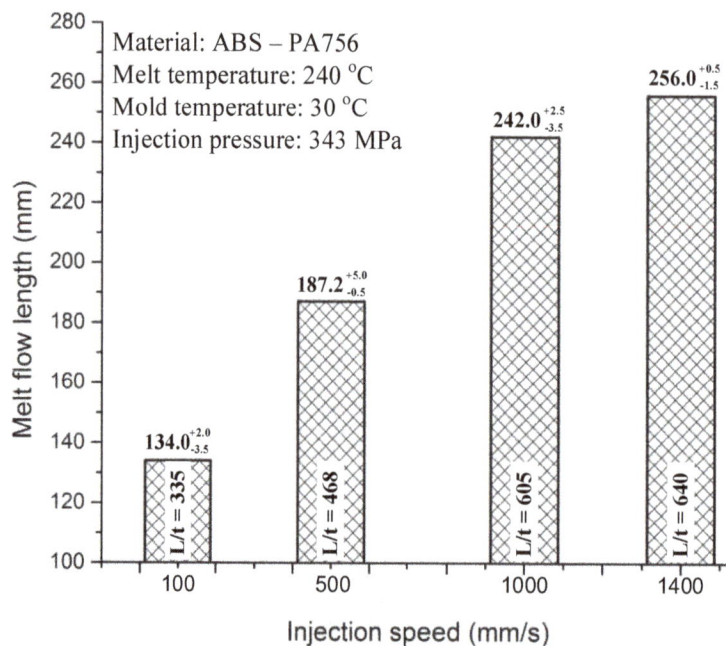

Figure 5. *Melt flow length comparison under different injection speed*

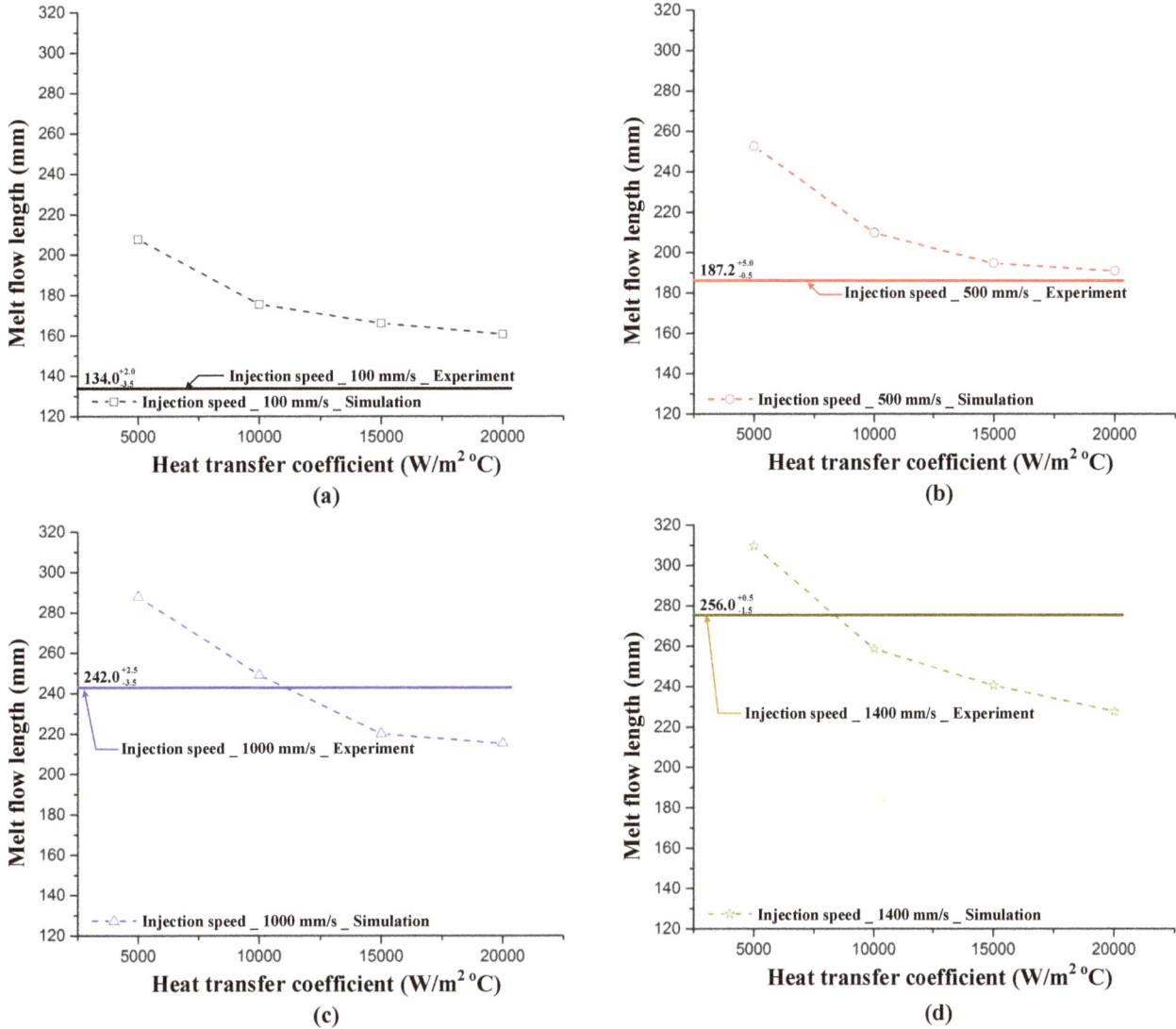

Figure 6. *Comparison of melt flow length between simulation and experiment under different heat transfer coefficient.*

4. Conclusion

In this research, by simulation and experiment, the effect of injection speed on the melt flow length was overseen. In addition, by varying the heat transfer coefficient in simulation cases, the accuracy of simulation was verified. Based on the result, these conclusions were obtained:

1. In injection molding process with the thin-wall parts, thermal energy is rapidly transferred from the melt to the mold material, it means that the HTC must be modified in the simulation based on the injection speed.

2. In the experiment, when the injection speed of 1400 mm/s was used, the melt flow length was improved 91% with the case of 100 mm/s, and the L/t ratios was increased from 335 to 640.

3. With higher HTC increased, the simulation results for L/t become shorter. When the injection speed exceeds 1000 mm/s, HTC should be adjusted to 10000 W/m2 °C, respectively

Acknowledgements

This research was supported by HCMC University of Technology and Education, Hochiminh City, Vietnam.

References

[1] S. C. Chen, W. H. Liao, J. P. Yeh, R. D. Chien, "Rheological behavior of PS polymer melt under ultra-high speed injection molding", Polym. Test., vol 31, pp. 864 – 869, October 2012.

[2] F. Yu, H. Deng, Q. Zhang, K. Wang, C. L. Zhang, F. Chen, Q. Fu, "Anisotropic multilayer conductive networks in carbon nanotubes filled polyethylene / polypropylene blends obtained through high speed thin wall injection molding", Polym., vol. 54, pp. 6425–6436, November 2013.

[3] M. Sortino, G. Totis, E. Kuljanic, "Comparison of injection molding technologies for the production of micro-optical devices", Procedia Eng., vol. 69, pp. 1296–1305, 2014.

[4]   C. C. Tsai, S. M. Hsieh, H. E. Kao, "Mechatronic design and injection speed control of an ultra-high-speed plastic injection molding machine", Mech., vol. 19, pp. 147–155, March 2009.

[5]   H. Bormuth, "High-speed injection moulding with Polypropylene", Mater. Design, vol. 5, pp. 137–138, June–July 1984.

[6]   D. Drummer, K. Vetter, "Expansion–injection–molding (EIM) by cavity near melt compression – About the process characteristic", CIRP J. Manuf. Sci. Technol., vol. 4, pp. 376–381, 2011.

[7]   B. Sha, S. Dimov, C. Griffiths, M.S. Packianather, "Micro-injection moulding: factors affecting the replication quality of micro features", 4M 2006 - Second International Conference on Multi-Material Micro Manufacture, pp. 269–272, 2006.

[8]   C.A. Griffiths, S.S. Dimov, D.T. Pham, "Micro injection moulding: the effects of tool surface finish on melt flow

behavior", 4M 2006 - Second International Conference on Multi-Material Micro Manufacture, pp. 373–376, 2006.

[9]   D. Annicchiarico, U. M. Attia, J. R. Alcock, "A methodology for shrinkage measurement in micro-injection moulding", Polym. Test., vol. 32, pp. 769–777, June 2013.

[10]   W. M. Yang, H. Yokoi, "Visual analysis of the flow behavior of core material in a fork portion of plastic sandwich injection molding", Polym. Test., vol 22, pp. 37–43, February 2003.

[11]   H. Yokoi, , N. Masuda, H. Mitsuhata, "Visualization analysis of flow front behavior during filling process of injection mold cavity by two-axis tracking system", J. Mater. Process. Technol., vol 130–131, pp. 328–333, December 2002.

[12]   http://www.boundarysys.com/attachments/098_Mold

Analysis and prediction of crack propagation in plates by the enriched free Galerkin method

Bui Manh Tuan[1, 2,*], Chen Yun Fei[1]

[1]School of Mechanical Engineering, Southeast University, Nanjing city, Jiangsu Province, China
[2]Faculty of Mechanical Engineering, Tuy Hoa Industrial College, Tuy Hoa City, Phu Yen Province, Vietnam

Email address:

buimanhtuan.tic@gmail.com (B. M. Tuan), yunfeichen@seu.edu.cn (Chen Yunfei)

Abstract: This paper presents a centre and edge crack analysis using meshless methods which is based on moving least squares (MLS) approximation. The unknown displacement function u(x) is approximated by moving least square approximation $u^h(x)$. These approximation are constructed by using a weight function which is based a monomial basis function and a set of non-constant coefficients. A subdivision that is similar to finite element method is used to provide a background mesh for numerical integration. An enriched EFG formulation with fracture problems is proposed to improve the solution accuracy for linear elastic fracture problem. The essential boundary conditions are enforced by Lagrange multipliers method. A code has been written in Matlab for the analysis of a crack tip. The obtained results of the developed EFG-code were compared to available experimental data and other numerical (exact methods and finite element method) methods.

Keywords: Crack, Stress Intensity Factor, EFG Method, Moving Least Squares Approximant, Crack Propagation

1. Introduction

There are many numerical methods applied for modeling cracks in mechanical problems such as the Boundary Element Method (BEM) [1], the Finite Element Method (FEM) [2], the extended Finite Element Method (XFEM) [3] and Meshless Methods (MMs). Among these methods, the element free Galerkin method and enriched element free Galerkin method has developed by Belytschko et al. [4,5] and it has been widely applied in fracture mechanics [6–8]. The meshless method works better than the traditional finite element method (FEM) in treatment of arbitrary evolving discontinuity. Due to the elements are independent of each other so the problems of crack analysis is simplified considerably which can be analyzed on a fixed mesh and unnecessary remeshing for crack development issues.

In this paper, we use a meshless method to analyze two-dimensional elastic problem by using Element-Free Galerkin (EFG) method that is based on moving least squares approximation (MLS) to construct the function approximation for the Galerkin weak form. These approximations are constructed by using a weight function that base on a monomial basis function and a set of non-constant coefficients. A subdivision similar to finite element method is used to provide a background mesh for numerical integration. The necessary boundary conditions are implemented by means of Lagrange multipliers. The method of enriched EFG that is enriched basis and nodal refinement is utilized to calculate and simulate crack growth. Therefore, the precision calculation is improved. The continuous crack propagation is modeled as a linear series of crack growth. In there, plate for cracks in the center and edge cracks is presented in this paper. The results obtained for two-dimensional problem with different of the number of nodes, crack length, load and dimensionless size of the support domain in the region of the crack tip compare to element finite method and exact method. In addition, Matlab code of EFG method also is offered in this paper. This Matlab code can be developed for meshfree application software or other meshfree method in the further.

2. MLS Approximations Functions

MLS functions were developed by Lancaster and Salkauskas in the literature [9] to approximate curves and surfaces, and then was used in EFGM method to generate shape functions [1,9]. In EFGM, a field variable u(x) is approximated by MLS approximation, $u^h(x)$ [22] which is given as

$$u(x) \cong u^h(x) = p^T(x).a(x) \quad \forall x \in \Omega_x \quad (1)$$

Where p(x) is a linearly independent basis of m functions

$$p^T(x) = [p_1(x) \, p_2(x) \ldots p_m(x)] \quad (2)$$

And a(x) collects the undetermined parameters of the approximation

$$a(x) = [a_1(x) \, a_2(x) \ldots a_m(x)]^T$$

The a(x) parameters are obtained by minimizing a weighted least square sum. The weighted least square sum denoted by L(x) can be written as follows:

$$L(x) = \sum_{I=1}^{n} \omega(x - x_I).[u^h(x_I, x) - u_I]^2$$
$$= \sum_{I=1}^{n} \omega(x - x_I).[p^T(x_I)a(x) - u_I]^2 \quad (3)$$

Where $\omega(x - x_I)$ is a weighting function which is nonzero on the influence domain of the node x_I; n is the number of points in the neighbourhood of x, and u_I is the nodal value of u at $x = x_I$. The dimension of the influence domain of each node and the choice of the weighting function are decisive parameters for the approximation by MLS [4].

Minimizing L(x) in order to the unknown parameters a(x) results in

$$A(x).a(x) = B(x).u \quad (4)$$

Or

$$a(x) = A^{-1}(x).B(x).u \quad (5)$$

With:

$$A(x) = \sum_{I=1}^{n} w(x - x_I).p(x_I).p^T(x_I) \quad (6)$$

$$B(x) = [w(x - x_1).p(x_1), w(x - x_2).p(x_2), \ldots$$
$$, w(x - x_n).p(x_n)] \quad (7)$$

Substituting the result (5) in the initial approximation (1), the MLS approximation is obtained as:

$$u^h(x) = \sum_{I=1}^{n} \Phi_I(x).u_I = \Phi(x)U \quad (8)$$

Where the shape function is defined by

$$\Phi_I(x) = \sum_{j=0}^{m} p_j(x)(A^{-1}(x)B(x))_{jI} = p^T A^{-1} B_I \quad (9)$$

Where m is the order of the polynomial p(x). To determine the derivatives from the displacement (8), it is necessary to obtain the shape function derivatives. The spatial derivatives of the shape functions are obtained by:

$$\Phi_{I,x} = (p^T A^{-1} B_I)_{,x} = p_{,x}^T A^{-1} B_I + p_{,x}^T (A^{-1})_{,x} + p^T A^{-1} B_{I,x} \quad (10)$$

Where

$$B_{I,x}(x) = \frac{d\omega}{dx}(x - x_I)p(x_I);$$
$$A_{,x}^{-1}(x) = -A^{-1} A_{,x} A^{-1}; \quad (11)$$
$$A_{,x} = \sum_{I=1}^{n} \omega(x - x_I)p(x_I)p^T(x_I)$$

It should be noted that EFG shape functions do not satisfy the Kronecker delta criterion: $\Phi_I(x)^1 \neq \delta_{ij}$, so $u^h(x_I) \neq u_I$, the nodal parameters u_I are not the nodal values of $u^h(x_I)$. Therefore, we use Lagrange multiplier method to enforce the essential boundary conditions.

3. Choice of Support Domain and Weight Function

The support domain usually used circular or rectangular. There is no difference if a circular or rectangular support domain is used in the EFG method [4,5]. A weight function needs properties as following:

- Compact support, i.e. zero outside the support domain.
- The values of all points in the support domain is positive.
- The value of its is maximum at the current point and decrease when moving outwards.

There are many kinds of function satisfying for these properties. In this paper, we used the quadratic spline function as follow

$$w(d_I) = \begin{cases} 1 - 6\left(\frac{d_I}{d_{mI}}\right)^2 + 8\left(\frac{d_I}{d_{mI}}\right)^3 - 3\left(\frac{d_I}{d_{mI}}\right)^2, & d_I \leq d_{mI} \\ 0 & d_I > d_{mI} \end{cases} \quad (12)$$

With: $d_I = \|x - x_I\|$
d_{mI} is the radius of influence domain x_I

$$d_{mI} = d_{max}.c_I$$

Where the scaling parameter d_{max} usually is chosen $1.5 \div 4$ for static analysis. The characteristic dimension parameter c_I represents the nodal spacing. If the nodes are uniformly distributed then c_I is the distance between two adjacent nodes.

4. Discrete Equations and Integration

In EFGM, the shape functions dissatisfied the Kronecker delta property. Therefore, we have to use Lagrange multiplier to invoke essential boundary. The Lagrange multiplier in [4] as follow as:

$$\tilde{L} = L + \int_{S_u} \lambda^T (u - \overline{u}) dS_u \qquad (13)$$

The Lagrange multiplier (λ) can be interpreted as the reaction forces needed to fulfill the displacement conditions at the boundary. The approximation given by:

$$u(x) \cong u^h(x) = \sum_{I=1}^{n} \Phi_I(x).u_I = \Phi(x)U \qquad (14)$$

By enforcing essential boundary conditions using Lagrange multiplier approach. Discretization of (14) results in [7].

$$\begin{bmatrix} K & G \\ G^T & o \end{bmatrix} \begin{Bmatrix} U \\ \Lambda \end{Bmatrix} = \begin{Bmatrix} f \\ q \end{Bmatrix} \qquad (15)$$

With K is stiffness matrix.

$$K_{IJ} = \int_{\Omega} B_I^T D B_J d\Omega \qquad (16)$$

f is the force vector.

$$f_I = \int_{\Omega} \Phi^T b d\Omega + \int_{S_T} \Phi^T t dS_T \qquad (17)$$

B_I is the partial derivatives of the shape function.

$$B_I = \begin{bmatrix} \Phi_{I,x} & 0 \\ 0 & \Phi_{I,y} \\ \Phi_{I,y} & \Phi_{I,x} \end{bmatrix} \quad N_K = \begin{bmatrix} N_K & 0 \\ 0 & N_K \end{bmatrix} \qquad (18)$$

$$G_{IK} = \int_{S_u} -N^T \Phi dS_u \quad q_K = \int_{S_u} -N^T \overline{u} dS_u \qquad (19)$$

For plane stress:

$$D = \frac{E}{1-v^2} \begin{bmatrix} 1 & v & 0 \\ v & 1 & 0 \\ 0 & 0 & (1-v)/2 \end{bmatrix} = D_e \qquad (20)$$

$$\sigma = \varepsilon.D \qquad (21)$$

5. The Stress and Displacement near Crack Tip

The analytical solution for the stresses of an infinite plate [10,11]

Mode 1:

$$\sigma_x = \frac{K_I}{\sqrt{2\pi r}} \cos\frac{\theta}{2} \left[1 - \sin\frac{\theta}{2} \sin\frac{3\theta}{2} \right] + O(r^0) \qquad (22)$$

$$\sigma_y = \frac{K_I}{\sqrt{2\pi r}} \cos\frac{\theta}{2} \left[1 + \sin\frac{\theta}{2} \sin\frac{3\theta}{2} \right] + O(r^0)$$

$$\tau_{xy} = \frac{K_I}{\sqrt{2\pi r}} \sin\frac{\theta}{2} \cos\frac{\theta}{2} \cos\frac{3\theta}{2} + O(r^0)$$

$$v = \frac{K_I}{4G} \sqrt{\frac{r}{2\pi}} \left[(2k+1) \sin\frac{\theta}{2} - \sin\frac{3\theta}{2} \right] + O(r^0) \qquad (23)$$

$$u = \frac{K_I}{4G} \sqrt{\frac{r}{2\pi}} \left[(2k-1) \cos\frac{\theta}{2} - \cos\frac{3\theta}{2} \right] + O(r^0)$$

Mode 2:

$$\sigma_x = -\frac{K_{II}}{\sqrt{2\pi r}} \sin\frac{\theta}{2} \left[2 + \cos\frac{\theta}{2} \cos\frac{3\theta}{2} \right] + O(r^0)$$

$$\sigma_y = \frac{K_{II}}{\sqrt{2\pi r}} \sin\frac{\theta}{2} \cos\frac{\theta}{2} \cos\frac{3\theta}{2} + O(r^0) \qquad (24)$$

$$\tau_{xy} = \frac{K_{II}}{\sqrt{2\pi r}} \cos\frac{\theta}{2} \left[1 - \sin\frac{\theta}{2} \sin\frac{3\theta}{2} \right] + O(r^0)$$

$$u = \frac{K_{II}}{4G} \sqrt{\frac{r}{2\pi}} \left[(2k+3) \sin\frac{\theta}{2} + \sin\frac{3\theta}{2} \right] + O(r) \qquad (25)$$

$$v = \frac{K_{II}}{4G} \sqrt{\frac{r}{2\pi}} \left[(2k-3) \cos\frac{\theta}{2} + \cos\frac{3\theta}{2} \right] + O(r)$$

$$w = 0$$

6. Enrichment Functions

In the EFG method, by adding extensions in Meshless methods to reflect the discontinuous displacement field generated by the crack. For the case of linear elastic fracture mechanics, two sets of functions are used: a Heaviside jump functions to capture the jump across the crack faces and asymptotic Branch functions that span the 2D asymptotic crack tip fields. The enriched approximation for fracture mechanics problems take the form [12-15]

$$u^h(x) = \sum_{I \in N} \Phi_I(x) u_I + \sum_{I \in N^b} \Phi_I(x) H(x) a_I$$
$$+ \sum_{I \in N^t} \Phi_I(x) B_k^{enr}(x) \beta_I \qquad (26)$$

In the equation (26), the first term is the standard approximations functions of EFG methods, the second term reflects discontinuous along the two sides of crack surface, and the third term reflects the crack tip singularity.

Where N is the entire set of particles in the domain, N^b is the set of particles whose domain of influence is completely bisected by the crack, N^s is the set of particles whose domain of influence is bisected by the crack tip as shown in Fig.1.

H(x) is Heaviside jump enriched function, and given by [16]:

$$H\left(f\left(\boldsymbol{x}\right)\right) = \begin{cases} +1 & f\left(\boldsymbol{x}\right) > 0 \\ -1 & f\left(\boldsymbol{x}\right) < 0 \end{cases} \quad (27)$$

B(x) is Branch enriched function, the crack tip extension function, and given by [17]:

$$B^{gr}(x) = \left[\sqrt{r}\cos(\frac{\theta}{2}) \quad \sqrt{r}\sin(\frac{\theta}{2}) \quad \sqrt{r}\sin(\frac{\theta}{2})\sin\theta \quad \sqrt{r}\cos(\frac{\theta}{2})\sin\theta\right] \quad (28)$$

The Branch enrichment is crucial to accurately locate the crack tip in enriched meshfree methods. The crack tip enrichment ensures that the crack is properly closed at the crack tip.

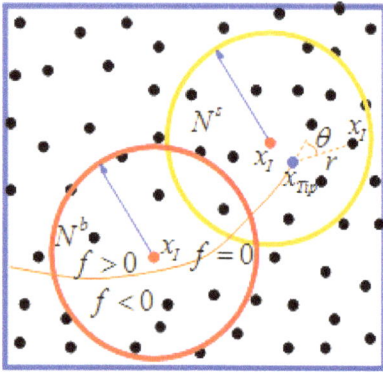

Figure 1. *Selection of the support domain near a crack face*

Where f(x) is the signed distance function from the crack line; x is the sample points, x_I is the distance from the surface cracks to the point x nearest; r and θ are polar coordinates with respect to the crack tip as shown in Fig. 1.

7. Crack Growth Direction

The propagation of crack requires a suitable criterion for a crack growth. The commonly used criteria are: the maximum principal stress criterion [18], the maximum energy release rate criterion [19], and the minimum strain energy density criterion [20].

In this study, we use maximum hoop stress theory, which assumes that the crack may grow in a direction perpendicular to the maximum principal stress [21].

$$\sigma_{r\theta} = \frac{1}{\sqrt{2\pi r}}\cos\frac{\theta}{2}\left[\frac{1}{2}K_I\sin\theta + \frac{1}{2}K_{II}(3\cos\theta - 1)\right] = 0 \quad (29)$$

Where K_I and K_{II} are the stress intensity factor of modes 1 and 2, respectively. The variables r and θ are as shown in Fig. 2. Therefore, the crack growth direction θ_0 for each crack increment is obtained by following

condition [22].

$$K_I \sin\theta_0 + K_{II}(3\cos\theta_0 - 1) = 0 \quad (30)$$

After solving the above equation, we obtain

$$\theta_0 = 2\arctan\left[\frac{1}{4}\left(\frac{K_I}{K_{II}}\right) \pm \sqrt{(\frac{K_I}{K_{II}})^2 + 8}\right] \quad (31)$$

According to this criterion, the equivalent mode I SIF is

$$K_e = K_I \cos^3(\theta_c / 2) - 3K_{II}\cos^2(\theta_c / 2)\sin(\theta_c / 2) \quad (32)$$

When equivalent stress intensity factor K_e is greater than fracture toughness of the material K_c, cracks began to expand. K_c is fracture toughness of the material.

The steps of calculations in Matlab code:

Step 1: Set up the nodal coordinates for a problem domain.

Step 2: Set up a background mesh for numerical integration.

Step 3: Determine the Gauss point, weight function and Jacôbi.

Step 4: Determine the domain of influence of each node in the model.

Step 5: Determine shape functions MLS and shape function derivatives.

Step 6: Enriched.

Step 7: Determine stiffness matrix K.

Step 8: Enforce essential boundary conditions using Lagrange multipliers.

Step 9: Assembled to form the master stiffness matrix and solve for equations.

Step 10: Solve for nodal parameters, solve for stresses.

Step 11: Solve for stress intensity factors and direction of crack propagation.

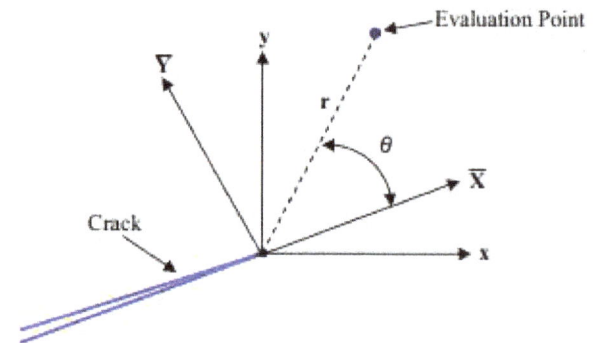

Figure 2. *Distance r and angle θ of a point x the crack tip*

8. Numerical Simulation

8.1. Rectangular Plate with a Center Crack under Tension Shown in Fig. 3.

The plate has an initial crack length of $2a$=40cm, a plate length of H=350cm, a plate width of b=175cm; Elastic modulus $E = 2\text{x}10^7\text{N/cm}^2$; Poisson's ratio v=0.3. Fracture toughness $K_c=140N/cm^{3/2}$; The dimensionless size of support domain d_{max}=1.75; a 4x4 Gauss quadrature is used and 6x6 nodes around the crack tip are selected for enrichment.

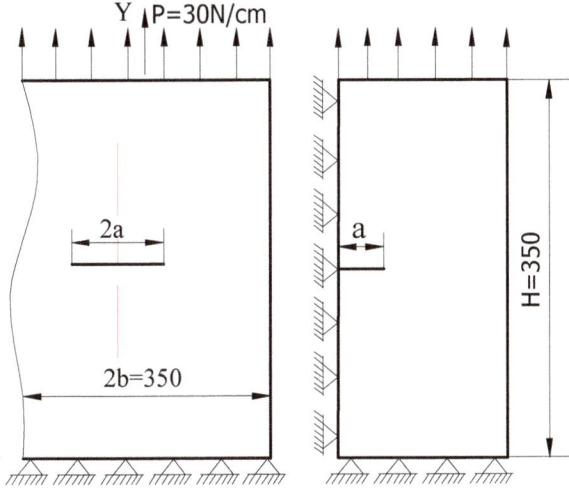

Figure 3. Mode I crack subjected to tensile load

Thanks to the symmetry, only half of the model calculations.

The reference mode I SIF is given by [11]

$$K_{Theory} = \sigma\sqrt{\pi a}F_1(a/b)$$

Where a is the crack length, b is the plate width and $F_1(a/b)$ is an empirical function given as

$$F_1(a/b) = \left[1 - 0.025(a/b)^2 + 0.06(a/b)^4\right]\sqrt{\sec\frac{\pi a}{2b}}$$

$$Error(\%) = \frac{K_{EFG} - K_{FEM}}{K_{EFG}}.100\%$$

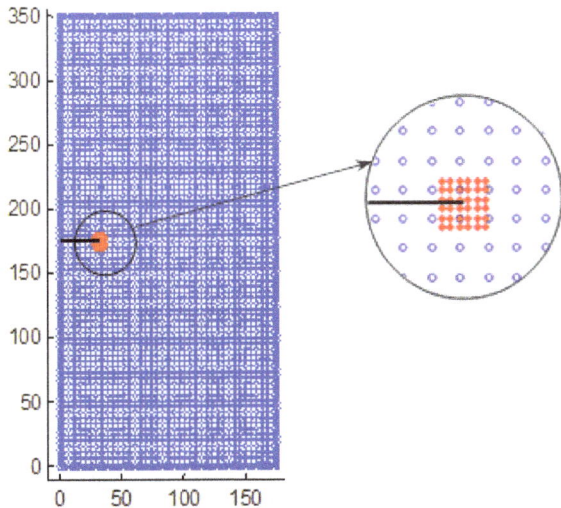

Figure 4. Nodal discretization and around the crack tips was refined with 6x6 nodes

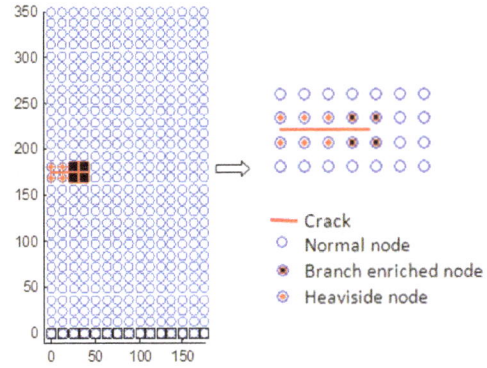

Figure 5. Normal and enriched nodes

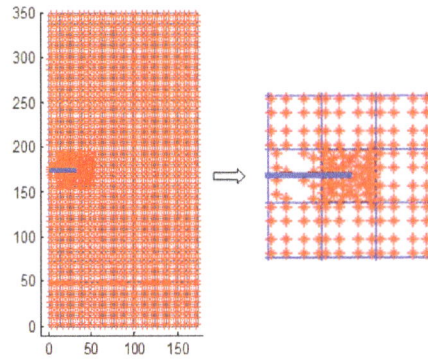

Figure 6. Enriched nodes and Gauss point distribution 4x4

8.1.1. Analysis of the Number of Nodes in the SIF Calculation

Table 1. Shows the mode I SIFs obtained using different of the number of nodes

Nodes	K_{EFG}	K_{Theory}	K_{FEM}[23]
10x20	247,2613	239,6579	255,9088
15x30	242,3825	239,6579	252,3577
20x40	241,0894	239,6579	245,3904
25x50	241,116	239,6579	243,7217
30x60	241,2555	239,6579	243,5303
35x70	241,1846	239,6579	243,511
40x80	241,0183	239,6579	243,6795
45x90	241,2739	239,6579	243,9402

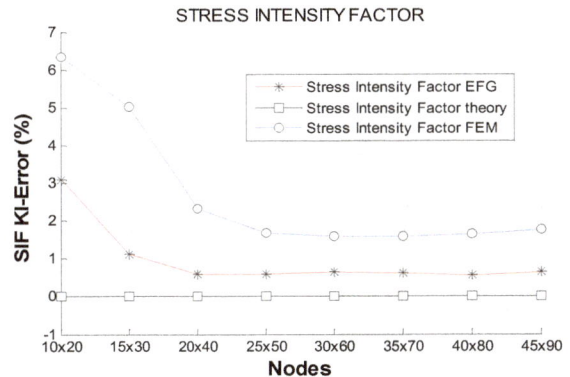

Figure 7. SIF % error of different nodes density

The results obtained in Table 1 show that the accuracy of the EFG method depend on the smoothing node density. If the node density is as smoothing as a high precision and vice versa. This article demonstrates that the problem of distributed nodes in EFG method great influence to the accuracy of calculations. To increase the accuracy level, we just simply add the button to the crack tip. Because no grid structure, hence the problem of adding nodes can be implemented and reduce computation time.

Using the results of EFG method compared to the results of element finite method and exact method, the values of error are mostly less than 1% is shown in Fig.7.

8.1.2. Analysis of the Crack Length in the SIF Calculation. With a=i*5(cm) (i=1:10)

Table 2. shows the mode I SIFs obtained using different of the crack length

a(cm)	Nel	K_{EFG}	K_{theory}	K_{FEM} [23]
5	40x80	119.5638	118.9573	118,09
10	40x80	169.3015	168.4756	167.1297
15	40x80	207.8535	206.8415	203.7212
20	40x80	241.0183	239.6579	237.8907
25	40x80	270.8392	269.1344	267.5569
30	40x80	298.436	296.4354	295.4391
35	40x80	324.9144	322.2807	322.2666
40	40x80	350.4446	347.1662	348.433
45	40x80	375.588	371.463	374.166
50	40x80	399.6855	395.4695	399.5562

From table 2, we can show that the different of the crack length under three different approaches. In Fig. 8 we also see that error of EFG method is small. When the crack length a = 45cm, the maximum error is about 1.0982%, the crack length a = 15cm, the minimum error of approximately 0.4868%.

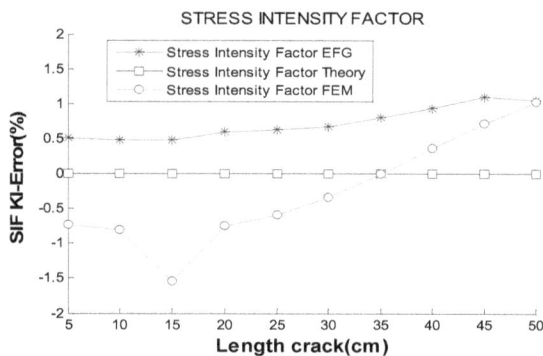

Figure 8. SIF% error of different crack length

8.1.3. Analysis Dimensionless Size of the Support Domain (d_{max}) in the SIF Calculation

Table 3. shows the mode I SIFs obtained using different of d_{max}

d_{max}	Nel	K_{EFG}	K_{Theory}	Error (%)
1.6	20x40	297,1386	306,9168	-3,2907
1.7	20x40	300,1515	306,9168	-2,2539
1.8	20x40	298,9854	306,9168	-2,6527
1.9	20x40	297,4066	306,9168	-3,1977
2.0	20x40	299,0257	306,9168	-2,6389
2.1	20x40	301,3948	306,9168	-1,8321
2.2	20x40	306,14	306,9168	-0,2537
2.3	20x40	317,113	306,9168	3,2153

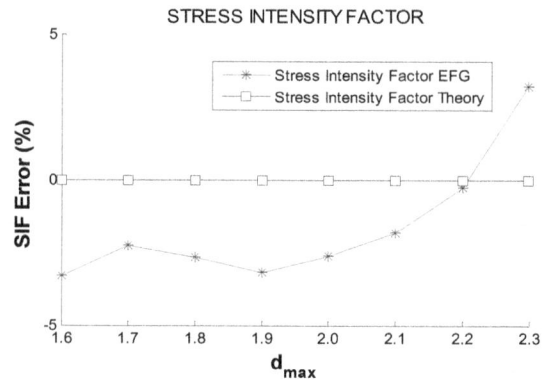

Figure 9. SIF% error of different d_{max}

Table 3 shows that the size of nodes in the influence regions have an important role in Meshless methods because the results obtained of Meshless method based on influence domain by using moving least squares to solve point within the variable. If the domain of influence parameters are chosen as $d_{max} = 1.7 \div 2.2$ then computational precision is high. Conversely, d_{max} is chosen outside this region then the effectiveness of approximation is not good.

8.1.4. Analysis of the Tension in the SIF Calculation: q=5*i (N/cm) (i=1:10)

From Fig. 10, we find that the calculation error is independent of the distribution of the load q. All computational errors always equal 0.8280%.

Table 4. shows the mode I SIFs obtained using different of load

q(N/cm)	Nel	K_{EFG}	K_{Theory}	K_{FEM} [23]
5	20x40	40,2263	39,9429	39,3728
10	20x40	80,4525	79,8859	78,745792
15	20x40	120,6788	119,8289	118,1186
20	20x40	160,9051	159,7719	157,4915
25	20x40	201,1314	199,7149	196,86448
30	20x40	241,3576	239,6579	236,2373
35	20x40	281,5839	279,6009	275,61027
40	20x40	321,8102	319,5438	314,9831
45	20x40	362,0365	359,4868	354,3561
50	20x40	402,2627	399,4298	393,7289

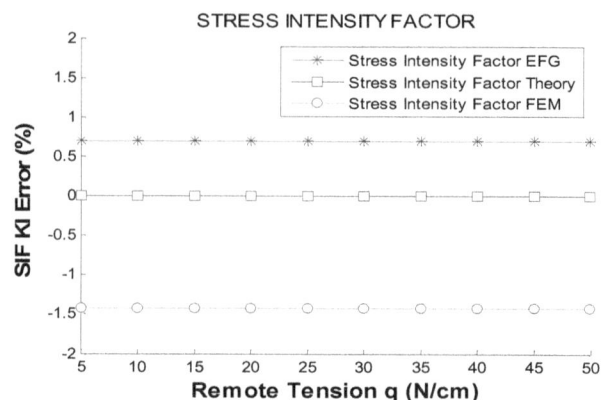

Figure 10. SIF % error of different load

8.2. Rectangular Plate with a Single Edge Crack under Tension Shown in Fig. 5.

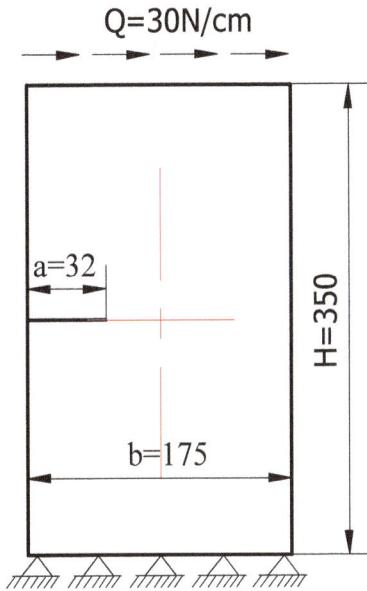

Figure 11. Single Edge Crack Computational model (mode2)

The plate has an initial crack length of a=32cm, a plate length of H=350cm, a plate width of b=175cm; Elastic modulus $E = 2 \times 10^7 N/cm^2$; Poisson's ratio $\upsilon=0.3$. Fracture toughness: $K_c=140N/cm^{3/2}$; The dimensionless size of support domain d_{max}=1.75;

A Gauss integration of 4x4 orders is used, without local refinement at the crack tip.

The reference mode II SIF is given by [11]

$$K_{II\ theory} = \sigma \sqrt{\pi a}.F_{II}\left(\frac{a}{b}\right)$$

Where a is the crack length, b is the plate width and $F_{II}(a/b)$ is an empirical function given as

$$F_{II}\left(\frac{a}{b}\right) = 1.122 - 0.231\left(\frac{a}{b}\right) + 10.55\left(\frac{a}{b}\right)^2$$

$$- 21.718\left(\frac{a}{b}\right)^3 + 30.382\left(\frac{a}{b}\right)^4$$

8.2.1. Analysis of the Number of Nodes in the SIF K_{II} Calculation

Table 5. Analysis different of the number of nodes in the SIF calculation

Nel	K_{II} EFG	K_{II} Theory	K_{II} [23] FEM
30x60	395,5081	401,1702	404,2531
35x70	396,2977	401,1702	403,3656
40x80	397,0112	401,1702	400,2353
45x90	397,8484	401,1702	399,7724
50x100	397,879	401,1702	399,4941
55x110	397,8911	401,1702	399,3723
60x120	397,9574	401,1702	399,3572
65x130	398,0053	401,1702	399,3854

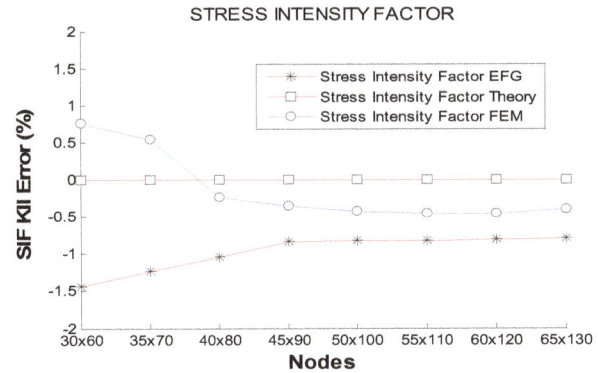

Figure 12. SIF K_{II} % error of different nodes density

In this section, computational analysis without smoothing at the crack tip.

The results obtained of Table 5 illustrates that the maximum error of EFG method and element finite method with the same number of nodes is 1.40% and 0.76%, respectively. When the mesh density is smoothing then computational precision is high. Thus, we can find that the computational precision depends greatly on the distribution of nodes.

8.2.2. Analysis of the Crack Length in the SIF K_{II} Calculation

Table 6. Analysis of the crack length in the SIF calculation. With a=i*10(cm) (i=1:8)

a(cm)	Nel	K_{II} EFG	K_{II} Theory	K_{II} [23] FEM
10	65x130	190,7816	191,6101	194,8391
20	65x130	285,7325	286,8245	290,5711
30	65x130	378,9788	381,316	384,4031
40	65x130	482,2136	485,6039	487,7887
50	65x130	602,4452	606,5331	609,3197
60	65x130	750,6046	752,7061	757,2961
70	65x130	933,2459	936,6675	942,2688
80	65x130	1172,198	1176,25	1179,348

From Table 6 and Fig. 3 show that the error of EFG method are small. When the crack length a = 40 cm, the maximum error calculation is 0.7030%. When the crack length a = 60 cm, the minimum error calculation is 0.2799%.

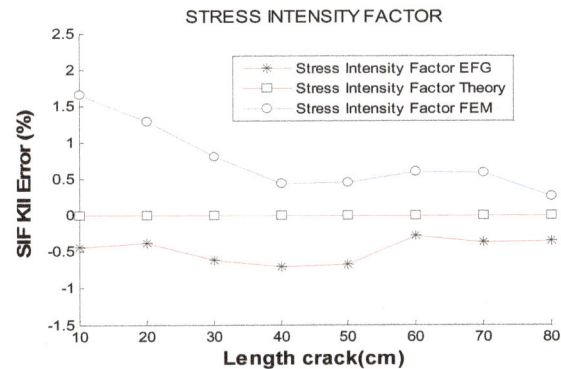

Figure 13. SIF KII % error of different crack length

8.3. Simulation Results of Displacement Fields and Stress Field

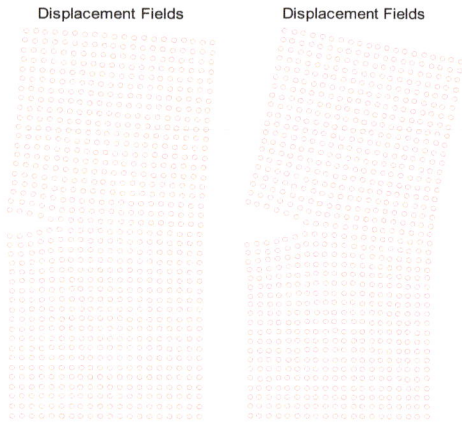

Figure 14. The node displacement fields after 2 steps

Fig. 14 presents the displacement field of edge crack finite plates. Where we can see that the node displacement fields after 2 steps is reflected under the effect of loads.

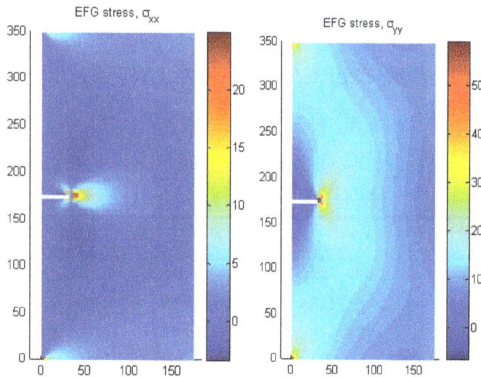

Figure 15. x, y direction stress field of center crack under tension

Fig. 15 and Fig. 16 present the stress contour plot of σ_{xx} and σ_{yy} direction of center and edge crack under tension. It also shows that the stress concentration at the crack, the singularity of the crack tip stress and and the stress field at the crack location outside is smooth.

All the problems have been simulated by EFG codes (algorithms) writting in MATLAB (R2010b).

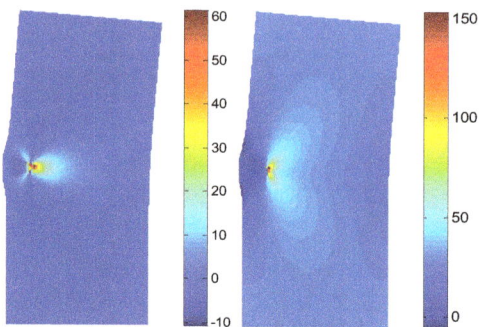

Figure 16. x, y direction stress field of single edge crack under tension

8.4. Simulation of Crack Propagation

Simulate the growth of crack in quasi-static is evaluated by equivalent stress intensity factor, and if its exceeds the fracture toughness of the material then the crack will be extended for some finite length (da) in a particular direction that is found by a suitable crack growth criterion. The step size (da) is a used to determine parameters, and da should be chosen enough minor to get an accurate crack growth path. The stress intensity factors are recalculated for the new crack geometry, and the next crack is extended according to a new direction.

In this paper, the following equation is calculated for each step of the cracking step [24]:

$$da_n = \frac{K_e^{(n)}}{K_e^{(0)}}.da_0$$

Where da_0 is initial cracking step, $K_e^{(0)}$ is initial equivalent stress intensity factor, da_n is step n for cracking step, $K_e^{(n)}$ is step n the equivalent stress intensity factor.

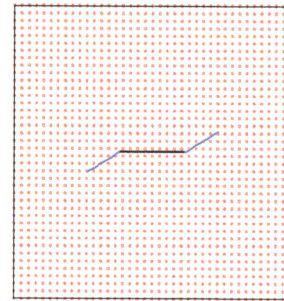

After 10 step

Figure 17. Crack growth path of centre crack

The obtained results of model crack propagation analysis by using a relatively coarse discretization of 40 × 40 nodes for centre crack and 20x40 nodes for edge crack. The crack growth increment is selected 0.1a (a=40) for this study, and the crack growth are simulated for 10 steps and 15 step. The result of crack path is shown as Fig. 17 and Fig. 18.

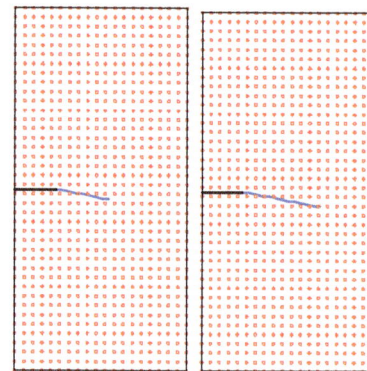

After 10step   After 15step

Figure 18. Crack growth path of single edge crack

9. Conclusions and Discussion

The Meshless computation has some outstanding advantages compare to the traditional finite element method (FEM) in treatment of arbitrary evolving discontinuity. Because of the independence of elements, the adaptive refinement can be easily achieved. Therefore, the crack-propagation analysis can be done easily and dramatically simplified. By introducing enriched functions, the extensions are added in the approximation of traditional Meshless method, the computation accuracy was improved. Through the computational results of stress intensity factors for edge and centre crack are compared to the results of element finite method and exact method, proving the Meshless method is very convenient for crack problem. In both cases the crack growth that is edge and centre crack are simulated for predicting fatigue crack propagation path. This method is used to allow crack growth without remeshing.

Through the analyses of numerical examples demonstrate, we can see that enriched EFG method which can solve the fracture problems is effective, and has practical merits for modeling crack growth problem. It is very promising in engineering application.

References

[1] S.T. Raveendra, P.K. Banerjee, "Boundary element analysis of cracks in thermally stresses planar structures," Int. J. Solids Struct, vol. 29, 1992, pp. 2301–2317.

[2] L.N. Gifford, P.D. Hilton, Stress intensity factors by enriched finite elements, Eng. Fract. Mech, vol.10, 1978, pp. 485–496.

[3] M. Duflot, "The extended finite element method in thermo-elastic fracture mechanics," Int. J. Numer. Methods Eng, Vol.74 , 2008, pp. 827–847.

[4] Belytschko T, Lu YY, Gu L, "Element-free Galerkin methods," International Journal for Numerical Methods in Engineering , 1994, vol.37, pp. 229–256.

[5] Fleming M, Chu YA, Moran B, Belytschko T, "Enriched element-free Galerkin methods for crack tip fields," Int J Numer Methods Eng, 1997, vol.40, pp. 1483–504.

[6] Belytschko T, Gu L, Lu YY, "Fracture and crack growth by element-free Galerkin methods," Modelling and Simulation in Materials Science and Engineering, 1994, vol. l2, pp. 519–534.

[7] Belytschko T, Lu YY, Gu L, Tabbara M, "Element-free Galerkin methods for static and dynamic fracture," International Journal of Solids and Structures 1995, vol.32, pp. 2547–2570.

[8] Belytschko T, Tabbara M, "Dynamic fracture using element-free Galerkin methods," International Journal for Numerical Methods in Engineering, 1996, vol.39, pp. 923–938.

[9] Lancaster P, Salkauskas K, "Surfaces generated by moving least squares methods," Math Comput, 1981, vol. 37, pp. 141–58.

[10] Nguyen P, Rabczuk T, Bordas S, Duflot M, "Meshless methods: a review and Computer implementation aspects," Math Comput Simul, 2008, vol.79, pp. 763–813.

[11] Tada Hiroshi, Paris Paul, Irwin George, "The Stress Analysis of Cracks Handbook [M]," Washington University, 1957.

[12] Dennis M Tracy, "Finite elements for determination of crack tip elastic stress intensity factors [J]," Engineering Fracture Mechanics, 1971, vol.3(3), pp. 255-266.

[13] Moes N, Dolbow J, Belytschko T, "A finite element method for crack growth without remeshing," International Journal for Numerical Methods in Engineering, 1999, vol.46(1), pp.131–150.

[14] Ma wen tao, Li ning, Shi jun ping, "Modelling crack growth by enriched Meshless method based on partition of unity," Chinese journal of computational Mechanics, 2013, vol.30, pp.28-33.

[15] N. Muthu, S. K. Maiti, B. G. Falzon, I. Guiamatsia, "A comparison of stress intensity factors obtained through crack closure integral and other approaches using Xtended element-free Galerkin method," Comput Mech, 2013, vol.52, pp. 587–605.

[16] N. Muthu, B.G.Falzon, S.K.Maiti, S.Khoddam, "Modified crack closure integral technique for extraction of SIFs in mesh free methods," Finite Elements in Analysis and Design, 2014, pp. 25–39.

[17] Sayyed Shahram Ghorashi, Soheil Mohammadi, Saeed-Reza Sabbagh-Yazdi, "Orthotropic enriched element free Galerkin method for fracture analysis of composites"' Engineering Fracture Mechanics, vol.78, 2012, pp. 1906–1927.

[18] Kong, X.M., Schluter, N. Dahl, W, "Effect of triaxial stress on mixed-mode fracture," Eng. Fract. Mech, 1995, vol.52(2), pp. 379–388.

[19] Khan, Sh.M.A, Khraisheh. M.K, "A new criterion for mixed mode fracture initiation based on the crack tip plastic core region," Int. J. Plast, 2004, vol.1, pp. 55–84.

[20] Himanshu Pathak, Akhilendra Singh, Indra Vir Singh, "Fatigue crack growth simulations of homogeneous and bi-material interfacial cracks using element free Galerkin method," Applied Mathematical Modelling, 2013, pp. 11-30.

[21] Erdogan F, Sih GC, "On the crack extension in plates under plane loading and transverse shear," J Basic Eng, 1963, vol.85, pp. 19–27.

[22] I.V. Singh, B.K. Mishra, S. Bhattacharya, R.U. Patil, "The numerical simulation of fatigue crack growth using extended finite element method," Int. J. Fatigue, 2011, vol.36, pp. 109–119.

[23] BuiManhTuan, Chenyunfei, "Determine stress intensity factors and stress distribution for a surface crack in plates" Journal of Southeast University, 2014, vol.44 (4), pp. 728-734.

[24] Marc Duflot, Hung Nguyen-Dang, "A meshless method with enriched weight functions for fatigue crack growth," Int. J. Numer. Meth. Engng, 2004, vol.59, pp. 1945–1961.

3

Simulation and experimental of oil-water flow with effect of heat transfer in horizontal pipe

Esam Mejbel Abed[*], Zahra'a Aamir Auda

Babylon University- College of Engineering, Mechanical Department, Babylon, Iraq

Email address:

dr.esammejbil@yahoo.com (Esam M. A.), zahraazahraa932@gmail.com (Zahra'a A. A.)

Abstract: There is a strong tendency for two immiscible fluids to arrange themselves so that the low viscosity constituent is in the region of high share. Therefore, it may be possible to introduce a beneficial effect in any flow of a very viscous liquid by introducing amount of a fluid lubricated as liquid-liquid oil-water flow. Two main classes of flows are seen, annular and small bubble in all experimental results. The pressure drop and mean heat-transfer coefficients were observed to depend strongly on the flow patterns. A correlation of the two-phase mean heat-transfer coefficients, based on a simple model of liquid flow, with a Reynolds number based on the actual mean velocity of the liquid mixture two-phase flow, were developed. An experimental rig facility has been designed and constructed, to enable measurements of local parameters in oil-water flow in the developing region of the flow in a 32 mm ID 6 m long pipe. The large discrepancies between model predictions and experimental data are reported in the literature review that the physics of oil-water flow is complex and not yet fully understood. The flow patterns that appear are classified in flow pattern maps as functions of either mixture velocity and water cut or superficial velocities. From these experiments a smaller number of annular flows are selected for studies of velocity and turbulence. The theoretical study was executed using software Fluent program, a modified turbulent diffusion model is presented. Simulation results carried out with the model show more physical predictions with respect to the particle deposition process and concentration profile. The theoretical results represent the pressure gradient distribution, velocity and mean heat transfer coefficient, pressure contours, velocity vectors, streamlines, and also velocity profiles. It was found that the methods with more restrictions (in terms of the applicable range of void fraction, liquid superficial Reynolds number) give better predictions.

Keywords: Two-Phase, Annular Flow, Flow Regime, Heat Transfer, CFD

1. Introduction

Oil–water two-phase flow widely exists in petroleum industry such as crude oil production and transportation through both horizontal and inclined pipes. During the simultaneous flow of two liquids there are several flow patterns. The occurrence of two and three phases flow in pipelines is very common in the petroleum industry.

The density difference between the two liquids has a substantial effect on the flow pattern. Two immiscible liquids of different densities tend to stratify when flowing in a pipeline. Thus, it is more difficult to produce a dispersed flow regime when the density difference is high. For low viscosities, the same flow patterns are observed when oils of different viscosities are used, but transition from one flow regime to another may appear at different superficial velocities. In general, the viscosity has a dual effect on the flow: (1) increasing viscosity can increase instability due to the different velocity profiles at the interface of the two layers, and (2) at the same time it helps to dissipate the energy that causes instability.

Lawrence and Panagiota studied liquid–liquid flows for stratified flow. The experimental data were obtained in a horizontal 14 mm ID acrylic pipe, for test oil and water superficial velocities ranging from 0.02 m/s to 0.51 m/s and from 0.05 m/s to 0.62 m/s, respectively. Using conductance probes, average interface heights were obtained at the pipe center and close to the pipe wall, which revealed a concave interface shape in all cases studied. A correlation between the two heights was developed that was used in the two-fluid model. In addition, from the time series of the probe signal at the pipe center, the average wave amplitude was calculated to be 0.0005 m and was used as an equivalent roughness in the interfacial shear stress model [1].

The effect of viscosity, studied by Russell, et al. and

Charles, et al., seems to have little or no effect on the observed flow patterns for oil-water flows [2],[3]. Glass carried out experiments for 1.2 meter long, 1 cm inner diameter tubing, in which the oil viscosity was varied from (10 to 30,000 cp), the oil gravity from (0.97 to 1.03), the volumetric water fraction from (0.09 to 0.8) and the superficial velocity of oil from (0.06 to 1.28) m/s, The indicated that the less viscous is the oil, the more is the core would tend to break up into globs of oil of various sizes [4]. Finally, Fujii, et al. studied in acrylic pipe 2.5 cm diameter ,the oil viscosity was 61.5 and the density ratio 0.98 .The additional measurements pressure gradient , holdup water were performed .The observed flow patterns were stratified flow ,stratified with mixing and with increasing in the flow rate the viscous core breakup to either large slugs and elongated bubble or spherical bubble [5].

The oil layer loses its continuity and moves as discrete droplets separated by the water continuous-phase. De Salve1, et al. used a wire mesh sensor, based on the measurement of the local instantaneous conductivity of the two-phase mixture has been used to characterize the fluid dynamics of the gas–liquid interface in a horizontal pipe flow. Experiments with a pipe of a nominal diameter of 19.5 mm and total length of 6 m, have been performed with air/water mixtures, at ambient conditions. The flow quality ranges from 0.00016 to 0.22 and the superficial velocities range from 0.1 to 10.5 m/s for air and from 0.02 to 1.7 m/s for water; the flow pattern is stratified, slug/plug and annular. A sensor with an inner diameter of 19.5 mm and a measuring matrix of 16×16 points equally distributed over the cross-section has been chosen for the measurements. From the analysis of the Wire Mesh Sensor digital signals the average and the local void fraction are evaluated and the flow patterns are identified with reference to space, time and flow rate boundary conditions [6].

Guzhov, et al. observed that the forward edge of the oil bodies is bent downward, probably by the action of the clockwise eddies. The water region turbulent energy tried to distribute larger oil droplets along the cross sectional area of the pipe, but the upward buoyancy prevails and a dispersion of oil in water over water layer is developed. With a further increase of the water superficial velocity, the frequency and intensity of water vortices increase, and more and smaller oil droplets are formed. Under these conditions, the dispersed oil in water flow pattern is formed [7].

The pressure drop in production pipelines has large impact on the design of a new field and on the operational costs. The pressure drop limits the maximum flow and is thus a critical parameter, both in terms of cost evaluations and production optimization. Panagiota, et al. studied the pressure drop measurement in oil and water pre wetted pipes for water and kerosene (Exxsol D80 with 801 kg/m3 and density and 1.6 MPa.s viscosity), founding that at the lower velocity the flows are either dispersed or separated with inter-entrainment at the liquid –liquid inter-phase [8].

There are three types of heat transfer modes namely, convection, conduction, and radiation. In pipelines and wellbores, convective heat losses occur between flowing fluids and the pipe wall. In a typical convective heat transfer, a hot surface heats the surrounding fluid, which is then carried away by fluid movement. As mentioned earlier, many separate studies have been carried out to predict convective heat transfer for pipe flow in two-phase flow. Zimmerman, et al. studied experimentally air–water flow with heat transfer in a 25 mm internal diameter horizontal pipe. The water superficial velocity varied from 24.2 m/s to 41.5 m/s and the air superficial velocity varied from 0.02 m/s to 0.09 m/s. The aim of the study was to determine the heat transfer coefficient and its connection to flow pattern and liquid film thickness. The flow patterns were visualized using a high speed video camera, and the film thickness was measured by the conductive tomography technique. The heat transfer coefficient was calculated from the temperature measurements using the infrared thermography method. It was found that the heat transfer coefficient at the bottom of the pipe is up to three times higher than that at the top, and becomes more uniform around the pipe for higher air flow-rates [9]. Hetsroni, et al. performed experiments to study the flow regimes and heat transfer in air–water flow in 8 θ inclined tubes of inner diameter of 49.2 and 25 mm. The flow regimes were investigated by using high-speed video technique and conductive tomography. The thermal patterns on the heated wall and local heat transfer coefficients were obtained by infrared thermograph. Under the conditions studied, disturbance waves of different forms were observed. The analysis of the behavior of the heat transfer coefficients, together with flow visualization and conductive tomography sowed that dry out took place in the open annular flow regimes with motionless or slowly moving droplets [10].

Roula and Dash studied Pressure drop through sudden contraction in small circular pipes have been numerically investigated, using air and water as the working fluids at room temperature and near atmospheric pressure. Two-phase computational fluid dynamics (CFD) calculations, using Eulerian–Eulerian model with the air phase being compressible, are employed to calculate the pressure drop across sudden contraction. The pressure drop is determined by extrapolating the computed pressure profiles upstream and downstream of the contraction. The larger and smaller tube diameters are 1.6 mm and 0.84 mm, respectively. Computations have been performed with single-phase water and air, and two-phase mixtures in a range of Reynolds number (considering all-liquid flow) from 1000 to 12000 and flow quality from $1.9 * 10^{-3}$ to $1.6 * 10^{-2}$. The contraction loss coefficients are found to be different for single-phase flow of air and water. The numerical results are validated against experimental data from the literature and are found to be in good agreement. Based on the numerical results as well as experimental data, a correlation is developed for two-phase flow, the pressure drop caused by the flow contraction [11].

Al-yaari and Abo-Sharkh investigated numerically, using commercial CFD package FLUENT 6.2 in horizontal pipe (0.0254), Oil-water stratified flow regime is simulated using Volume of Fluid (VOF) multiphase flow approach. RNG k–ε turbulence model is adopted. Mesh independent study has been achieved to decide on the mesh size. The phase separation is investigated for the tested stratified flow points. CFD Numerical simulation predicted the stratified flow pattern and smoothness and the type of the interface. On the other hand, while the oil layer was clearly predicated by the CFD model, water layer was not clearly predicted as a clear segregated layer [12].

2. Experimental Apparatus and Procedure

The analysis performed on experimental laboratory data; provides the main source of information about specific two-phase flow regimes. This research presents a detailed description of the experimental rig used to study the oil-water annular flow with heat transfer in horizontal pipe. A liquid-liquid flow facility has been built to study phenomenon of two- phase flow with heat transfer in pipe in the experimental rig. The experimental rig is build up within the fluid laboratories of the Mechanical Engineering Department at University of Babylon. The physical properties of the system, the criteria of experimental design and the operational principle of each instrument are explained. The experimental facility consists of a main pipe flow test section made from 32mm inner diameter and 6m length. A 2m Perspex manufactured from methyl methacrylate monomers. A 2m pvc, Polyvinyl chloride (C4H8) is the most widely used of any of thermoplastics, the specification of pvc is (20-630)mm OD (6)m length (2-30)mm thickness, and 2m copper pipe of 32mm diameter with a circular cross section area. The test section is a 0.6m long at a distance of 4m from entry. For imaging the flow of clear use lighting system consists of a two fluorescent lamp number two in beside camera. The system is equipped with a diffusive white surface in front of the lamp for greater light uniformity. The rigid steel frame supporting the test pipe section is constructed to fix the pipes with no vibration as shown in Fig.(1a,b). The experiments are carried out at ambient laboratory conditions of approximately 25℃ temperature and pressure of 1bar. The experimental work includes two flow case liquid-liquid with and without heat transfer effect. The physical properties of the fluids used in the experiments are as shown in Table (1). The experiments two-phase liquid-liquid horizontal flow system are explained to investigation annular liquid-liquid flow through flow with heat(36 EXP.).

Table (1). *Fluid properties*

Product name	Oil	Water
Density	820 Kg/m³	1000 Kg/m³
Viscosity	1.52 *10⁻³ Kg/s.m	0.89*10⁻³ kg/s.m
Surface Tension	27.6 mN/m at 25℃	71.99 N/m
Oil-Water interfacial tension	44.69 m.N/m at 25℃	44.69 N/m

3. Experiments Limitation

A flow can be laminar in the beginning for the passage of

water only and then the flow convert to the turbulent case at passage the oil fluid. It is found that a flow in a pipe is laminar if the Reynolds number based on diameter of the pipe is less than 2000 and is turbulent if it is greater than 4000. Transitional flow prevails between these two limits. Reynolds number for turbulent flow is [13].

$$\text{Re} = (\rho_m . U_m . D)/\mu_m \qquad (1)$$

The entrance length required for fully developed velocity profile to form laminar and turbulent flow respectively:

$$\text{For laminar flow Le} = 0.06D \text{ Re} \qquad (2)$$

$$\text{For turbulent flow Le} = 4.4 \text{ D Re}^{1/6} \qquad (3)$$

For present work and according to the pipe diameter, higher velocity of the liquid-liquid phase, the entrance length is 0.8723 m as equation (3). Therefore, the design of the experimental set-up carried out for pipe according to the maximum entrance length. The most physically based explanation is that:

Superficial Velocities and flow rates, the flow rate and superficial velocity of the fluid flow was measured as follows. The liquid flow rate is read directly from the float flow meter in (l/min), while the superficial velocity determined from the following equation as:

$$Q = Us * A \qquad (4)$$

When A is the cross section area (the diameter of the entrance pipe for water equal to (Dw= 11.2 mm) and the entrance pipe for oil equal to (Do=23.2 mm)) and Us is the mixture velocity equal to the summation of the water and oil velocity.

The inlet water cut is defined by dividing the flow rate of water, Q_w, by the sum of flow rate of water and oil, Q_w+Q_g, as:

$$\lambda = \frac{w}{Qw+Qo} \qquad (5)$$

The mean heat-transfer coefficients (h_{TP}) for the two-phase flow were calculated experimentally as reported by reference [14]:

$$hTPEXP = \frac{1}{L}\int h \, dZ = \frac{1}{L}\sum_{K=1}^{NST} hK \, \Delta ZK \qquad (6)$$

$$\text{Where } Q = h \, A \, \Delta T \text{ ,then} \qquad (7)$$

$$hTPEXP = \frac{1}{L}\sum_{K=1}^{NST}\left(\frac{Q}{A \, (Tw-Tb)}\right)\Delta ZK \qquad (8)$$

Where h is the local mean heat transfer coefficient, L is the length of the test section, k is the index of the thermocouple stations, NST is the number of the thermocouple station, and Δz is the distance between each thermocouple and equal to 1 m. Q is heat flux supply. A is the cross section area. Tb is the bulk temperature in each thermocouple station, which equal the temperature recorded by temperature recorder with SD card data logger with time. Tw is the wall temperature of test section.

The wall shear stress, the mean shear stress at the wall for Newtonian and non-Newtonian fluids and for all flow regimes is given by reference [15]:

$$\tau_w = \frac{D}{4} \frac{\Delta p}{\Delta x} \tag{9}$$

4. Simulation Work

Two-phase oil-liquid annular flow is simulated in this work using the commercial CFD code FLUENT 6.3.26 in order to solve the governing equations. The problem considers the transient tracking of a oil-liquid interface. The working fluids are water and oil, and in order to develop an applicable analysis of the flow field on fluent 6.3.26 inclusively in ANSYS CFD.

4.1. Mesh Generation

The partial differential equations of fluid flow are not usually amenable to analytical solutions, except for very simple cases. Therefore, in order to analyze fluid flows, flow domains are split into smaller sub domains called elements or cells and the collection of all elements is known as mesh or grid. The governing equations are solved inside each of these portions of the domain. Care must be taken to ensure proper continuity of solution across the common interfaces between two sub domains, so that the approximate solutions inside various portions can be put together to give a complete picture of fluid flow in the entire domain. Grid generation is often considered as the most important and most time consuming part of CFD simulation as shown in fig. 2.

4.2. Governing Equations

The fundamental governing equations of fluid dynamics in the theoretical work are continuity and momentum Equations in three dimensional.

4.2.1. Conservation of Mass

$$\frac{\partial}{\partial t} \alpha_q \rho_q + \nabla . \alpha_q \rho_q \vec{u}_q = \sum_{p=1}^{n} \dot{m}_{pq} \tag{10}$$

4.2.2. Conservation of Momentum

$$\frac{\partial}{\partial t} \alpha_q \rho_q \vec{u}_q + \nabla.(\alpha_q \rho_q \vec{u}_q \otimes \vec{u}_q) = -\alpha_q \nabla p + \nabla . \alpha_q \bar{\bar{\tau}}_q + \alpha_q \rho_q \vec{F}_q + \sum_{p=1}^{n}(R_{pq} + \dot{m}_{pq} \vec{u}_{pq}) \tag{11}$$

4.2.3. Conservation of Energy

$$\frac{\partial}{\partial t}(\alpha_q \rho_q h_q) + \nabla . (\alpha_q \rho_q \vec{u}_q h_q) = - \alpha_q \frac{dpq}{dt} + \bar{\tau}_k : \nabla \vec{u}_q - \nabla . \vec{q}_q + s_q + \sum_{p=1}^{n}(Q_{pq} + \dot{m}_{pq} h_{pq)} \tag{12}$$

4.3. Boundary Condition Application

The boundary domain in the present problem is dependent on flow variables at the domain boundaries Specify fluxes of mass, momentum, energy, etc. into the domain. Defining boundary conditions involves: Identifying the location of the boundaries (e.g., inlets, walls, symmetry). Also, the

turbulence kinetic energy k and its dissipation rate ε initial guess those are estimated with the following equations according to Launder and Spalding [16]:

$$K = 3/2 \, I^2 \, Uin2 \tag{13}$$

$$\varepsilon = 2K_{in}^{3/2}/d \tag{14}$$

The turbulence intensity for fully developed pipe flow is:

$$I = 0.16 Re^{-1/8} \tag{15}$$

5. Results and Discussion

The experimental and theoretical work results are illustrated in the present section, which include the results obtained for pressure reading, inlet water cut, shear stress, viscosity and the effect of heat on the pressure as well as calculate the mean heat transfer coefficient with applied three thermal loads by using thermal heaters "eight finger shape" with a capacity of 2000 watt per heater for oil-water for different flow rates is presented.

5.1. Experimental Result

5.1.1. Flow Visualization

The first and simplest approach to study two-phase flow behavior in deviated pipes is to visualize the flow. Flow patterns play very important roles in two-phase flow to explain the phenomena of two-phase flow. Each regime has certain hydrodynamic characteristics, occurrence in nature and many applications in industries. It is clarify the form of flow with the effect of the heat, Fig. (3) shows an instantaneous side view of oil-water flow into the pipe, obtained by high speed video camera (AOS imaging studio v3). The flow is from the left to the right, the distance in the flow direction, shown in this image is L = 60 mm, to show the effect of heat transfer on the flow fluids applying power of the amount of 8000 watt and the rate of flow of the oil ranges from 5 to 10 liters per minute and the rate of flow of water ranging from 5 to 15 liters per minute by using the video system. Noting change the type of flow-type annular to type small bubbles due to decrease viscosity, which decreased with increasing temperature as shown in Fig.(3) bubbles of oil mediates the flow of water.

5.1.2. Pressure Gradient

The pressure drop in oil-water flow is dependent on flow pattern conditions. After applying the thermal heater loading with various value of heat flux from (4000) watt to (12000) watt for three values of water flow rate and four values of oil flow rate. Applying the second thermal load with the heat flux equal 8000 watt. Fig.(4) represents the relation between the pressure gradient and oil superficial velocity for various water superficial velocity. The pressure gradient increases with increasing the mixture velocity. Fig.(5),(6) and (7) show the pressure versus tap locations (x1, x2, x3,x4 and x5). The pressure decreasing gradually until reach minimum value at the end of the pipe. Fig.(8) represents the pressure gradient fluctuation along the pipe with heat transfer effect, for oil flow rate equal 40 l/min and water flow rate equal 10,15 and 20 l/min. It is show that the pressure gradients reading fluctuate with time due to two-phase effect. The pressure sensor recorded the pressure with time for five taps located along the pipe, and the pressure increased-decreased with time.

5.1.3. Mean Heat Transfer Coefficient

Fig.(9),(10) and (11) show variation of the overall mean heat transfer coefficient with the mixture Reynolds number. The mean heat transfer coefficient increase with increasing the mixture Reynolds number at the same power load . So, it's increased until reach maximum value when Reynolds number equal to 65125 for water superficial equal 20 l/min and power 12000 watt. It's noted that the mean heat transfer coefficient increased with increased the power until reach maximum value at 80.234 kw/m^2.k when power equal 12000 watt.

Fig.(12) shows the bulk temperature readings along section for thermocouple stations. It's noted that the bulk temperature increase with increasing distance along the pipe until reach maximum value at the end of the pipe, for the same mixture flow rate. Also increase with increasing mixture flow rate for the same position.

Fig.(13) describes the relationship between the overall mean heat transfer coefficient values in all thermocouples station versus oil superficial velocity, These were calculated through every run for water superficial velocity from 1.69 to 3.38 m/s, and oil superficial velocity from 0.39 to 1.58 m/s. The mean heat transfer coefficient increases with increased oil due to decreasing the bulk and wall temperature as a result of increased oil velocity and thermal insulator.

5.1.4. Effect Inlet Water Cut λ

Average inlet water cut at least gradually from 0.2 to 0.5 when the speed of the water equal to 1.69 m/s., also from 0.2727 to 0.6 at the speed of water equal 2.54 m/s finally from 0.3333to 0.6666 at the speed of water equal 3.38 m/s. Fig.(14) shows the water cut profiles measured for the experimental matrix conditions.

As can be seen, the water tends to settle at the bottom of the pipe, resulting a small difference in the measured water cut. At inlet water cut indicating slow settling of water droplets. The fact that the water cut is not zero at the oil rich layer at any data set indicates that the flow in the developing region.

5.1.5. Shear Stress (τ_w)

Fig.(15) presents the wall shear stress versus mixture Reynolds number for different mixture superficial velocity with variable heating load (4000, 8000 and 120000) watt respectively, it's noted the wall shear stress increases with increasing the power applied and increasing the mixture velocity.

5.1.6. Viscosity μ

The quantity of " μ" is called the viscosity, which is a property of the fluid. It is an indication of how much internal friction is present. Some fluids, such oils, have high viscosity, and a substantial applied stress is required to cause these fluids to flow. Other fluids, such as water, have lower viscosity and flow more easily for the same applied stress. In general, liquid viscosity decreases exponentially with increasing temperature, but gas viscosity increases with temperature. It's often varies considerably with temperature, and that effect must be considered in calculations. Fig.(16) shows the dynamic viscosity versus different temperature that resulted from applied three heating loads for different water velocity a fixed oil velocity.

5.2. Simulation Results

The brief literature review presented at the beginning of the chapter two reveals that additional work is required for establishing computational procedures leading to reliable predictions of oil-water annular flow. Available oil-water annular pipe flow experimental data are used to test the simulations obtained by FLUENT to study distribution phase, pressure, speed and shear stress. This shows that the Fluent depends on the mixture velocity and void fraction. These trials were used to match precisely with the terms of the annular flow that has been studied, which was extracted in practice.

Fig.(17) represents the mesh that has been applied to the geometry and the number of cells, 5000.

Fig.(18) displays the small bubble that obtained after practice the condition and complete the iteration which lasted for more than 24 hours after the reduction of the system to reduce the iteration required and the time needed.

5.2.1. Mean Heat Transfer Coefficient and Bulk Temperature

Fig.(19) shows the heat transfer coefficient when the amount of power load 8000 watts. Note the increased heat transfer coefficient gradually due to increasing temperatures over time and this increase also produces a proven rate of ability and proven a flow rate of water used a 20 l/m with the gradient of the values of the rate of oil, which ranges from 10 to 40 l/m so the oil is heated at a higher rate of water and thus increases coefficient heat Transfer.

Fig.(20) represents the bulk temperature change versus oil superficial velocity, it's be the highest value when the flow rate of the oil equal to 20 l/m and then begin a downward as result of temperature stability to refer again overpaid when flow rate of oil increases to 40 l/m.

6. Comparison Experimental and Theoretical Results

Comparing the results is important to determine the percentage of error between them and the reasons for the difference between the two results. The comparison also shows the existence of the causes leading to the inaccuracy of the practical results, including leaking pipes, flow meter or oil rate may be re-used when each experiment contains a proportion of the water. This comparison between the theoretical and practical results will be interpreted in the following section.

6.1. Heat Transfer Effect

Comparisons are presented on mean heat transfer coefficient and bulk temperature. Fig.(21) presents comparisons between the model predictions and the measured data for the effect of oil superficial velocity on the mean heat transfer coefficient along the pipe at power load equal to 8000 watt and water superficial velocity equal to 3.38 m/s. It can be observed that the model predictions follow trend of the data fairly well and the theoretical mean heat transfer effect coefficient has similar behavior as the experiments. Fig.(22) illustrates the bulk temperature versus oil superficial velocity, it can be observed that bulk temperature changed continuously with dependent on oil superficial velocity until reach maximum to (24.91) °C at Uo equal to (0.79)m/s.

The percentage theoretical heat transfer coefficient decreases over experimental heat transfer coefficient are 6%.

7. Conclusion

1. The work reveals that the annular flow regime exists over a wider range of phase flow rates. As a result, regime maps and transition equations available for gas-liquid cases cannot be used as such to predict the patterns in liquid –liquid flows.
2. Pressure drop along the pipe was direct proportional for changes in oil-water superficial velocities.
3. Mean heat transfer coefficient increases with increasing heat flux that used in experimental work and its decreases with increasing in mixture superficial velocity. Finally, it's increased with wall temperature.
4. Void fraction has reverse behavior to that of hold up to wards changes in liquid superficial velocities and towards other flow characteristics.
5. CFD calculations using Fluent 6.3.26 were performed to predict the oil-water annular flow.
6. A model for the calculation of fully-developed, turbulent-turbulent oil-water annular flow in horizontal pipe is presented. The model is based on a numerical solution of the basic governing differential equations using a finite-volume method in a bipolar coordinate system, applying a simple mixing-length turbulence model. The moving wall assumption was implemented for the prediction of the interface behavior.
7. Volume of fluid (VOF) multiphase model with RNG-k-ε two equations turbulent model was selected among other different multiphase and turbulent models based on the convergence, prediction of the oil-water annular flow pattern and the smoothness of the interface.
8. Care should be taken while initializing the CFD solver to obtain convergence.
9. Mesh independent study has been achieved to decide on the optimum mesh size to be used in the simulation process.
10. The volume fraction value specified the phase inversion point which determined which of the two liquids dispersed in the other.
11. In this study water dispersed in oil for volume fractions less than 0.6 which represents the inversion volume fraction at which the oil began to dispersed in water.
12. All simulations give good agreement with the expected flow regime annular.
13. The numerical model solves the resulting set of algebraic equations in an iterative way, simultaneously for both oil and water layers. The pressure gradient is calculated based on the condition that the velocity field in both layers must satisfy the total flow rate.

Fig. 1a. *The experimental rig.*

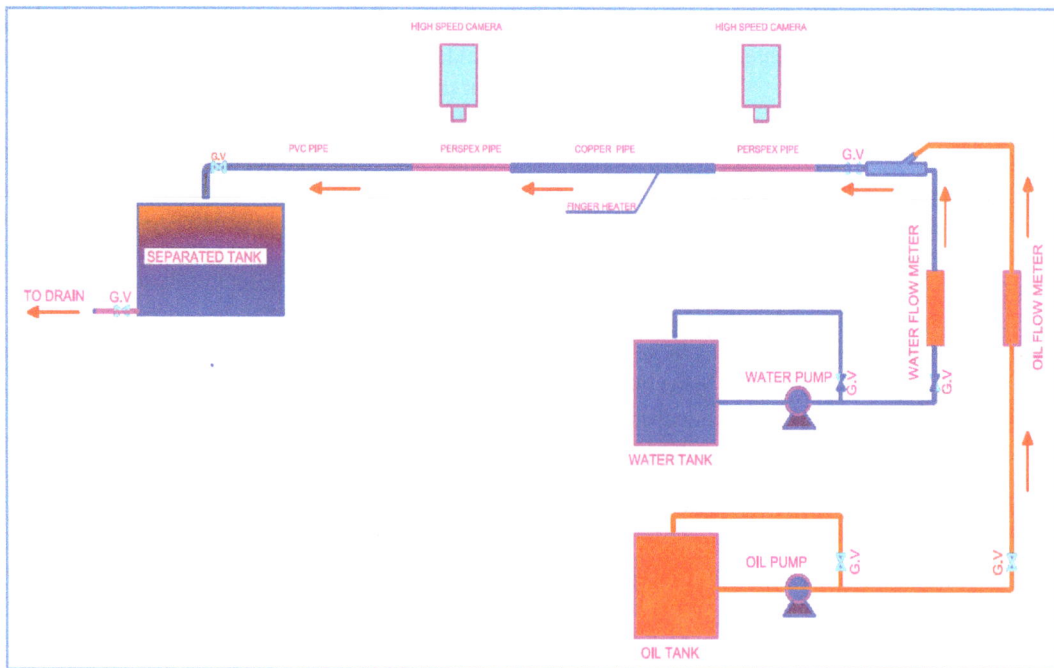

Fig. 1b. *The schematic of the experimental rig*.

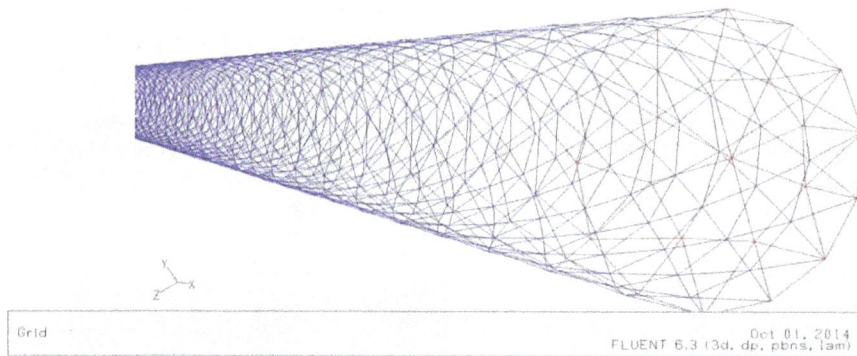

Fig. 2. *the approach mesh used to create the pipe geometry*.

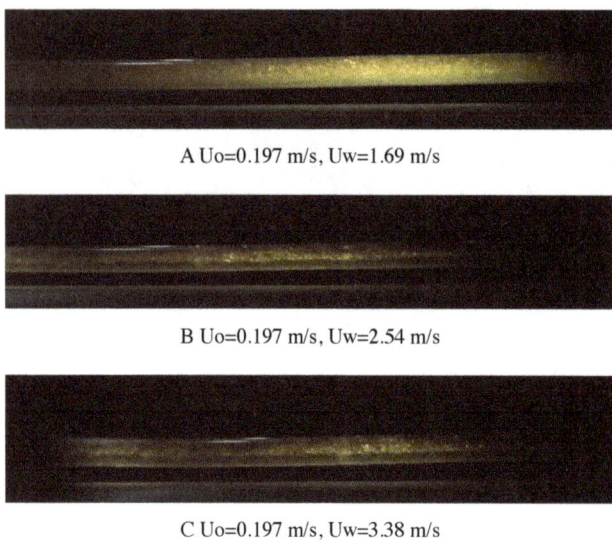

A Uo=0.197 m/s, Uw=1.69 m/s

B Uo=0.197 m/s, Uw=2.54 m/s

C Uo=0.197 m/s, Uw=3.38 m/s

Fig. 3. *Flow visualization at Qo=5 l/m and Qw=10,15,20 l/m at power load equal to 8000 watt*.

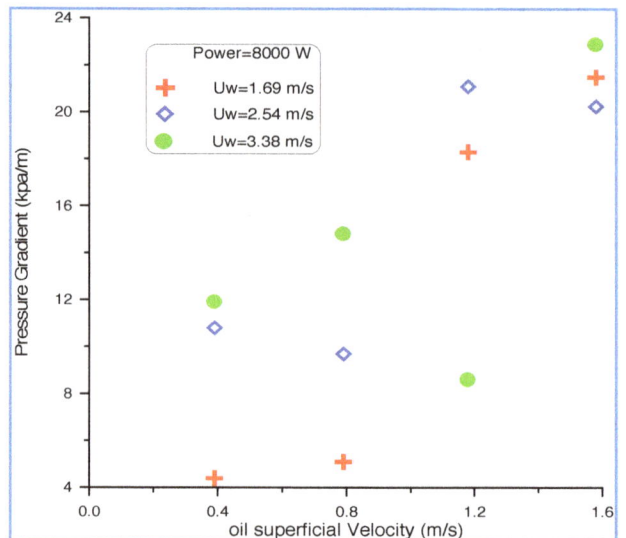

Fig. 4. *shows the relation between the pressure gradient and oil velocity at power =8000 watt*.

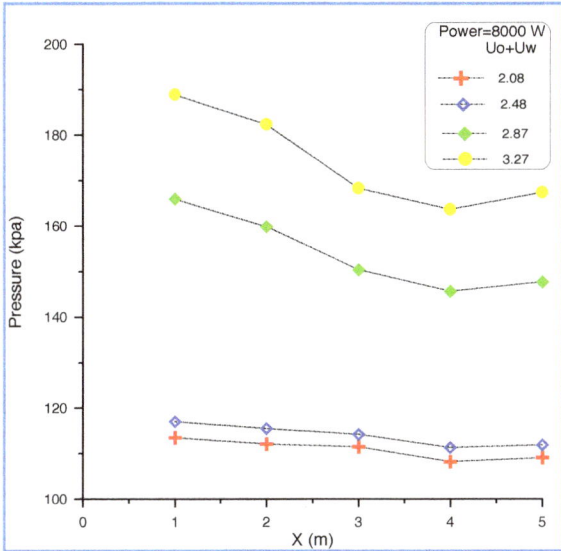

Fig. 5. *Pressure reading of the different sensors along the pipe.*

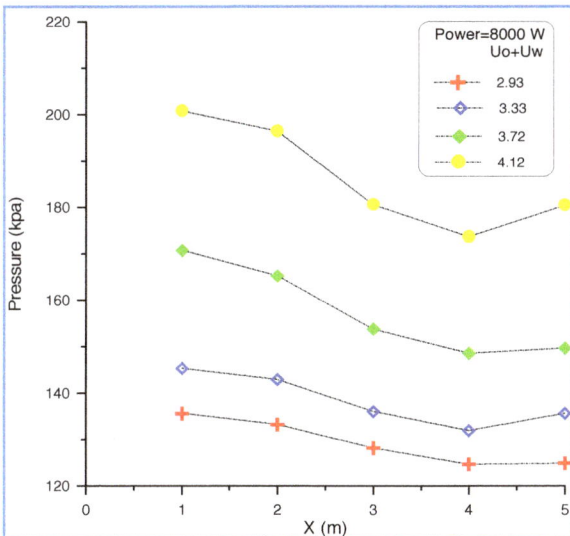

Fig. 6 *Pressure reading of the different sensors along the pipe.*

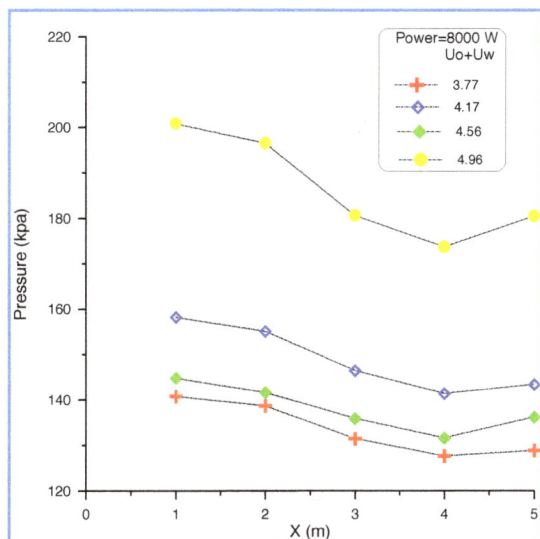

Fig. 7. *Pressure reading of the different sensors along the pipe from.*

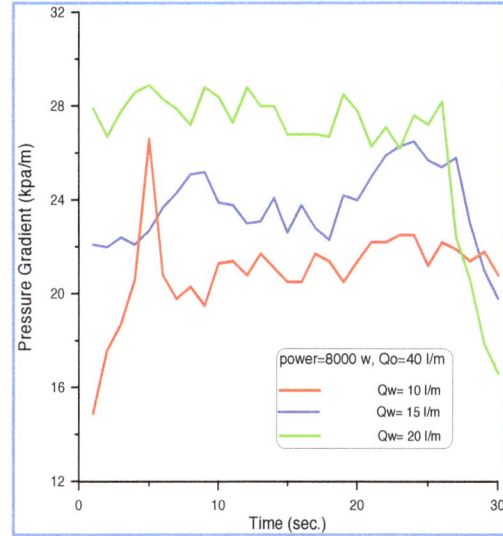

Fig. 8. *Effect of Time evolution of Pressure obtained by Experimental.*

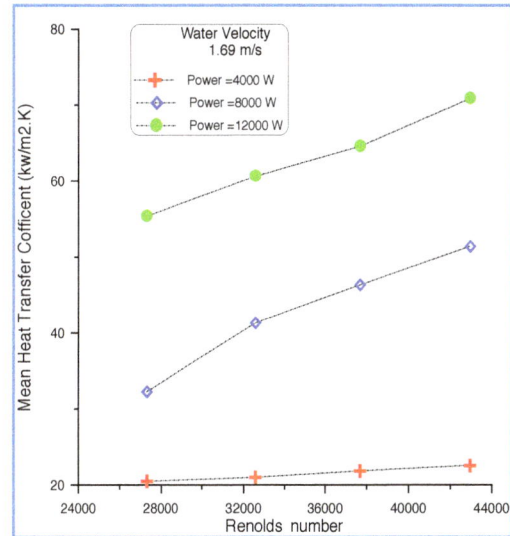

Fig. 9. *Variation of the overall mean heat transfer coefficient with the mixture Reynolds number for various power.*

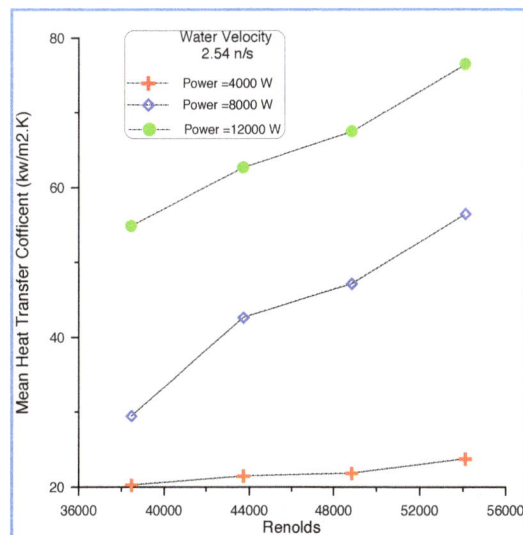

Fig. 10. *Variation of the overall mean heat transfer coefficient with the mixture Reynolds number for various power.*

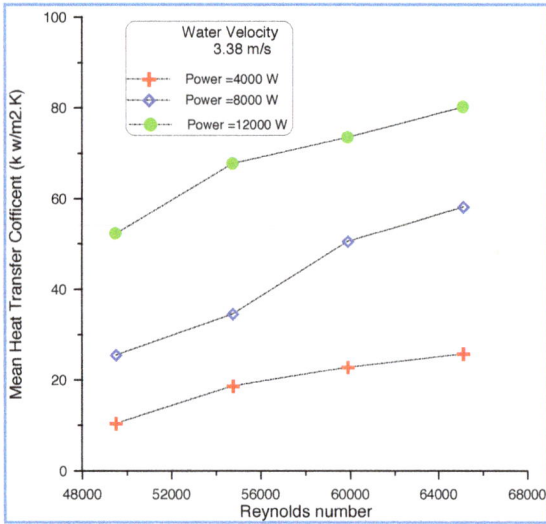

Fig. 11. *Variation of the overall mean heat transfer coefficient with the mixture Reynolds number for various power.*

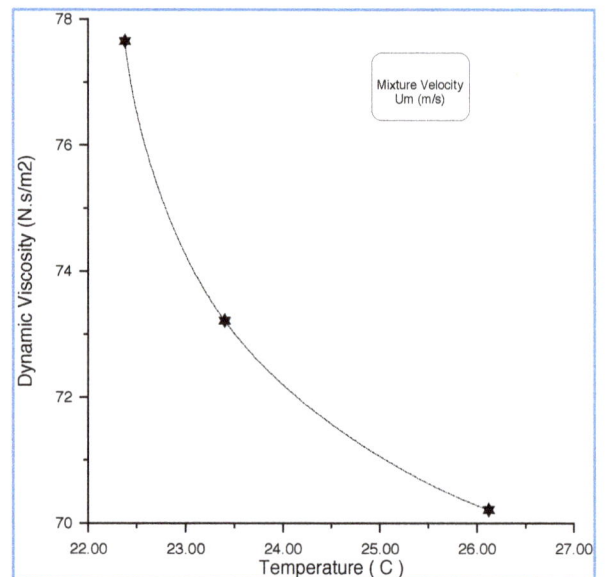

Fig. 12. *Bulk Temperature distribution along the pipe.*

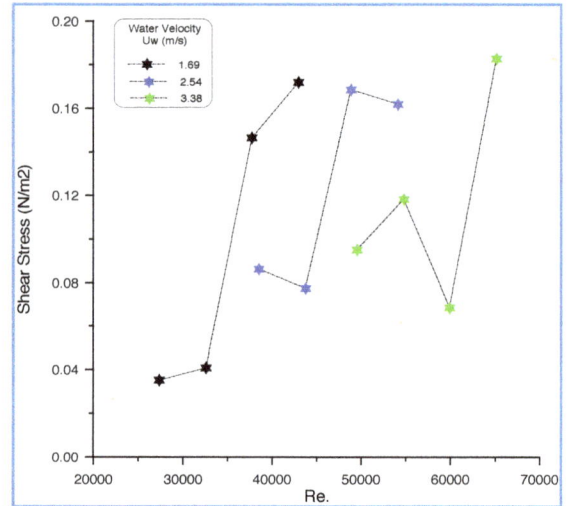

Fig. 13. *Effect the mean heat transfer coefficient with oil-water two-phase flow.*

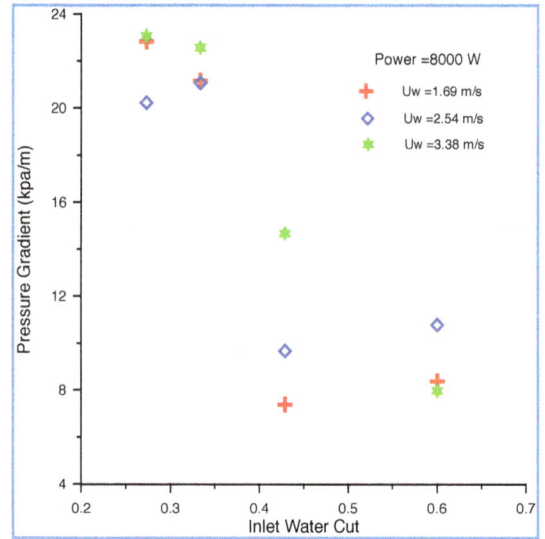

Fig. 14. *Effect inlet water cut on the pressure gradient*

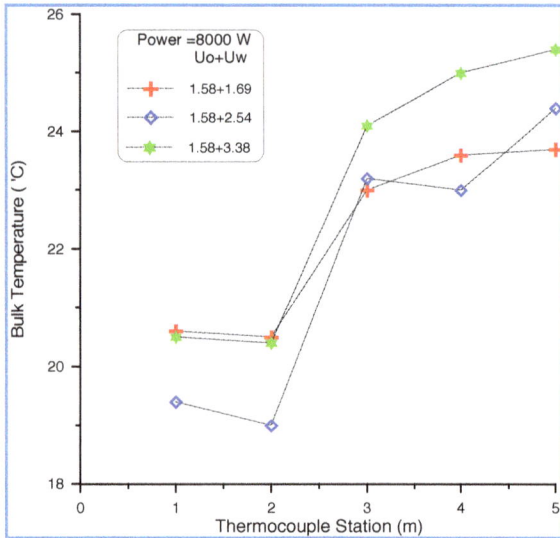

Fig. 15. *Effect of Reynolds number on shear stress at wall for different water superficial velocity at power 8000 watt.*

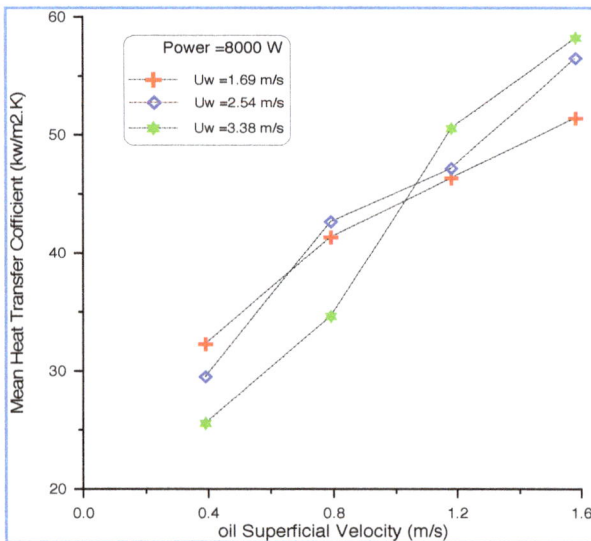

Fig. 16. *Effect the temperature on dynamic viscosity.*

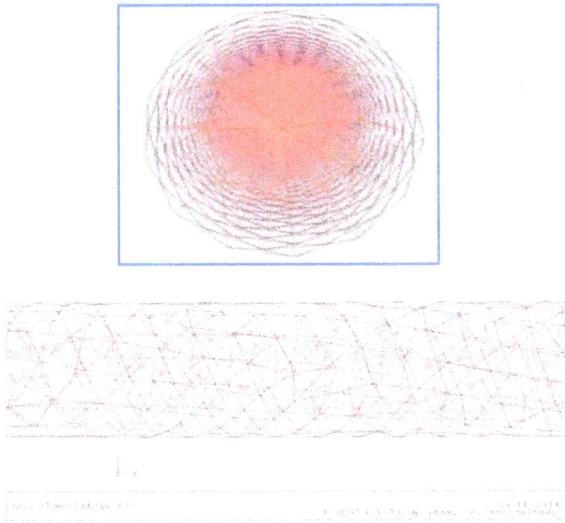

Fig. 17. Represent the pipe geometry of computational flow domain.

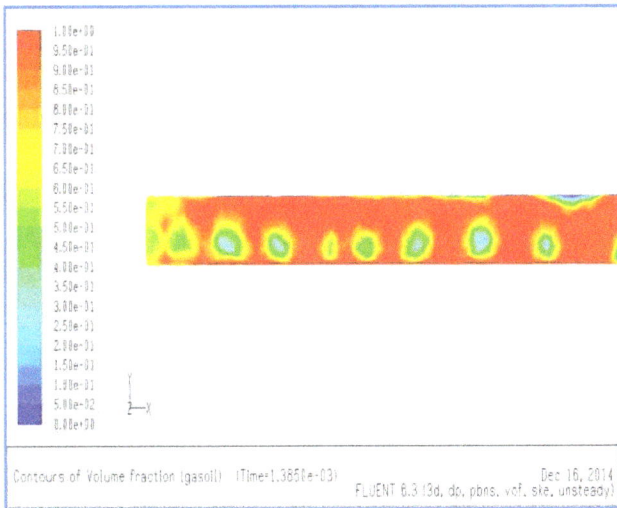

Fig. 20. Bulk Temperature versus oil superficial velocity.

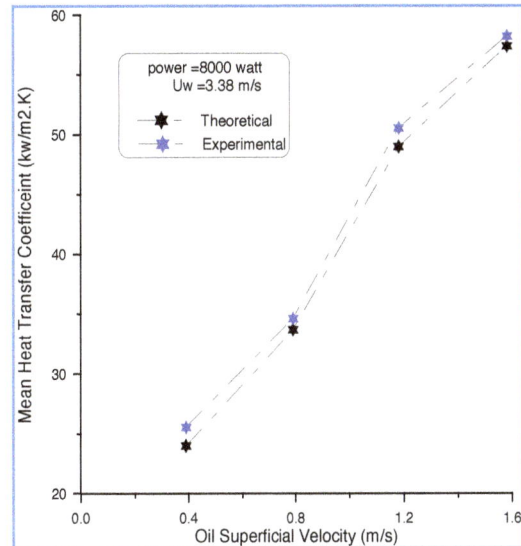

Fig. 18. Flow distribution V.F 0.6, Um 4 m/s.

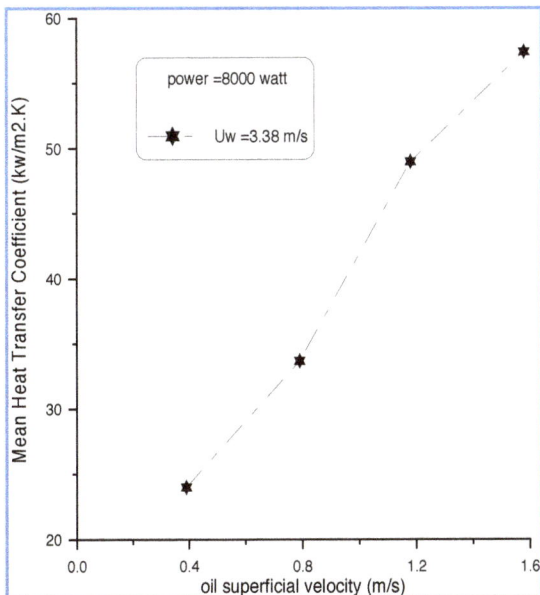

Fig. 21. Experimental and theoretical Mean Heat Transfer Coefficient comparison for oil-water flow at Uw=3.38 m/s.

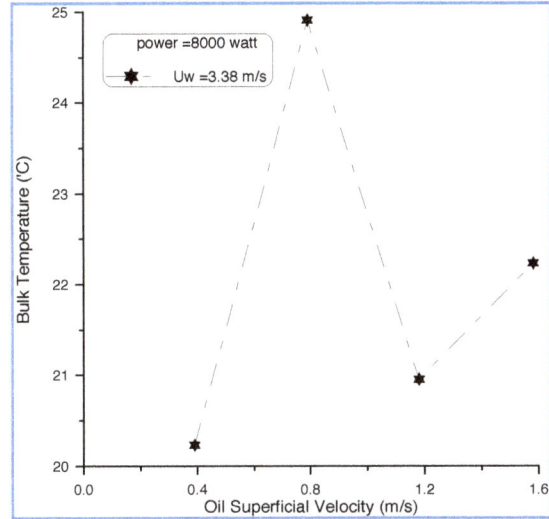

Fig. 19. Mean Heat Transfer Coefficient versus oil superficial velocity at power equal to 8000 watt.

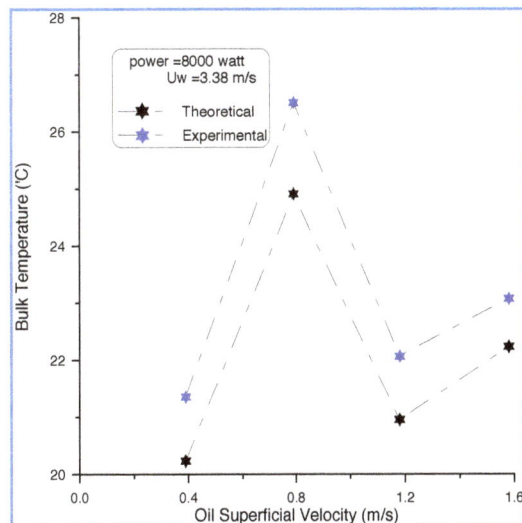

Fig. 22. Experimental and theoretical Bulk Temperature comparison for oil-water flow at Uw=3.38 m/s.

Nomenclature

ΔP	Pressure Gradient. (pa/m^2) A		Cross section area. (m)
D	Diameter. (m)	Do	Oil Diameter. (m)
Dw	Water Diameter. (m)	HL	Holdup. (---)
L	Length. (m)	Le	Entrance Length. (m)
Qo	Oil Flow Rate. (m^3/s)	Qw	Water Flow Rate. (m^3/s)
Re	Reynolds no. (---)	Us	Superficial Velocity. (m)
L1	Lighter liquid.	L2	heavier liquid
CFD	Computational Fluid Dynamic. QCV		Quick-closing valve.
VOF	Volume of Fluid.		RGB:- Red, Green, and Blue Image.
K	Turbulence Kinetic Energy		
σ_k, σ_ε	Constants in the $k - \varepsilon$ model.		
ε	Dissipation Rate.	λ	Input water cut.
ρ	Density. (Kg/m^3)	α	Input volume fraction.
μ	Viscosity. (kg/m.s)	I	Turbulence Intensity.

References

[1] Lawrence C. Edomwonyi-Otu, Panagiota Angeli∗, " *Pressure drop and holdup predictions in horizontaloil–water flows for curved and wavy interfaces*", chemical engineering research and design, Cherd-1615, 2014.

[2] Russell, T.W.R. Hodgsen, G.W. and Govier, G.W., "*Horizontal Pipeline Flow of Mixtures of Oil And Water,*" Can. Jornal Chem. Eng., Vol.37, pp.(9-17), 1959.

[3] Charles, M.E. and Redberger, P.J., "*The Reduction of Pressure Gradients in Oil Pipelines by the addition of Water; Numerical Analysis of Stratified Flow*" Chem. Eng., Vol.40, pp.(70-75), 1962.

[4] Glass, W., "*Water Addition Aids Pumping Viscous Oils*". Chem. Eng. Prog., Vol.57, pp.(116-118).

[5] Fujii, T., Otha, J., Nakazawa, T., and Morimoto, O., "*The Behavior of an Immiscible Equal-Density Liquid-Liquid Two-Phase Flow in a Horizontal Tube*", JSME Journal Series B,

Fluids and Thermal Engineering, Vol.30, No.1, PP.(22-29), 1994.

[6] M De Salve1, G Monni1 and B Panella1, "*Horizontal Air-Water Flow Analysis with Wire Mesh Sensor*", Politecnico di Torino, Energy Department, C.so Duca degli Abruzzi 24, 10129 Torino, IT,pp.(177-182),2012.

[7] Guzhov, A., Grishin, A.D., Medredev, V.F. and Medredeva, O.P.: "*Emulsion Formation during the Flow of Two Immiscible Liquids*" Neft. Choz., Vol.8, pp.(58-61), 1973.

[8] Panagiota Angeli, Geoffrey F. Hewitt, "*Pressure Drop Measurements in Oil and Water Prewetted Pipes*", Department of Chemical and Biochemical Engineering, University College London, London, UK, 2011.

[9] R. Zimmerman, M. Gurevich, A. Mosyak, R. Rozenblit, G. Hetsroni," *Heat transfer to air–water annular flow in a horizontal pipe*", International Journal of Multiphase Flow, Vol.32, pp.(1-19), 2006.

[10] G. Hetsroni∗, R. Rozenblit," *Thermal patterns on a heated wall in vertical air-water flow*", International Journal of Multiphase Flow, Vol.26, pp.(147-167), 2000.

[11] Manmatha K. Roula, Sukanta K. Dash." *Flow of Single-Phase Water and Two-Phase Air-Water Mixtures through Sudden Contractions in Mini Channels*", International Journal of Engineering Research and Applications (IJERA) ISSN, Vol. 2, Issue 5, pp.757-763,September- October 2012.

[12] Mohammed A. Al-Yaari∗ and Basel F. Abu-Sharkh," *CFD Prediction of Stratified Oil-Water Flow in a Horizontal Pipe*", Asian Transactions on Engineering (ATE), Vol.1, pp. (2221-4267), 2011.

[13] Holman, J.P., Gajda, W.J. "*Experimental Methods for Engineers*", McGraw-Hill Book Company, Fourth Edition, 1984.

[14] Ghajar, A. J., Tang, C.C, 2007, "*Heat Transfer Measurements, Flow Pattern Maps, and Flow Visualization for Non-Boiling Two-Phase Flow in Horizontal and Slightly Inclined Pipe*", J. Heat Transfer Engineering, Vol. 28, No. 6.

[15] Sylvester, N.D., Dowling, R.H., Mino, H.P., and Brill, J.P., " *Drag Reduction in Two Phase Gas-Liquid Flow*", Paper No. L11477e, University of Tulsa, January, 1997.

[16] Launder, B. and Spalding, D. "*The Numerical Computation of Turbulent Flows*", Computer Methods in Applied Mechanics and Engineering, Vol. 3, pp. (269-289), 1974.

4

Experimental study by visualisation of behavioural properties of vortex structures on the upper surface of an ogive of revolution ß 52°

Abderrahmane Abene

Université de Valenciennes, Laboratoire d'Aérodynamique, d'Energétique et de l'Environnement, I. S. T. V, Valenciennes, 59300 Aulnoy lez valenciennes France

Email address:

a.abene@yahoo.fr

Abstract: A large number of studies of flow visualisations, developed on the upper surface of delta or gothic wings and on that of ogives of revolution, have been carried out in the wind tunnel of the Valenciennes University aerodynamics and hydrodynamics laboratory (LAH). These studies have provided a better understanding of the development and the positioning of vortex structures and have enabled, in particular, the preferential nature of intervortex angles, thereby defined, to be determined on a wide range of Reynolds. This paper concerns in particular the study by visualisations of the behavioural properties on the upper surface of an ogive of revolution having an apex angle of 68.6° at a low angle of attack and conducted at variable speeds. It has been noted that variations in speed have no influence at all on the behavioural properties of the development of vortex structures whereas, by contrast, changes to the angles of incidence do indeed strongly influence that development. The study of the ascent of the vortex breakdown at high angles of attack has revealed original behavioural properties which find expression notably in the discontinuous evolution, in terms of the apex angle, of those angles of attack which define the beginning and the end of the ascent of this vortex breakdown. These properties undoubtedly reflect those already observed in similar studies carried out on delta and gothic wings and on cones. However, no current theory seems to be able to provide a straightforward explanation of these phenomena.

Keywords: Vortex, Visualization, Structures, Reynolds, Speed

1. Introduction Reynolds

The concept of a preferential angle was introduced for the first time in 1972 by M. LE RAY and his colleagues and stems from their studies of liquid helium [1, 2]. These angles, which have the following relatively simple analytic formula :

$$\operatorname{Cos} \theta_{\ell m} = \frac{m}{\sqrt{\ell(\ell+1)}} \quad \text{where } \ell \geq m \qquad (1)$$

can be classified by their most frequent values into two groups :

1st group :

($\ell = m$) : $\theta_{11} = 45°$; $\theta_{22} = 35.3°$, $\theta_{33} = 30°$, $\theta_{44} = 26.6°$, $\theta_{55} = 24.1°$, $\theta_{66} = 22.2°$, $\theta_{77} = 20.7°$, $\theta_{88} = 19.4°$, $\theta_{99} = 18.4°$

2nd group :

($m = 2$) : $\theta_{22} = 35.3°$, $\theta_{32} = 54.7°$, $\theta_{42} = 63.4°$, $\theta_{52} = 68.6°$, $\theta_{62} = 72°$, $\theta_{72} = 74.5°$, $\theta_{82} = 76.4°$, $\theta_{92} = 77.8°$

A particularly high number of studies have been carried out to date into delta or gothic wings and into some combinations of such components that form more or less simple slender bodies; they have dealt as much with the development of approximate theories as with the definition of models specifying vortex lift per unit area.

Visualisations of hyperlifting vortex structures, mainly those carried out by H. WERLE [3 to 9], the analysis of pressure fields and of speeds created by these vortices, with or without their bursting, notably J.L. SOLIGNAC's analysis [10, 11], also provide quite outstanding studies that are the standard works in their field. Already described fully in such papers as those by W. STAHL and A. ASCHAR [12 to 13], these studies offer today in their entirety, a thorough knowledge of the properties of various types of slender bodies.

However, given that the character of most of the aspects referred to remain empirical and limited to this or that angle of incidence [3 to 18] or to a numeric field, the way lies open, starting out from experimental data and various factors of analysis [19 to 21], for new attempts to be undertaken to examine the fundamental problems related to the positioning of vortices created by such slender bodies.

A large number of photographic and videographic visualisations concerning vortex flows developed on the upper-surface of delta or gothic wings and of cones [33, 34] have been carried out within the Valenciennes University laboratory with a view to providing a better understanding of the development and the positioning of vortex structures under the influence of apex angles and angles of incidence. These visualisations have enabled priority to be accorded to the study of examples of the most elementary shaped bodies, *i.e.* delta and gothic wings. The results obtained, and already fully described in previous papers [19 to 23], have been expressed by some very simple mathematical formulae; that very simplicity perhaps calls for some essential explanations.

The angles between the vortices have, in fact, been found, on a wide range of Reynolds, to have a preferential nature, thereby underlining a simple angular characterisation of the positioning relative to the vortex torques. Moreover, it is worth recalling here that vortices appeared as single or double structures depending on the nature of the apex angle. The studies carried out on such slender bodies have enabled the conditions of the existence of secondary vortex torques to be determined. We have shown in particular that the vortex structure developed on the upper surface of delta wings is a double one for non preferential or preferential angles but only for those higher than 45°. In the same way, the vortex structure developed on the upper surface of gothic wings has been shown to be a double one for both non preferential and preferential angles but only when they are higher than or equal to 68.6°.

However, in either case, we have been able to formulate a simple law, called the law of filiation, which expresses a notable angular relationship between the main vortex structure and the leading edges characterised by an apex angle from the 1st group [22 , 23, 37 and 38]. Simultaneously, studies carried out on the cone enabled the existence of preferential intervortex angles to be confirmed [22].

Vortex torques developed on the upper surface of the ogive of revolution were generally shown to have double structures, a finding which raises the delicate fluid equivalence question of the leading edges of the delta wing. We have also, by analogy with the law of filiation referred to above, been able to monitor an angular characterisation of similar simplicity but one which expresses a notable relationship between the main and secondary vortex structures. This new behavioural property, equally called the law of filiation, without doubt concerns the ogive of revolution as a whole. Limited as it is to the case of an ogive of revolution having an apex angle of 68.6°, our present study can consequently only be taken as an example, albeit a detailed one, of this phenomenon.

2. Models Used in the Experiment

The various models subjected to trials in the wind tunnel are of the circular-base cone-shaped type.

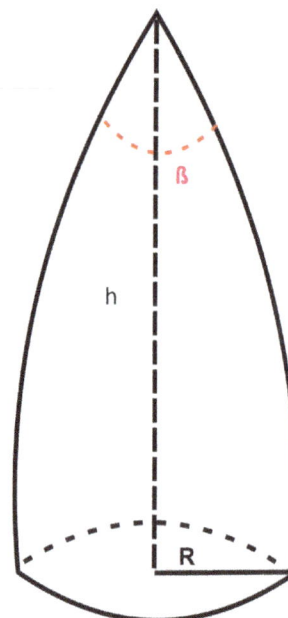

Diagram n° 1. The principles of constructing an ogive of revolution

β : summit (apex) angle
h : height of the ogive of revolution
R : radius of circular base

3. Visualisation Techniques

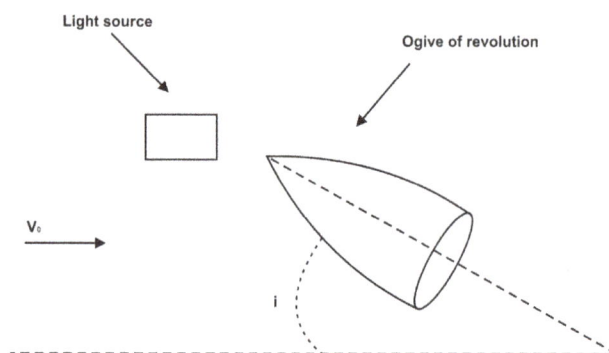

Diagram n° 2. Diagrammatic representation of the experimental device

i : incidence
v_0 : speed of flow to infinity upstream

The visualisations in their entirety were carried out within the test section of the wind tunnel {a 45 x 45 cm² section} at the LAH. The vortex structures developed on the upper surface were made visible by providing for emissions of white smoke at the ogive of revolution tip; this was obtained by injecting oil under pressure through a tube of small diameter and then, by means of an integrated electrical system, vaporising the oil immediately as it left the probe. The ogive of revolution was fixed onto an axis relayed to a cursor graduated from 0° to 360° which enabled the incidence of inclination to be varied. The visualisations were captured on photographs and videofilm which are today stored in the department's database.

4. Conditions under which the Experiments were Carried Out

Visualisations, employing smoke, were carried out at low speed (v = 3 to 5 m/s) which gives the flow a Reynolds' number ranging from 19000 to 80000 {h: length of reference}. The height h and the apex angle β of the ogive of revolution are respectively 110 mm and 68.6°. The incidence i varies over a range of from 25° to 65°.

5. Results of the Experiments

In order to avoid rendering this paper unduly lengthy, we summarise the findings by stating simply that visualisations at increasing incidence enabled the progressive birth of vortex structures to be carefully monitored; at mean incidence, these structures become concentrated and stable and thus pass from a structure of flow with raised edges to the standard vortex tube structures.

Those vortices derived from the apex and perfectly described by H. WERLE [8], result in fact from the "cornet-like" spiral coiling of the flow which detaches itself from each side of the streamlined body. Their axes are rectilinear and are cut near the apex depending on the angles {respectively interior and exterior ω_1 and ω_2} while the pace of the pseudo-flow in a transversal plane is probably close to that which figures in diagram n° 3 where the external vortices are closer to the wall than are the interior ones.

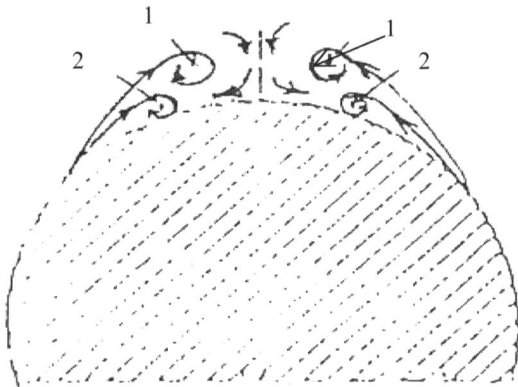

Diagram n° 3. *Vortex to turn your exterior and interior inverse sense and variable speed [22to33].*

1: interior vortex
2: exterior vortex

Whatever the incidence, these intervortex angles have shown themselves to have a preferential nature and have proved to be especially stable since they do not depend, in the conditions applied to the experiment, on speed {or on Reynolds}. However, it is worth noting that such angles, contrary to what takes place on the upper surface of delta or gothic wings, even so depend on the incidence, a fact which

bestows on them a discontinuous evolution since they have a tendency to conserve their preferential aspect {*see graph n° 1*}.

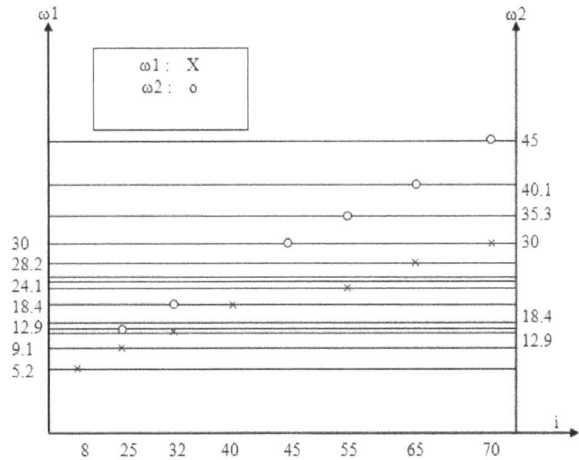

Graph n° 1. *The evolution of interior and exterior intervortex angles in relation to the incidence 19000<Re<80000*

Moreover, we noticed that there existed, in every case, a notable correspondence between both the interior and exterior intervortex angles. It has in fact been confirmed that the intervortex angles ω_2, in their entirety, belong to the first group of preferential angles {those with the notation $\theta_{\ell\ell}$} and consequently infer, in accordance with the law of filiation, that the intervortex angles ω_1 are also of the same first group {*see Table n° 1*} but with the notation

$$\theta_{2\ell+1, 2\ell+1} \qquad (2)$$

$$\omega_2 = \theta_{\ell\ell} \rightarrow \omega_1 = \theta_{2\ell+1, 2\ell+1} \qquad (3)$$

Table n° 1. *Angular relationship between interior and exterior intervortex angles (in degrees) 19000<Re<80000*

i	w_2		w_1	
8	8.7	$q_{39.39}$	5.2	$q_{45,45}$
25	12.9	$q_{29.29}$	9.1	$q_{39.39}$
32	18.4	$q_{19.19}$	12.9	$q_{29.29}$
35	19.4	$q_{9.9}$	13.6	$q_{21.21}$
40	26.6	$q_{8.8}$	18.4	$q_{19.19}$
45	30	$q_{4.4}$	20.7	$q_{17.17}$
50	35.3	$q_{3.3}$	24.1	$q_{9.9}$
55	35.3	$q_{2.2}$	24.1	$q_{7.7}$
58	40.1	$q_{2.2}$	28.2	$q_{5.5}$
62	45	q_{11}	30	$q_{3.3}$

This law, established in the particular case of the ogive of revolution, inevitably refers back to the quite analogous results we observed [22 and 23] from our studies of gothic wings and cones and of delta wings even if, in the case of gothic wings, the notable angular correspondence was described between the apex angles β of the first group and the associated intervortex angle α.

$$\beta = \theta_{\ell\ell} \rightarrow \alpha = \theta_{2\ell+1, 2\ell+1} \qquad (4)$$

Be that as it may, the description of the behavioural law

may be taken a little further if we make a point of referring again to the first definition of preferential angles, namely :

$$\cos \theta_{\ell\ell} = \frac{\ell}{\sqrt{\ell(\ell+1)}} \qquad (5)$$

In these conditions,

$$\sin^2\theta_{\ell\ell} = 1 + \cos^2\theta_{\ell\ell} = \frac{1}{\ell+1} \qquad (6)$$

$$\sin^2\theta_{2\ell+1,\,2\ell+1} = \frac{1}{2(\ell+1)} \qquad (7)$$

which can be simplified to :

$$\sin^2\theta_{2\ell+1,\,2\ell+1} = \frac{\sin^2\theta_{\ell\ell}}{2} \qquad (8)$$

and thus :

$$\sin^2\omega_1 = \frac{\sin^2\omega_2}{2} \qquad (9)$$

Some of the visualisations obtained, where incidences were increasing from 25° to 55°, provide a better appreciation of the evolution of vortex structures developed on the upper surface of an ogive of revolution having an included angle of 68.6° {see Views n° 1, 2, 3, 4 and 5}. However, it is worth noting that even at a higher incidence of 65°, we can notice the existence of two flows separated by an interval not fed by smoke and thus constituting a non vortex passage {see View n°6}.

Incidence of An angle of 32°

View n° 1. An angle incidence of 8°; the flow is uniform on the upper side of the wing. The boundary layer is observed but there is non flow separation as yet.

β = 52°   i = 32   ω1 = 12.9°   ω2 = 18.4°
19000<Re<80000

Incidence of An angle of40 °

View n° 2. When the incidence is increased, the main vortices and the secondary vortex torque are now observed, the speed remaining constant. However, variation in the speed have no effect at all on the formation of vortex angle of 32°

β = 52°   i = 40°   ω1 = 18,4°   ω2 = 26,6°
19000<Re<80000

Incidence of An angle of 45°

View n° 3. At 45 ° incidence, I see the phenomenon of pulse bursts outside vortex that rises to the apex and the angle between ''he outer vortices and ω2 26, 6 and the angle between the main vortices is ω 12.9.

β = 52°   i = 45°   ω1 = 20,7°   ω2 =30°
19000<Re<80000

Incidence of An angle of 50°

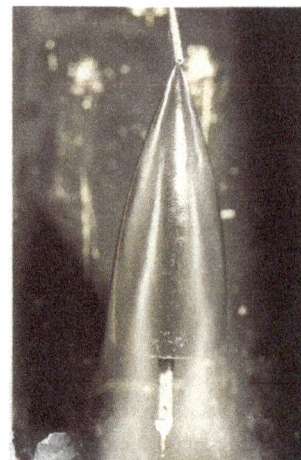

View n° 4. The breakdown phenomenon of the secondary vortices first appears as from an angle of incidence of 40°.

β = 52°   i = 50°   ω1 = 24,1°   ω2 =35.3°
19000<Re<80000

Incidence of An angle of 58°

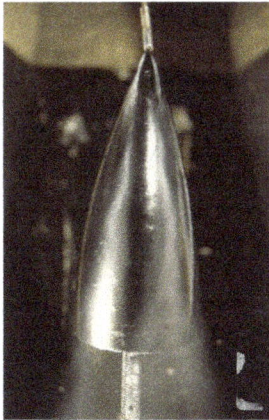

View n° 5. The formation of two vortex torques is detected, separated by an air corridor (a non-vortex passage) while the breakdown of the both main and secondary vortices begin. Angle of incidence of 55°.

$\beta = 52°$ $i = 58°$ $\omega1 = 28,2°$ $\omega2 = 40,1°$
$19000 < Re < 80000$

Incidence of An angle of 62°

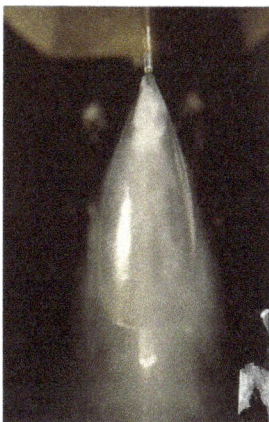

View n° 6. By increasing the incidence to 70°, the breakdown phenomenon of vortex structures becomes apparent, the flow of which, according to the incidence, swirl towards the apex of the give of revolution.

$\beta = 52°$ $i = 62°$ $\omega1 = 30°$ $\omega2 = 45°$
$19000 < Re < 80000$

The whole of these two air flows and of this passage is contained in a global angle of 45° whereas the angle formed between the left- and right-hand edges of each of the two air flows is 30°. At higher incidences, the turbulence causes considerable diffusion of the smoke and therefore does not enable clear visualisations to be made of the vortex structures; the latter subsequently deteriorate by bursting into torch forms that are initially stationary and then are subjected to alternate movements before those movements become thereafter chaotic.

6. Conclusion

While, on a wide range of Reynolds, the preferential nature of the intervortex angles present on the upper surface of delta and gothic wings and of ogives of revolution would seem to be fully catalogued, the very existence of the law of filiation relative to such slender bodies expresses a certain universality of behaviour and reveals the fundamental feature of our study.

At the present time, however, no complete theoretical approach would seem to be capable of providing a straightforward explanation of the simplicity of these results. The progressive evolution from elementary vortices of the sheer flow before take-off towards a particularly stable vortex system, wherein spatial positioning reveals an original organisation, still remains today an enigma. It is, of course, difficult to prejudge the lines along which one or more future studies may follow, studies which could lead to a theoretical explanation. However, perhaps we may be permitted to note that the phenomena, in which the sine squared of an angle also plays a part, are created by the simple structures of stationary and unsteady fluid mechanics.

It is in this way that the flow - emitted by an ogive of revolution having a demi-span of θ at its summit on a tridimensional dipole with the same axis as that of the cone, or with its summit at the centre of a vortex ring equivalent to the dipole - is proportional to $Sin^2\theta$. This is how energy - emitted by an oscillating electromagnetic dipole (with properties analogous to those of the oscillating fluid dipole that plays a part in aerodynamics or in hydrodynamics) in a θ direction with regard to the axis of the dipole - is, energy too, proportional to $Sin^2\theta$ or to $cos^2\theta = 1- sin^2\theta$ in the case of the acoustic dipole. It is also a law in $Sin^2\theta$ that gives the dependence, with regard to the angle of attenuation of the second sound thermal waves in liquid helium, by rectilinear vortices which form the θ angle with the direction of the propagation of this wave.

It is, moreover, by coupling this law with the concept of that preferential angle, formed by helicoidally vortices with their axes, that the authors of the papers referred to in [1] and [2] interpreted the discontinuous angular behaviour of these vortex systems in liquid helium [1, 2 and 19].

Finally, and this ultimate remark is probably not the least important one, the suction force, to which a profile - an infinitely thin and localised plane, let us remember - is subjected in the immediate vicinity of its leading edge, is, that force, too, proportional to the sine squared of an angle, in this case the angle of incidence.

As concerns the possible links, of the phenomena we have described, with the properties of an emission or of an absorption of a flow or of a wave - whose source may be dipolar or multipolar - it is perhaps interesting to note that the range of speeds of a tridimensional dipole {characterised by the angle between the radius vector of a point of the fluid and the speed of this fluid} is linked to the angular positioning of this point, with regard to the dipole {characterised by the polar angle between the axis of the dipole and the radius vector}, by a series of striking correspondences between the most simple preferential angles. Moreover, if we now consider the force of interaction between two dipoles of relatively simple orientation - the most simple is one with two parallel dipoles, but numerous other layouts give equally curious results - the range of interaction forces {characterised by the angle between the radius vector joining the two dipoles with this interaction force}, possesses, in its turn, together with the two other ranges of relative positioning and of speeds {characterised as described above}, two new entireties of quite striking correspondences E. TRUCKENBRODT [24].

Where the notable orientations of interaction forces between two parallel dipoles are concerned, and with regard to the common direction of the axes of these dipoles, we have extracted a few particular cases from the general calculations made by W. KÖNIG [25] and from his final result given in J.W-C. RAYLEIGH's very famous book on acoustics [26].

These references contain expressions of : the components of the force exerted by one sphere on another in the presence of a uniform wind pattern to infinity; where the fluid flow is perfect; the line joining the centres of spheres forming the θ angle with the direction of the wind to infinity - this force is the same as that exerted by one sphere on another when those spheres are moving parallel to each other and at the same speed in an immobile fluid to infinity - but it is known that each of these spheres is equivalent to a tridimensional dipole.

The sole particular cases commented on by W. KÖNIG [25] and J.W-C. RAYLEIGH [26] are those where the centres of the spheres are aligned, {i.e. in the direction of the wind}, and where these spheres therefore exert on each other a repulsion force {i.e. perpendicularly to the wind} with, in this case, a gravitational interaction force which explains the formation of very fine powder ridges, perpendicular to the axis of a sound tube, within its antinodes (ventral segments) of vibration.

But a whole series of other consequences from the general formulae found in references [25] and [26] seem, they too, to be very significant. One of the most important of these particular cases seems to us to be that where the interaction force between two parallel dipoles is itself parallel to them. Formula n° 4 on page 47 of reference [26] immediately shows that this case corresponds to the θ angle, cancelling the Legendre polynomial 1-5cos²θ, i.e. at

$$\cos\theta = \frac{1}{\sqrt{5}} = \frac{2}{\sqrt{4(4+1)}} = \cos\theta_{42} \qquad (10)$$

according to the defining formula of preferential angles given at the beginning of this paper.

This angle $\theta_{42}=63.4°$ is, moreover, the angle between the diagonals of the famous "Golden Rectangle" discovered by architects and employed by them from time immemorial [21].

In this same train of thought, it is striking to note, in references [27 to 29] the role played systematically by the angle $\theta_{32}=54.7°$ (cancelling the Legendre polynomial 1-3cos²θ) in the sound emission of an axisymetric jet and of two interaction forcing vortex rings or of one forcing vortex ring in the presence of a sphere.

In short, many other well-known hydrodynamic and aerodynamic phenomena are rich in preferential angles, the theory of which has at some or other been fully elaborated. This is the case found in the very subtle and elegant theory described in particular in the works of H. LAMB [30] and of

J. LIGHTHILL [31]. In the wake, the crests of waves, in a curvilinear triangle form, will in fact each disappear at two counter flow points, the alignment of which, along two right-hand sides, determines a total span of the wake at twice 19.4° here and there of the axis of this wake, axis with which the counterblow tangents, associated with the crests, form an angle of 54.7° while also forming with each corresponding edge of the wake an angle of 35.3° {i.e. 54.7° - 19.4°}.

It is there where the following relation is to be found, never interpreted before now, in terms of preferential angles :

$$\theta_{32} = \theta_{22} + \theta_{88} \qquad (11)$$

$$54,7° = 35,3° + 19,4° \qquad (12)$$

The link between the wake of a ship, being the result of the combination of bidimensional surface waves shed in various directions, and the phenomena described above may appear at first sight to be very mysterious. We may, however, be permitted to reason that the paper by E. LEVI [32] under the title "An oscillating approach to turbulence", so suggestively illustrated by figure n° 1 on page 352 of his study {an illustration which represents the frontier of a wake or of a maximum layer as a swell induced by the emission of vortices} perhaps provides the starting point of a profitable line of further research which could lead to a better understanding of the omnipresence of preferential angles and of their filiations in tridimensional flows, and in particular in those developed around slender bodies.

The essential question to be pursued, and one which remains as yet to be entirely addressed, seems to us primarily to lie with structures and wave propagation. What is required is the explanation of the link between those preferential angles, which appear in structures exterior to the borderline layer, and structures, which would also probably need to be termed preferential angles, to be found in the forms and modes of wave propagation. The latter have recently been the subject in a close study of "the coherent structures of turbulence", structures that are, in particular, present in laminar-turbulent transition zones or in zones of anisotropic turbulence, especially where they relate to layers.

Notations

$\theta_{\ell m}$: preferential angle associated with the whole numbers ℓ and m

ℓ, m : whole numbers such as $\ell \geq m$

β : apex (included) angle

h : height of the ogive of revolution

R : radius of the base of the ogive of revolution

I : incidence

V_o : speed of the flow to infinity upstream

Re : Reynolds' number

ω_1 : the main or interior intervortex angle

ω_2 : the secondary or exterior intervortex angle

α : the main intervortex angle present on the upper surface of delta or gothic wings

References

[1]   M. LERAY and M. FRANCOIS Stability of a vortex lattice in the presence of a superflow. *Physics letters 34A 14 April 1971, pp 431-432.*

[2]   J.P. DEROYON, M.J. DEROYON and M. LERAY. Experimental evidence of macroscopic spatial quantification of angular momentum in rotating. Hélium. *Physics letters, A45, September 1973, pp 237-238.*

[3]   H. WERLE Structure des décollements sur les ailes cylindriques . *La recherche aérospatiale, n° 3, Mai 1986, pp 13-19.*

[4]   H. WERLE Tourbillons d'ailes minces très élancées. *La recherche aérospatiale, n° 109, Novembre-Décembre 1965, pp 3-12.*

[5]   H. WERLE Interaction tourbillonnaire sur voilures en delta fixées ou oscillantes (visualisation hydrodynamique). *La recherche aérospatiale, n° 2, Mars-Avril 1980, pp 43-68.*

[6]   H. WERLE et M. GALLON Etude par visualisation hydrodynamique de divers procédés de contrôle d'écoulements décollés . *La recherche aérospatiale, n° 2, Mars-Avril 1976, pp 75-94.*

[7]   H. WERLE Le décollement sur le corps de révolution à basse vitesse. *La Recherche aéronautique n° 90, Septembre-Octobre 1962, pp l-14.*

[8]   H. WERLE Exploitation quantitative des visualisations d'écoulements obtenues dans les tunnels hydrodynamiques de l'ONERA. *La recherche aérospatiale, n° 6, Novembre-Décembre 1990, pp 49-72.*

[9]   H. WERLE Structure des décollements sur les ailes en flèche. *La recherche aérospatiale, n° 2, Mars-Avril 1980, pp 85-108.*

[10] J.L. SOLIGNAC, D. PAGAN et P. MOLTON Examen de certaines propriétés de l'écoulement à l'extrados d'une aile delta . *Rapport technique 37/1147 AN, Septembre 1988.*

[11] J.L. SOLIGNAC, D. PAGAN et P. MOLTON Etude expérimentale de l'écoulement à l'extrados d'une aile delta en régime incompressible . *La recherche aérospatiale n° 6, Novembre-Décembre 1989, pp 47-65.*

[12] W. STAHL Experimental investigations of asymmetric vortex flow behind elliptic cones at incidences. *AIAA Journal Vol 3, n° 5.1993, pp 966-968.*

[13] W. STAHL, A. ASGHAR Suppression of vortex asymmetry behind circular cones. *AIAA Journal, Vol 28, n° 6, 1990, pp 1138-1140.*

[14] W. STAHL, M. MAHMOOD and A. ASGHAR Experimental investigations of the vortex flow on delta wings at high incidence. AIAA Journal, Vol .30, n° 4, 1992, pp 1027-1032.

[15] A. AYOUB, STANFORD UNIVI SITY, C.A. and B.G Mc LAGHLAM, NASA, Research Center, Mofliet Field CA. Slender delta wing at high angle of attack, flow visualisation study . *AIAA Journal , June 8-10, 1987, Honolulu, Hawaï.*

[16] O.K. REDINIOTIS, H. STAPOUNTZIS and D.P. TELIONIS Periodic Vortex Shedding over Delta Wings. *AIAA Journal , Vol 31, n° 9, September 1993, pp 1555-1562.*

[17] M. ROY Caractère de l'écoulement autour d'une aile en flèche accentuée. *C.R.Acad. Sci. Paris, t. 234, 1952, pp 2501-2503.*

[18] R. LEGENDRE Ecoulement au voisinage de la pointe avant d'une aile à forte flèche aux: moyennes incidences. *La recherche aéronautique n° 30, 1952 et n° 31, 1953 .*

[19] M. LERAY, J. P. DEROYON, M.J. DEROYON et C. MINAIR Critères angulaires de stabilité d'un tourbillon hélicoïdal ou d'un couple de tourbillons rectilignes, rôle des angles privilégiés dans l'optimisation des ailes, voiles, coques des avions et des navires. *Communication à l'association technique maritime et aéronautique (ATMA). Publication dans le bulletin de l'ATMA, Session 85, Paris 1985, pp 511-529.*

[20] C. MINAIR Les angles privilégiés, grands invariants et universaux : une approche par la dynamique des fluides, l'esthétique et physio-biologie. *Thèse de doctorat d'état, Université de Valenciennes, Octobre 1987.*

[21] M. LERAY Dialogue du physicien et de l'esthète : les Angles Privilégiés *Communication et langage, n° 45, 1980, pp 49-69.*

[22] A. ABENE Etude systématique des positions et de la stabilité des structures tourbillonnaires au dessus d'ailes ogives et de cônes. *Thèse de doctorat. Université de Valenciennes, Juillet 1990.*

[23] M. BENKIR Persistance et destruction des structures tourbillonnaires concentrées ou partielles au dessus des ailes delta. *Thèse de doctorat. Université de Valenciennes, Avril 1990.*

[24] E. TRUCKENBRODT Fluidmechanik band Elementare strömungsvorgänge dichteversanderlicher fluide sowie potential und gnenzschichtströmungen. *Springer Verlag, Berlin, 1980, § 5.3,2.6 d Räumliche Dipole Strömung, pp 167-168.*

[25] W. KÖNIG *Wied Ann, Tome XLII 1891, pp 353-549.*

[26] J .W-Ç. RAYLEIGH Theory of sound. *Volume II, second edition, § 253b, Striafion in Kundt's tubes 1896, Dover Publications. New York, 1945, pp 45-47.*

[27] W. MÖHRING Modelling low Mach number noise in mechanics of sound generation in flows. *IUTAM/ICA/ALAA Symposium, Göttingen, Germany, August 28-31 1979, Edited by E.A. Müller, Springer Verlag, Berlin. 1980, pp 85-96.*

[28] C.D. MÙLLER, F. OBERMEIER Vortex sound, in vortex motion. *Edited by H. Hasimoto and T. Kambe, North Holland, Amsterdam, 1988, pp 43-51.*

[29] T. MINOTA, T. KAMBE and T. MUKARAMI Acoustic emission from interaction of a vortex ring with a sphere, in vortex motion. *Edited by H. Hasimoto and T. Kambe, North Holland, Amsterdam, 1988, pp 357-362.*

[30] H. LAMB Hydrodynamics. *Sixth Edition 1932, reprinted by Dover Publications, New York 1945, § 256, Ship Waves, pp 433-440.*

[31] J. LIGHTHILL Waves in fluids. *Cambridge University Press, UK, 1978, § 3.10, Ship Waves, pp 269-279.*

[32] E. LEVI An oscillating approach to turbulence in unsteady turbulent shear flows. *IUTAM Symposium, Toulouse, France, May 5-8 1981, Edited by R.Michel, J. Cousteix and R. Houdeville, Springer Verlag, 1981, pp 348-358.*

[33] A. ABENE La détermination des positions relatives des structures tourbillonnaires au-dessus des ailes ogives et des cônes. *Habilitation à diriger des recherches, 9 Octobre 2002.*

[34] V. DUBOIS Etude détaillée des structures tourbillonnaires à l'extrados de corps élancés simples, complexes et de révolutions. *Thèse de doctorat le 16 Décembre 2005.*

[35] A. ABENE and V. DUBOIS Fundamental aspects of the vortex flow on cones. Angular characterisation of hyperlifting vortex torques; the law of filiation *Canadian Journal of Physics, n° 86, Aôut 2008.*

[36] A. ABENE, V. DUBOIS et M. SI YOUCEF Etude expérimentale sur des capteurs solaires à air : application pour le séchage de la prune. *Revue internationale d'héliotechnique, n° 33, pp 2-18, Printemps 2006.*

[37] A. ABENE and V. DUBOIS, « A experimental study of air heating plane solar panels as used in drying mangoes », International Journal of Energy Research, ref ER-09-1016, Juin 2009.

[38] A. ABENE "Visualisation of Vortex Structures Developed on the Upper Surface of Double-Delta Wings", J Aeronaut Aerospace Eng 2:118, doi:10.4172/2168-9792.1000118. Septembre 2013

5

Performance Assessment of HFC Group Refrigerants in Window Air Conditioning System

Boda Hadya[1], A. M. K. Prasad[1], Suresh Akella[2]

[1]Mechanical Engineering Department, U. C. E., Osmania University, Hyderabad, Telangana State, India
[2]Sreyas Institute of Engineering and Technology, Affiliated to J. N. T. U., Hyderabad, Telangana State, India

Email address:
hadya.ou@gmail.com (B. Hadya)

Abstract: This paper discusses the different operating conditions of three refrigerants i.e. R22, R410A and R32. For analysis 1 Ton of window air conditioner has been chosen. Basically the operating cycle is a simple vapour compression refrigeration cycle with hermitically sealed compressor. The HFC groups of refrigerants do not have any Ozone Depletion Potential and refrigerants like R32 has the advantage of lower Global Warming Potential (GWP:675) with compare to R22 (GWP:1700) and R410A(GWP:2100) refrigerants. As per the indication of Montreal and Kyoto protocol, the world community has decided to phase down the HCFC group refrigerants. The refrigeration and air conditioning industry required to evaluate new alternative refrigerants to HCFC-22 also known as R22. The performance parameters like, pressure ratio, condensing temperature, discharge temperature of the compressor, power consumption and COP of the system were analyzed and compared.

Keywords: ODP, GWP, Ton of Refrigeration (TR), COP, Discharge Temperature, Pressure Ratio

1. Introduction

The science and practice of creating a controlled climate in indoor space is called air conditioning. The refrigeration cycle can also be applied for air conditioning system, now days there are various methods of cooling systems are available, but vapour compression refrigeration system is better and efficient method for air conditioning purpose. Vapour compression refrigeration cycles have two advantages [1-3]. First, a large amount of thermal energy is required to change a liquid to a vapor, and therefore a lot of heat can be removed from the air-conditioned space. Second, the isothermal nature of the vaporization allows extraction of heat without raising the temperature of the working fluid to the temperature of whatever is being cooled [4-5].

2. Literature Review

To search for (HCFCs group) alternative to refrigerant 22, a comprehensive literature study has been carried out for various alternate refrigerants which are being used with vapour compression refrigeration system. James M.Clam [6] in this research article the author stated the important of the next stage environmental safe refrigerants as per the Montreal and Kyoto protocol agreements. The author conclude that there no ideal refrigerant which suitable for all the applications

Cavallini A, [2] conducted experiments by using ozone friendly refrigerants and compared the performance characteristics in a vapour compression refrigeration system by four different refrigerants like, 125, R134a, R143a and R152a in his investigation the R152a coefficient of performance is good with compare to the other three refrigerants and also r152a has zero ODP and very low global warming potential.The world community has been searching for alternative refrigerants, the major compressor manufacturer's plans to convert 50% in R410A and 50% in R32 by 2015, for commercial sectors for Room Air-Conditioning 85% in R410 and 15% Indonesia plans to prohibit HCFC import 2013 and HCFC production in 2015 [7-13]. Also decided to use R32 instead of R410A.Kuwait are interested in R32 and requested a contact with Daikin. Kuwait is doing a joint evaluation for R32 performance under high ambient temperature condition with Daikin as the representative of Middle East countries. In many of the research papers deal with performance comparison for different parameters as the R22 phase down from the existing air conditioning system [14-16]

3. Experimental Methodology

A window air conditioner is basically an enclosed assembly designed to be a compact unit primarily for mounting in a window, through the wall. The test unit under investigation is placed in the opening according to the specifications such that the heat rejection section is in the outdoor room and the heat absorption section is in the indoor room for a cooling test. The test unit will be powered in either the indoor and outdoor test room. Both test rooms are separately enclosed in another insulated room called controlled air space.

The basic components in a window air conditioner are as follows:

1. Compressor, 2. Condenser, 3. Evaporator and 4. Expansion valve

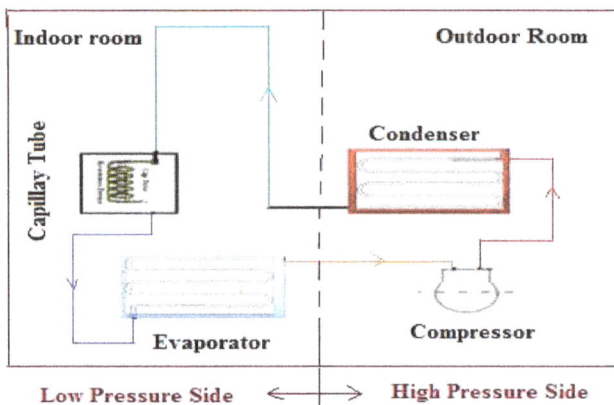

Figure 1. *System Design considerations.*

Figure 1 shows, the importance of proper system design when hermetic compressors are used on appliances, Compressors cannot over emphasized, because the motor and compressor assembly in the hermetic compressor necessitate holding mechanical, electrical and thermodynamic variables within the limits specified for safe and trouble free operations.

The P-h diagram (Moeller diagram) for refrigeration cycle which is shown in figure 2 with four basic processes are frequently used in the analysis of Vapour Compression Refrigeration cycle, process 1 to 2 is Compression, process 2 to 3 heat rejection in the Condenser, process 3 to 4 Expansion (Throttling) and process 4 to 1 is Evaporation i.e. heat absorbed in the evaporator. The performance characteristics are can be computed for Compressor work (Wc), Refrigeration Effect (Q_E) and Coefficient of Performance (COP). COP = (h_1-h_4 / h_2-h_1)

Where,

h_1 and h2 are Enthalpies of Refrigerant at the inlet and outlet of compressor (kJ/kg).

h_3= h_4 are Enthalpies of Refrigerant at the inlet and outlet of expansion valve (kJ/kg).

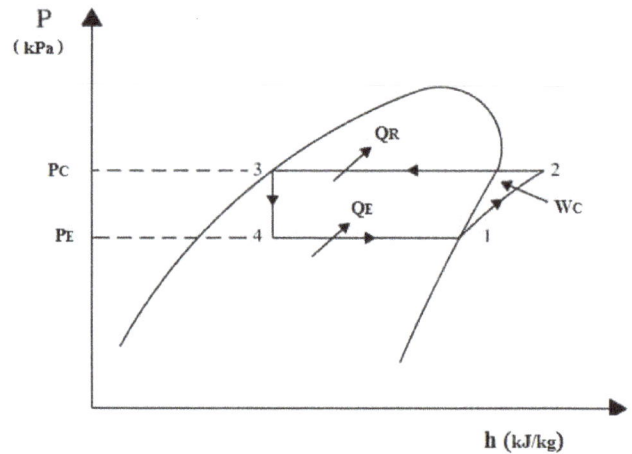

Figure 2. *P-h Diagram for refrigeration cycle.*

Figure 3. *Compressor with thermocouples.*

Figure 3, shows the measurement of temperature by attaching the thermocouple at various positions like Compressor shell (Top, Middle and Bottom locations) the selection of compressor for particular application is depends on the operating pressure and temperatures. The different types of hermitically sealed compressors are (RN, RR and RK are Rotary Compressors) and (AW–Reciprocating compressors). The temperature of the of locations in refrigeration system has been measured using with high precision RTD's and also the flow of refrigerant measured using mass flow meter device (Micro-Motion). Following will cover the key measurement required and the sensors we would typically use to acquire those measurements:

1. Temperature, 2.Pressure, 3.Power and 4.Flow

Table 1. *Standard instrument details used for calibration of temperature.*

Nomenclature	Make & odel	SI.No.	Uncertainaty	Traceable to
RTD sensor	PT-100	TIC/RTD/2	±0.25˚C	STQC

Table 2. *Standard instrument details used for calibration pressure.*

Nomenclature	Make&model	SI.No.	Uncertainaty	Traceable to
Pressure Controller With sensor	Ajay sensor 289 series	ASP 781/5	±0.018 bar	Nagman/Cert.No.2007-08/CFC/19100

Table 1&2 shows the standard instruments with their   specification, which were used for measurements and error for measurement uncertainty is ± 0.13% at 95 confidence level.

Resistive Temperature Detector (RTD); Thermocouple (T-type or J-Type

4. Results and Discussions

The following conclusion has been drawn, after the comparason of both simulated and experimental results. With compare to other two refrigerants R32 refrigerants performance characteristics were found to be better.

Table 3. Pressure Ratio.

| Refrigerant | Suction pressure | | | | |
	SET 1	SET 2	SET 3	SET 4	SET 5
R22	620.2	619.9	619.9	619.9	620
R410A	988.2	988.2	988.4	988.3	988.4
R32	1016	1016.1	1015.8	1015.5	1015.7
Refrigerant	Discharge Pressure				
R22	2147	2148	2145	2146	2146
R410A	3385	3386	3387	3386	3386
R32	3385	3386	3387	3386	3386

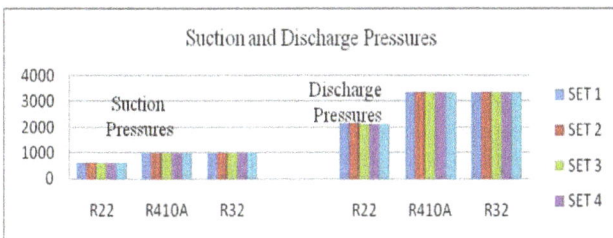

Figure 4. Compressor with thermocouples.

Figure 4 shows the experimental suction and discharge pressures at evaporator temperature of 7.2°C and condenser temperature of 54.5°C, the results obtained with respect to the pressure, from the plot for both the cases i.e suction and discharge pressure are as follows for R22 suction and discharge pressures 619.9 kPa and 2145 kPa. R410A suction and discharge pressures for 410A 988.4and 3387 kPa and for R32 the suction and discharge pressures are 1015.8 kPa and 3387 kPa, it is observed that; for all the cases the R32 attains higher suction and discharge pressures.

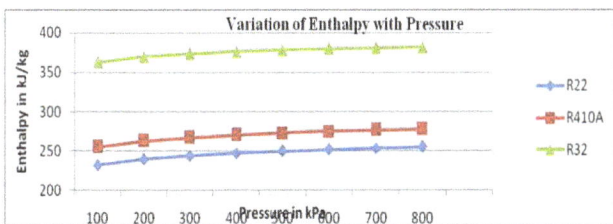

Figure 5. Variation of enthalpy with Pressure。

Figure 5 show the variation of enthalpy, with pressure the enthalpy is very high for the R32vand very low for R22 refrigerants. The thermodynamic concept involves vapour pressure and latent heat which is called enthalpy of evaporation.It is the indication of specific internal energy and specific flow work. It is observed from the figure that for the operating pressure of 800 kPa the enthalpy for R32 is 382

kJ/kg for R410A 280kj/kg and for R22 is 255 kJ/kg.With compare to R22 and R410A the R32 enthalpy is 33% and 27% more, this may be a advantage to R32 refrigerant.

Table 4. Discharge Temperature for different set values.

Refrigerants	SET 1	SET 2	SET 3	SET 4	SET 5
R22	89.8	89.7	89.5	89.7	89.5
R410A	89	89	89.2	89.2	89.2
R32	112	111.9	112	112.1	112.1

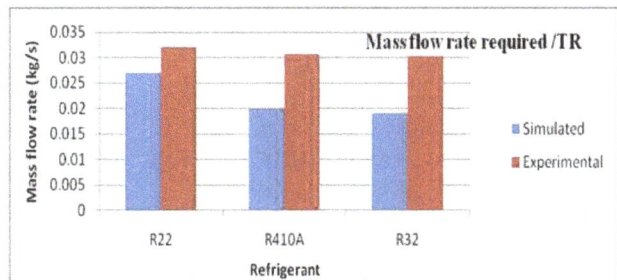

Figure 6. Mass flow require per TR.

Figure 5 shows the amount of mass flow required for 1 "Ton of Refrigeration" (TR) for the R32 required amount of mass flow rate is 0.0302 kg/s, R4410A is 0.0307 kg/s and R22 is 0.032 kg /s. For the same cooling capacity with compare to R22 and R410A refrigerants R32 is 5.6% and 4%is less.

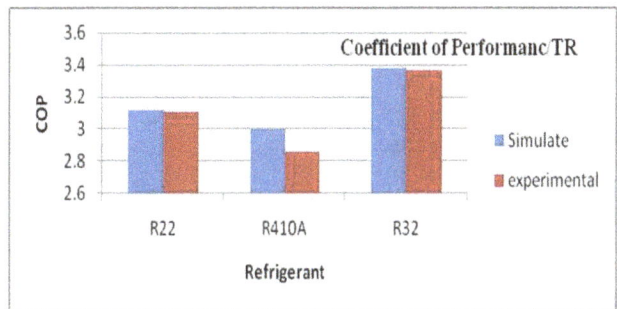

Figure 7. COP of the selected refrigerants.

The measured COP for the refrigerants are plotted and shown in figure 7 for condenser temperature of 54.5°C and evaporator temperature of 7.2°C, the R32 performs better than R410A.COP for R32 is 3 where as R410A COP is 2.89, consumption of power also low for R32 refrigerant. As R32 is having more refrigeration effect, the mass flows rate of refrigerant for R32 also less with compare to R410A.

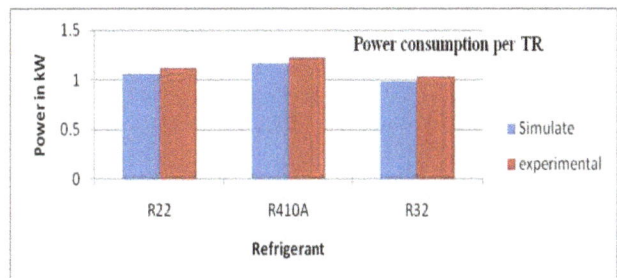

Figure 8. Power Consumption for selected refrigerants.

Figure 8 show the Power consumption for selected refrigerants for constant condenser and evaporator temperature i.e. Tc = 54.5°C and Te = 7.2°C. In all the cases the simulated power required is less than the experimental values. As per experimental analysis per 1 TR capacity the power consumption three refrigerants R22, R410A and R32 are 1.122kW, 1.225kW and 1.038 kW, with compare to R22 and R410A, the refrigerant R32 requires 7.84% and 15% less power.

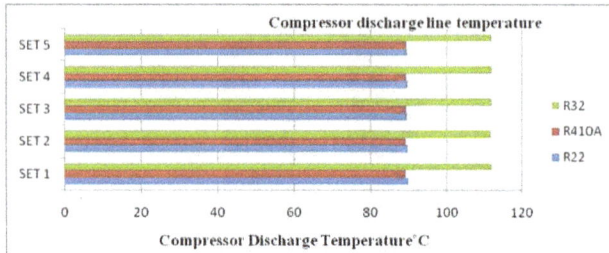

Figure 9. *Compressor discharge temperatures.*

From the figure 9 the discharge line temperature for R32 is 112°C where as R22 is 89.7°C and R410A is 99.3°C, with compare to other two refrigerants R32 discharge temperature 19% and 12% higher. While designing the system it should be seen that the compressor discharge temperatures should not be exceed 130°C.

5. Conclusions

1. R32 is the most balanced solution, not depleting the Ozone layer, smaller global warming compared to R410A.The global warming effect is 69 % and 61.7% less with compare to R410A and R22. It may be considered as alternative to R22 as well as 410A because environment friendly refrigerant.
2. An ideal refrigerant should have high latent heat of vaporisation which results higher refrigerating effect, hence coefficient of performance increases. R32 is having high latent heat i.e. 337 kJ/kg which is 36% higher than the R22.hence the advantage of this refrigerant is refrigerant effect is more per kg of refrigerant circulation in refrigeration cycle. The refrigerant R410A is having the latent of 238kJ/kg.
3. The refrigerant R32 possess higher pressure in both the cases i.e. suction and discharges pressures. The experimental suction and discharge pressures at evaporator temperature of 7.2°C and condenser temperature of 54.5°C, for R22 suction and discharge pressures 619.9 kPa and 2145 kPa. R410A suction and discharge pressures for 410A 988.4and 3387 kPa and for R32 the suction and discharge pressures are 1015.8 kPa and 3387 kPa.
4. From the experimental analysis per 1 TR capacity the power consumption three refrigerants R22, R410A and R32 are 1.122kW, 1.225kW and 1.038 kW, with compare to R22 and R410A, the refrigerant R32 requires 7.84% and 15% less power.

5. High compression ratio, consequently higher gas discharge gas temperature, if not controlled by proper system, may result in oil and refrigerant break down, forming carbon deposits on the valve plate which results choking of compressor.

References

[1] S.Devotta, A S Padalkar, N.K.Sane, "Performance assessment of HCFC 22 window Air conditioner retrofitted with R407 C", *Applied Thermal Engineering.* 25 pp: 2937 -2949. (2005)

[2] Cavallini A., "Working fluids for mechanical refrigeration" *International Journal of Refrigeration"* 19, pp; 485-496. (1996)

[3] Samuel, F.,Yana Motta, Pioter,A.Domanski, "Performance of R-22and Its alternative working At high outdoor Temperatures" *Eighth International Conference at Purdue university.*PP 47-54 (2000).

[4] E Halimic, D Ross, B Agnew, A Anderson, I Potts "A comparison of the operating performance of alternative refrigerants"Applied Thermal Engineering, 23,(12), Pp 1441-1451.

[5] S. Devotta,, A.V. Waghmare, N.N. Sawant,B.M. Domkundwar, "Alternatives to HCFC-22 For air conditioners" Applied Thermal Engineering 21(6) pp: 703-715. (2001)

[6] James M. Calm, Donald J. Wuebbles, Atul k. Jain; "Impacts on global ozone and climate from use and emission of 2, 2-dichloro-1, 1, 1-trifluoroethane (HCFC-123) Climatic Change 42: 439–474, *Kluwer Academic Publishers. Printed in the Netherlands.* (1999)

[7] Hung Pham, "Next Generation Refrigerants:Standards and Climate Policy Implications of Engineering Constraints"*American Council for an Energy-Efficient Economy*, PP;282-294, (2010).

[8] Chinnaraj, C., R. Vijayan, P. Govindarajan, "Analysis of Eco friendly Refrigerants Usage in Air-Conditioner" *American Journal of Environmental Sciences* 7 (6): 510-514, (2011).

[9] Piotr A. Domanski, David Yashar, "Comparable Performance Evaluation of HC and HFC Refrigerants in an Optimized System" *National Institute of Standards and Technology*,100 Bureau Drive, Stop 8631,Gaithersburg, MD 20899-8631, USA.

[10] Shanwei, M., Z. Chuan, C. Jiangping and C. Zhiujiu, "Experimental research on refrigerant mass flow coefficient of electronic expansion valve", Applied *Thermal Eng.*, 25, pp: 2351-2366, (2005).

[11] M.W.Spatz,Y.Motta, AN, "evolution of option for replacing HCFC-22 in medium temperature refrigeration systems," *International Journal of Refrigeration* 27 475-483, (2004).

[12] Vance Payne and Piotr A. Domanski "A Comparison of an R22 and an R410A Air Conditioner Operating at High Ambient Temperatures" Proceedings of *International refrigeration and Air-Conditioning Conference*,2002.

[13] Kontomaris, K., 2012. A zero-ODP, low GWP working fluid for high temperature heating and power generation from low temperature heat: DR-2. In: Proc. *JRAIA International Symposium* 2012, pp. 212-216. (2012)

[14] Lemmon, E.W., Huber, M.L., McLinden, M.O., 2010. Reference Fluid Thermodynamic and Transport Properties - REFPROP Ver. 9.0. *National Institute of Standards and Technology, Boulder,* CO, USA. (2010)

[15] Kayukawa, Y., Tanaka, K., Kano, Y., Fujita, Y., Akasaka, R., Higashi, Y., 2012. Experimental evaluation of the fundamental properties for low-GWP refrigerant HFO-1234ze(Z). *In: Proc.*

Int. Symp. New Refrigerants and Environmental Technology, Kobe, Japan, p. 231. (2012)

[16] Matsukura, N., Okuda, S., Nagai, K., Ueda, K., 2012. Study of application of HFO-1234ze(E) to hot water centrifugal heat pump e evaluation of low GWP refrigerant HFO-1234ze(E) in high temperature region. In: *Proc. JSRAE Annual conf., Sapporo, Japan,* pp. 1-4(2012).

6

Highly Turbulent Flow Laminarized by Hairy Pipe Walls

Bo Anders Nordell[1, *], Ragnar Oskar Gawelin[2]

[1]Department of Architecture and Water, Luleå University of Technology, SE-97187 Luleå, Sweden
[2]Enskilda Gymnasiet, SE-11161 Stockholm, Sweden

Email address:
bon@ltu.se (B. A. Nordell), ragnar.gawelin@enskildagymasiet.se (R. O. Gawelin)

Abstract: Nature has found ways to laminarize turbulent flows, as demonstrated by the high swim speed of dolphins and the silent flight of owls. Owls locate their prey by hearing and need to fly silently. In both cases it has something to do with the soft pliable surface of the moving body and the wavy pattern that occurs on the dolphin skin and the owl feathers. Our objective was to investigate whether a pipe lined with a hairy soft carpet would "laminarize" air flows. The degree of laminarization was determined by the velocity profile. Manual pressure measurements were done to determine the air velocity at cross-sections along the pipe. Varying flow rates were tested before the hair was cut increasingly shorter. It was found that for some hair lengths the velocity profile approached the parabolic form of laminar flow at very high Reynolds number.

Keywords: Flows in Pipes and Nozzles, High-Reynolds-number Turbulence, Interactions with Surfaces, Laboratory Studies, Laminarization, Stability of Laminar Flows, Velocity Measurements

1. Introduction

Many attempts have been made in reducing turbulence and to laminarize flows in order to influence friction and heat transfer. Reference [1] made experiments on the effects of fluid injection of an initially fully developed, low Reynolds (Re) number (from 3090 to 6350) turbulent pipe flow. They used a porous-walled pipe lined by a Rayon cloth. By injecting air through the porous wall at a Re ≈100 a local instant laminarization was observed close to the wall. Even in the core region the flow was slightly laminarized. However, in the downstream portion of the porous tube the entire flow undergoes a re-transition to fully developed turbulence. It is also known that accelerated turbulent flows show variations in heat transfer and friction that are interpreted as laminarization of the flow. This laminarization occurs only in the boundary layer close to the pipe wall [2]. Laminarization has also been observed and studied in unsteady turbulent flows [3] and in accelerating open channel flows [4]. The main problem with the different methods to laminarize pipe flows is that they only work locally and that the flow returns back to chaotic flow pattern after some time or distance from the occurrence. A recent literature review shows little or no progress.

However, nature has developed various ways to reduce turbulence or at least the consequences of turbulence. Some examples were looked into as part of current work, e.g. polymer additives for friction reduction in water [5,6], wave pattern in bronchial cilia in the human air ways [7], the silent flight of owls [8,9,10,11,12] and the super-efficient swimming ability of dolphins [13,14].

There are many different types of owl but it is normally a rather big bird that flies at a speed of 20 m/s. The seemingly soundless flight of owls is necessary as it depends on its hearing to localize its prey [15]. It is generally accepted that its hairy feathers and the softness and pliability of the whole body enable its soundless flight [10,12]. The dolphin swims at 15-20 m/s and its motion through water has been the subject of numerous studies. In 1938 James Gray found that its maximum muscle power of approximately two horse powers is not in proportion to the high speed at which it can swim. This observation, Gray's paradox, concludes that the power of a dolphin was a factor of ten too small to overcome its drag forces in water [14]. Gray assumed that dolphin skin must have special anti-drag properties. Coustaeu observed that longitudinal folds or patterns occur on the dolphin's skin when it is swimming at high speed [16]. This is partly explained by a fluid fat layer just beneath the dolphin's permeable outer skin. The resilient skin seems to dissolve eddies, thereby reducing friction between the dolphin and the water.

So, it seems as if the owl can fly and the dolphin can swim with little or no turbulence. The owl and the dolphin have very soft and resilient surfaces that diminish occurring swirls in air and water. Such soft wavy surfaces seem to unravel occurring eddies.

2. Objectives

The main objective was to investigate if the turbulence of an air flow would be reduced by a pressure equalizing pipe wall material. The lining fabric should be soft, hairy and pliable enough to change shape when subjected to the pressure of an eddy. The rate of laminarization was analyzed by the shape of the velocity profile. Current study is based on measurements that were carried out by the authors already in 1979 [17,18].

2.1. Method and Test Set-up

The test setup consisted of a plastic (PVC) pipe, Length=5.16 m; Diameter=0.15 m, which was lined with a hairy carpet. The pipe was cut in two half pipes, to make it easier to glue the carpet onto the pipe wall, and then attached back together again, and carefully sealed. The carpet consisted of a 100% polyester fur fabric (fake fur) with long hair, initially=80 mm. Air was pumped through the pipe while the static and total pressures of the air flow were measured at 15 sections along the pipe. These pressure measurements were made by a Prandtl probe [17,18], vertically and horizontally at each 5 mm of the diameter, to determine the velocity profile. Since the objective was to study how the velocity profile of the air flow was influenced by the soft walls (the hairy carpet) of the pipe, the hair was cut before a new series of measurements started. Finally only the fabric of the carpet remained.

The total pressure of a flow is the sum of the static pressure and dynamic pressure, i.e. the "wind force" of the flow. Since the latter depends on air density the dry and wet temperature of the air were also measured while the ambient air pressure was obtained from the local weather station. The test equipment used was standard laboratory equipment for studies of ventilation systems and related problems (Fig.1). The air flow rate was determined in the rectangular section (114 mm x 126 mm), before it reached the entrance to the test pipe, according to the guidelines stated by the Nordic Ventilation Group [19].

2.2. Theoretical Background

The theoretical background applied here can be found in introductory books on fluid dynamics see e.g. [20]. The state of a flow is characterized by its Reynolds Number (Re), which for a pipe flow this is given by:

$$Re = \frac{u_{mean} \cdot D}{\upsilon} \qquad (1)$$

which is a function of mean flow velocity u_{mean} (m s^{-1}), fluid viscosity ν (m^2 s^{-1}), and diameter of the pipe D (m). The size of the dimensionless Re means that the state of the flow is laminar for Re<2300 and turbulent for Re >4,000. In between these values the flow is in an intermediate state where the flow is neither laminar nor turbulent. For even greater Re the flow becomes increasingly turbulent. In laminar pipe flow the fluid velocity is described by a parabolic profile while the turbulent velocity profile is blunter. The laminar flow velocity u(r) is stable while the turbulent is not. The volume flow rate is obtained after integration over the cross sectional area, A.

$$Q = \int_0^R 2\pi \cdot r \cdot u(r) \cdot \partial r = U \cdot A \qquad (2)$$

The flow rate Q (m^3 s^{-1}) is for a laminar flow confined by a parabolic curve of height u_{max}, which is the maximum flow velocity.

$$Q = A \frac{u_{max}}{2} \qquad (3)$$

In fluid flows three different pressures are defined; total pressure (p_t), static pressure (p_s) and dynamic pressure (p_d). The static pressure is the one that the fluid medium exerts on the pipe wall, perpendicular to the flow direction. The dynamic pressure is the pressure that the fluid is creating in its flow direction, while the sum these two pressures is the total pressure.

$$p_t = p_{d+}p_s \qquad (4)$$

The dynamic pressure equals the kinetic energy of a flow divided by its volume, which means that it is the volumetric kinetic energy. So, by starting with the classical equation for kinetic energy E, we get;

$$E = \frac{m \cdot u^2}{2} \qquad (5)$$

Here, m is mass (kg), u is velocity (m/s) and ρ is density (kg m-3). Now, by dividing this energy by the volume flow rate $\dot{V}$ (m3 s-1) the hydro dynamic pressure (Pd) is given by

$$p_d = \frac{E}{\dot{V}} = \frac{m \cdot u^2}{2 \cdot \dot{V}} = \frac{\rho \cdot u^2}{2} \qquad (6)$$

The pressure drop, i.e. the friction loss, along the pipe was measured for each studied case of flow rate and hair length. In the case of a flow through a straight pipe with constant diameter the pressure drop is given by:

$$\Delta p = \frac{\lambda \cdot L \cdot p_d}{D_h} \rightarrow \lambda = \frac{\Delta p \cdot D_h}{L \cdot p_d} \qquad (7)$$

where Δp = pressure drop (Pa); λ= friction number (-); L = pipe length (m); D_h = hydraulic diameter (m);

Performed laboratory studies on turbulence reduction in pipe flow meant that 5600 manual pressure measurements were carried out to determine the flow velocities at different cross

TEST SETUP. Ventilation system by Airflow Developments Ltd, High Wycombe, UK.

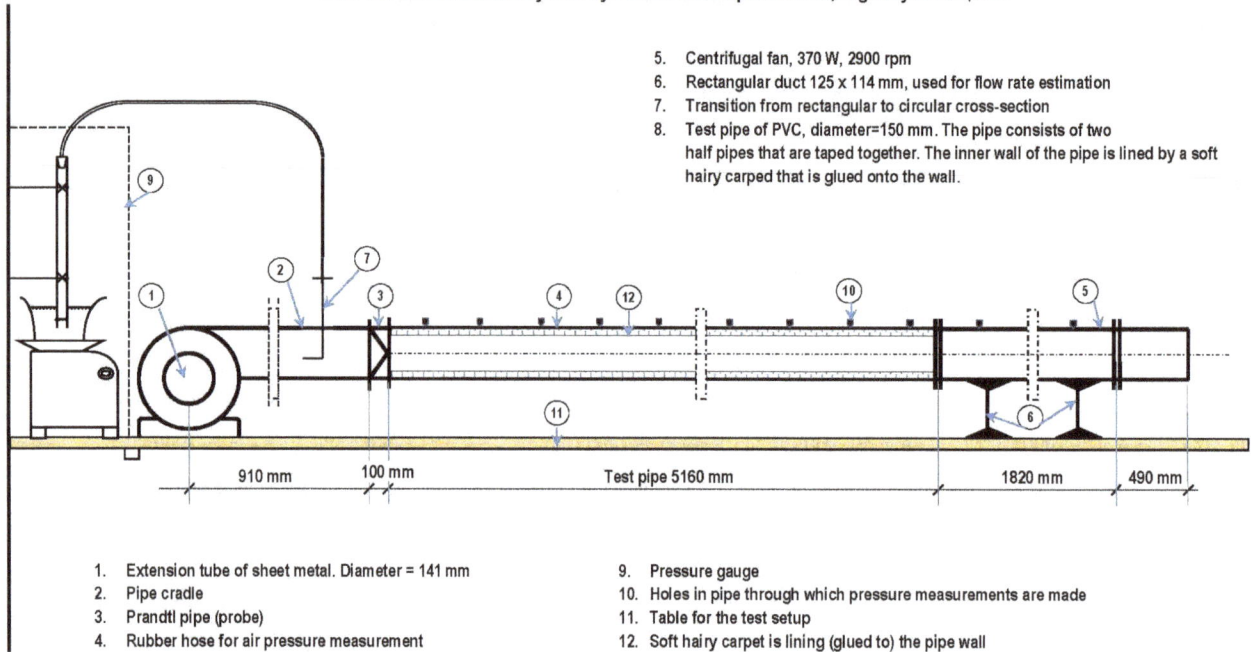

5. Centrifugal fan, 370 W, 2900 rpm
6. Rectangular duct 125 x 114 mm, used for flow rate estimation
7. Transition from rectangular to circular cross-section
8. Test pipe of PVC, diameter=150 mm. The pipe consists of two half pipes that are taped together. The inner wall of the pipe is lined by a soft hairy carped that is glued onto the wall.

910 mm   100 mm   Test pipe 5160 mm   1820 mm   490 mm

1. Extension tube of sheet metal. Diameter = 141 mm
2. Pipe cradle
3. Prandtl pipe (probe)
4. Rubber hose for air pressure measurement
9. Pressure gauge
10. Holes in pipe through which pressure measurements are made
11. Table for the test setup
12. Soft hairy carpet is lining (glued to) the pipe wall

Figure 1. *Description of the test setup that was used for the laminarization experiment in 1979 at LTU, Sweden.*

sectional areas of the pipe. Further pressure measurements were made in the section before the test pipe in order to determine the flow rate. The measurements were made during May and June 1979 and are available on the Internet [17].

The measured static pressure drops linearly along the test pipe as expected. One notable observation is that this pressure becomes negative for hair lengths greater than 15 mm. The negative static pressure was built up within a few seconds after the start of the air flow.

In performed laboratory test the flow velocity was determined by the dynamic pressure (p_d) of the air over cross-sectional areas of the test pipe. In order to do this the total pressure (p_t) and the static pressure (p_s) were measured, and the dynamic pressure was determined by eq. (4).

These measurements were made manually with, at the time, a common method for reliable air pressure measurements.

Rubber hose

Glass pipe
D=31.8 mm

Prandtl pipe

$H = P_t, P_s$

$P_t, P_s \rightarrow$

Figure 2. *Reliable pressure measurements were obtained by connecting a Prandtl probe to a water-filled vessel placed on an accurate balance.*

Actually the air pressure was transferred through a rubber pipe into a glass pipe that was partly submerged into a water filled container. The pressure was then seen as a reduced water level (H) in the glass pipe. Since the water-filled vessel was placed on a balance the mass of the displaced water was measured.

The mass of the displaced water was very accurately measured and a correspondingly accurate pressure was determined, see Fig. 1 and Fig. 2.

$$p_d = \frac{\rho_a \cdot u^2}{2} = H \rho_w g \text{ where } H = \frac{m_d \cdot 4}{\rho_w \cdot \pi d^2} \qquad (8)$$

The densities (kg m-3) of air and water are denoted ρ_a and ρ_w. The gravitational constant is g (m s^{-2}) and H is the displaced water i.e. air pressure in mm water column, see Fig. 2. The mass of displace water (m_d) is what is measured by the balance and recalculated into dynamic pressure and corresponding air velocity u is then given by eq. 9.

3. Results

3.1. Wave Pattern

A frequent observation in our studies of related flow phenomena was the presence of waves and wave pattern. When a dolphin swims at high velocity its skin vibrates and a wave pattern can be seen on its body. A flying owl has a blurry surface because its body surface, which is very flexible and resilient for vortices, shows a wavy movement. Even the cilia covered bronchial surface shows such wavy patterns.

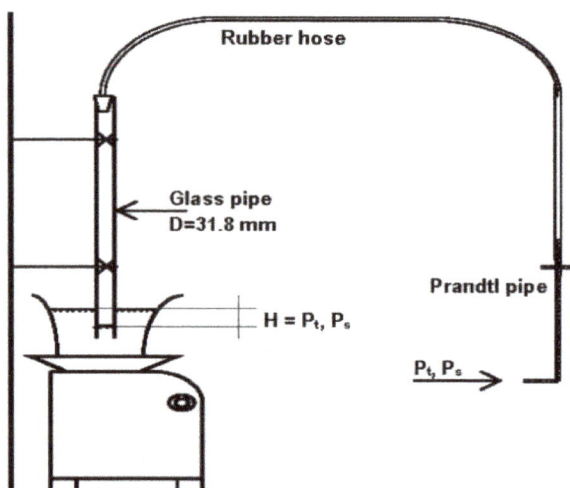

Based on these observations we expected to see a wave pattern also occur in the hairy test pipe. In order to investigate possible waves, the carpet was combed before each test run. After each test, the test pipe was opened "for a haircut". It was then observed that the mat was scruffy and disordered at the entrance of the pipe (~0.2 m), while a soft wave pattern had been formed along the rest of its length. The synthetic carpet retained the property of the waves even after the fan stopped. This pattern was ripple-like with a wave length of 0.1 – 0.2 m.

3.2. Friction Loss

The friction number was determined for the different hair lengths of the wall carpet, Fig 3. It is seen that the friction number was about the same for the 80 mm and the 50 mm hair length, and thereafter reduced for each haircut. The friction number is almost constant as a function of flowrate while it increases with increasing hair length. In laminar flows the friction number is $\lambda = 64/Re$ i.e. the friction reduces linearly with the increasing Re for Re < 2500. However, here with Re > 50000, this linear function for laminar flow is not valid.

3.3. Laminarization of Velocity Profile

The velocity profile of the pipe air flow was determined for various hair lengths of the wall carpet and for flows with Re between 30,000 and 150,000. In all tests the flow had an obvious turbulent velocity profile at the pipe entrance, which for some hair lengths the profile is almost perfectly laminar was changed along the pipe until it in some cases formed a rotated parabolic shape. Performed measurements show that even at very high Re. The Fig. 4a and Fig. 4b show the air velocity profiles along the pipe at Re~60000 for 8 different hair lengths. Measurements on other Re numbers are documented in [17].

The turbulent velocity profile in the smooth pipe (case 1:4) does not change along the 5 m test pipe. For the 80 mm hair length (case 2:4) the shape of the velocity profile is close to laminar at the end of the pipe and for 50 mm hair length (case 3:4) it is even better. Case 4:4 (25 mm hair length) is not as good as the 50 mm and 15 mm hair length (case 5:4). We therefore assume that our measurements were less accurate in case 4:4. It has most likely to do with an error in determining the flow rate. For increasingly shorter hair length (10 mm, 2 mm i.e. case 6:4 and case 7:4) the velocity profiles tend to change towards a more turbulent shape. In case 8:4 which means that all hair was removed, the hair was burned away so that only the mat canvas remained, the velocity profile is again fully turbulent.

In Fig. 5 the velocity profiles for the maximum Re of all eight cases are shown. Here, the profiles for all sections along the pipe are shown in the same diagram to show how the profiles are developing along the pipe. Also the ideal laminar profile is included in all graphs. The hair lengths of 50 mm and 15 mm (case 3 and case 5), are very close to laminar velocity profile. The reason why 25 mm hair is not as close is probably because of some measurement error, maybe in determining the flow rate. It is also obvious that a hair length of 10 mm or shorter is not able to laminarize the flow. For a hair length of 2 mm the velocity shows an almost turbulent velocity profile. Measured data and all velocity profiles are available [17].

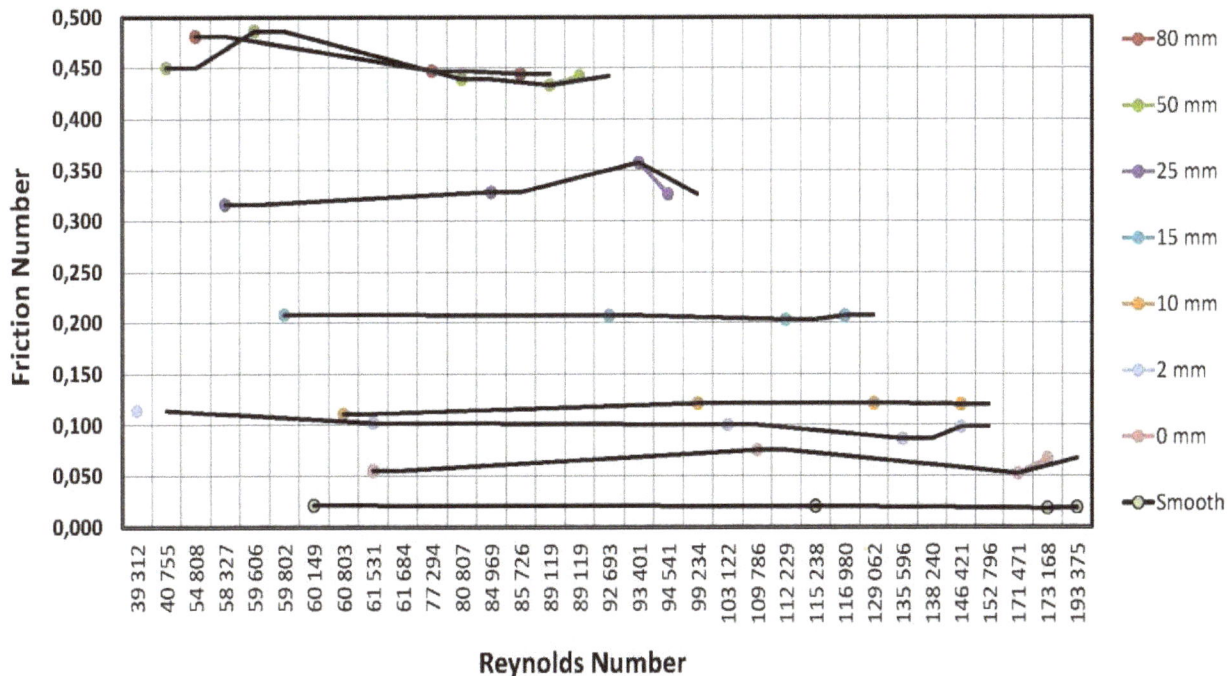

Figure 3. *Friction number for various hair lengths of the carpet lining the test pipe.*

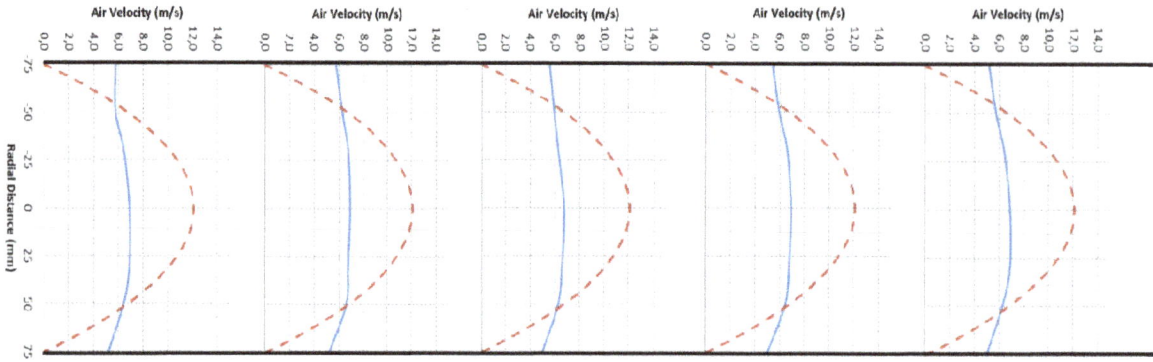

Case: 1:4, Re=60149, Smooth Pipe (no hair), Velocity profiles at 1.0, 2.0, 3.0, 4.0, 5.0 m from pipe entrance

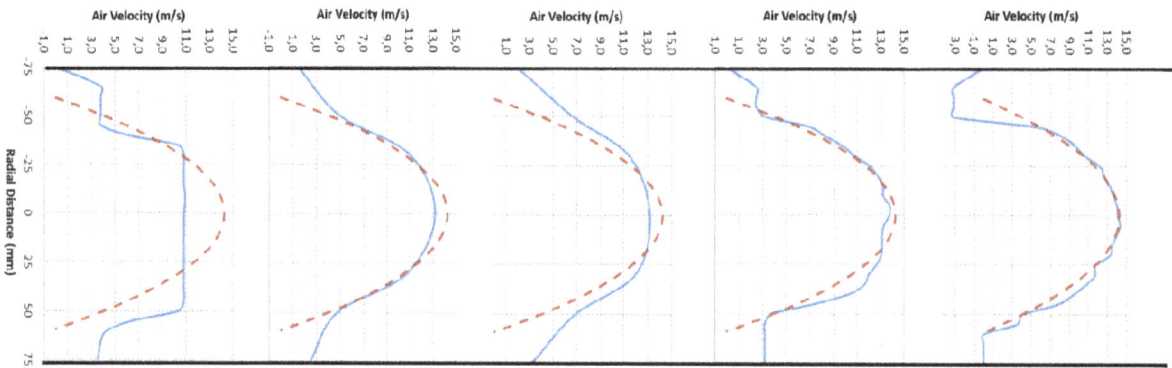

Case: 2:4, Re=55737, Hair= 80 mm, Velocity profiles at 0.15, 1.0, 3.0, 4.0, 5.0 m from pipe entrance

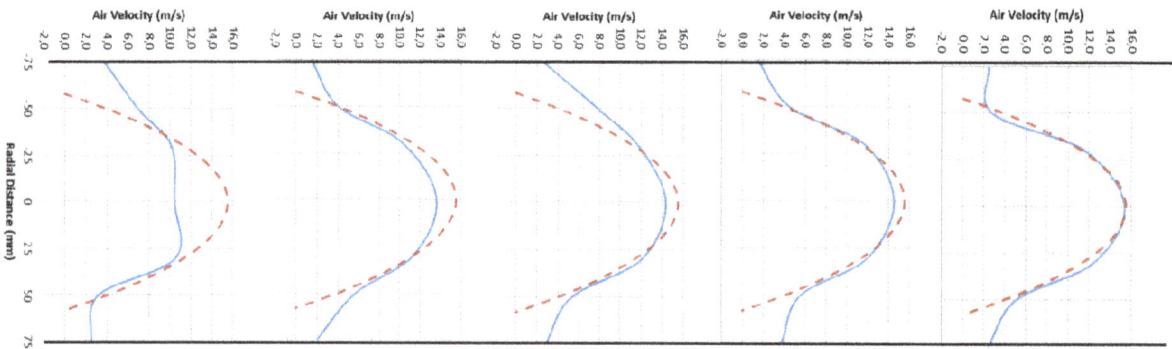

Case: 3:4, Re=59606, Hair= 50 mm, Velocity profiles at 0.15, 2.0, 4.0, 4.5, 5.0 m from pipe entrance

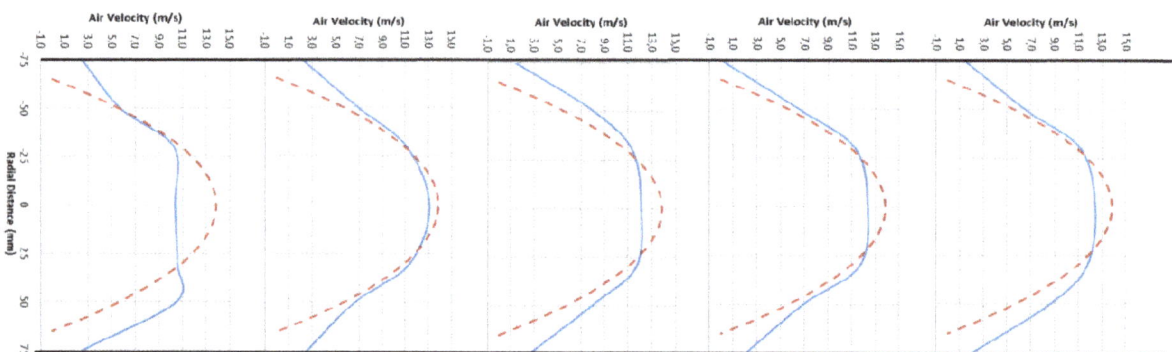

Case: 4:4, Re=59673, Hair= 25 mm, Velocity profiles at 0.15, 2.0, 4.5, 4.8, 5.0 m from pipe entrance

Figure 4a. Profiles of measured air velocity (solid line) and the ideal laminar velocity profile (broken line) at different sections along the pipe. Each case represents different Re numbers and hair lengths of the polyester carpet that is lining the pipe. The 8 cases presented in here are all for air flows with Re ~ 60000.

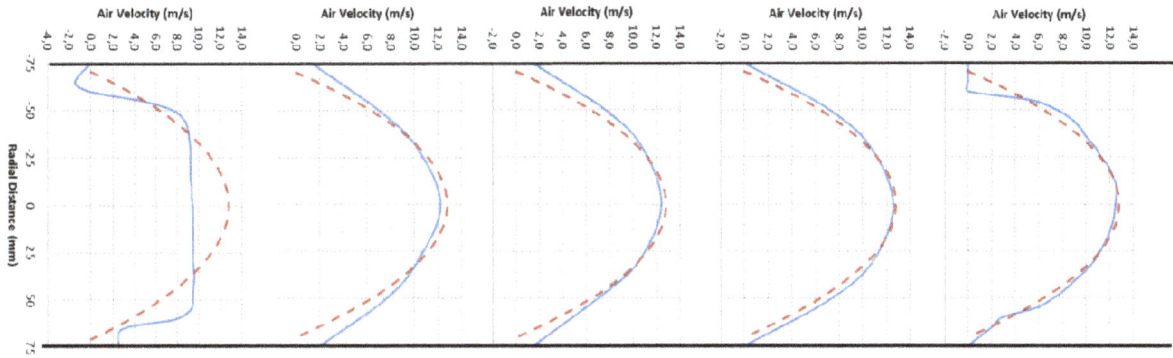

Case: 5:4, Re=59802, Hair= 15 mm, Velocity profiles at 0.15, 2.0, 4.7, 4.8, 5.0 m from pipe entrance

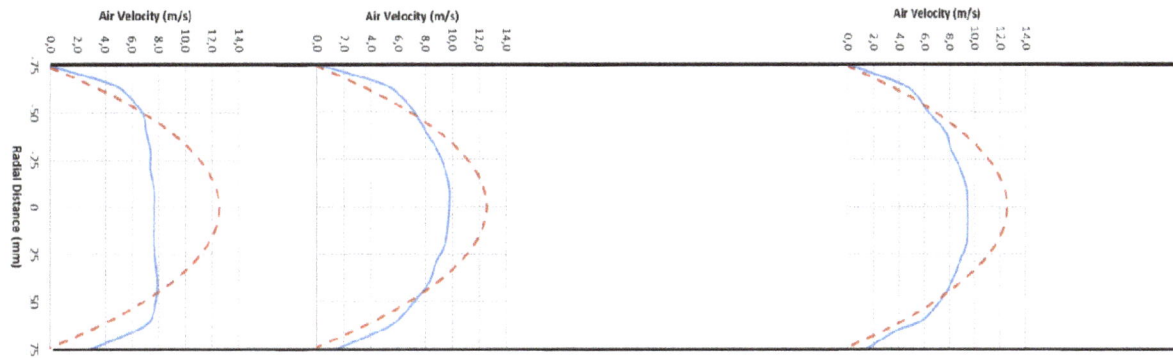

Case: 6:4, Re=60229, Hair= 10 mm, Velocity profiles at 0.15, 2.0, 5.0 m from pipe entrance

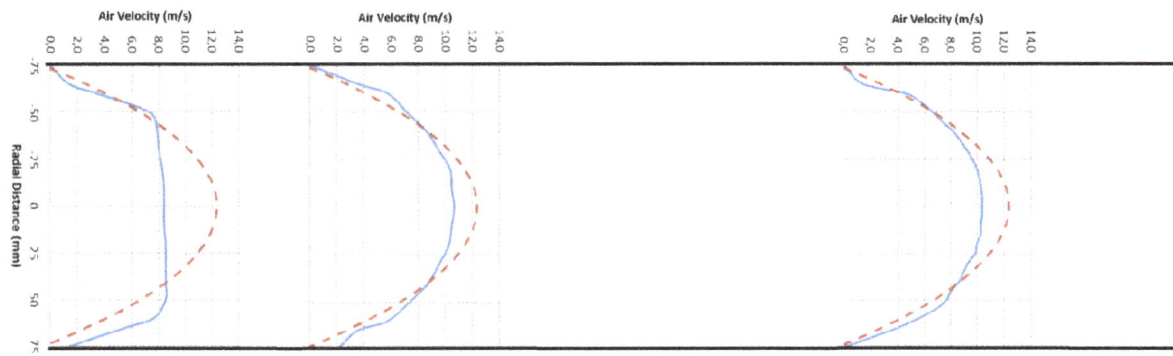

Case: 7:4, Re=61531, Hair= 2 mm, Velocity profiles at 0.15, 2.0, 5.0 m from pipe entrance

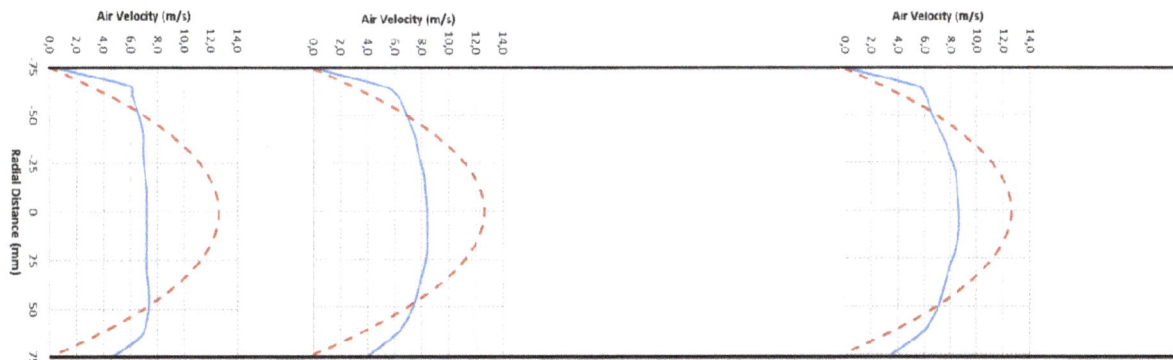

Case: 8:4, Re=62250, Hair= 0 mm (only canvas), Velocity profiles at 0.15, 2.0, 5.0 m from pipe entrance

Figure 4b. Profiles of measured air velocity (solid line) and the ideal laminar velocity profile (broken line) at different sections along the pipe. Each case represents different Re numbers and hair lengths of the polyester carpet that is lining the pipe. The 8 cases presented in here are all for air flows with Re ~ 60000.

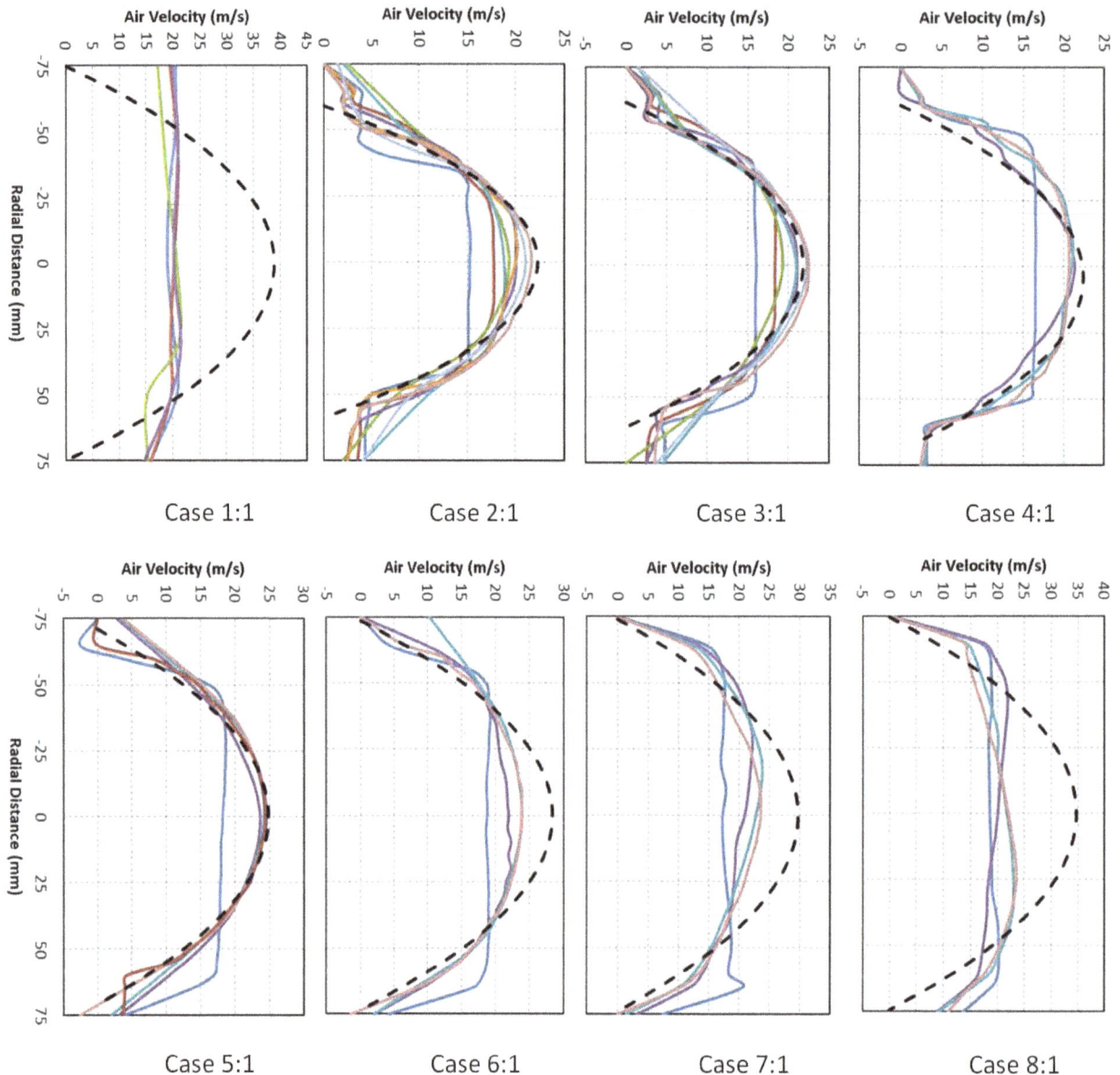

Figure 5. *The measured velocity profiles, which are tubulent at pipe entrance, are for some cases changed toward a laminar shape as it flows along the pipe.*

Case 1 is for a smooth pipe while the pipe is lined with a hairy carpet in the other cases. The hair is cut shorter for for each case, starting with 80 mm in Case 2. Case 1:1, Smooth pipe, Re =193375; Case 2:1, Hair length = 80 mm, Re = 85726; Case 3:1, Hair length = 50 mm, Re = 89119; Case 4:1, Hair length = 25 mm, Re = 94541; Case 5:1, Hair length = 15 mm, Re = 116980; Case 6:1, Hair length = 10 mm, Re = 138240; Case 7:1, Hair length = 2 mm, Re = 146421; Case 8:1, Hair length = 0 mm, Re = 171471.

4. Conclusions

Our objective was to investigate whether a pipe lined with a hairy soft carpet would laminarize air flows. The degree of laminarization was determined by the change of velocity profile along the test pipe. Varying flow rates were tested for each hair length before it was cut. It was shown that highly turbulent air flows (Re>100,000) seem to be laminarized by the hairy walls of the test pipe. If the hair of the wall mat is not too short the initially turbulent velocity profile changes towards a more laminar velocity profile. Some hair lengths worked better than others. An expected wave pattern occurred on the pipe wall. The evolving velocity profile is

certainly less random than the typical turbulent profile at such high Re numbers. The friction number is considerably greater than that of an ideal laminar friction number. However, the ideal laminar friction should not be applicable at such high Re. The linear pressure drop along the pipe, in spite of the increasingly laminarized flow, indicates that that there is cost of energy in maintaining the laminarized profile. Performed measurements are available [17].

Acknowledgement

We are grateful for the encouraging interest in our work that at the time was shown by late Prof. Mårten T. Landahl (1927-1999) at Aeronautics and Astronautics, MIT, and

Mechanical Engineering, KTH, Stockholm. We acknowledge the careful work of Katrina Nordell in transferring the handwritten measurements into digital format.

Nomenclature

A	cross sectional area of pipe	m^2
D	pipe diameter	M
D_h	hydraulic diameter	M
E	kinetic energy	$kg\ m^2\ s^{-2}$
g	gravitational constant	$m\ s^{-2}$
H	height of displaced water (i.e. air pressure in m water column)	M
L	pipe length	M
λ	friction number	–
m	mass	kg
m_d	mass of displaced water	kg
ν	dynamic viscosity	$m^2\ s^{-1}$
Δp	drop in static pressure	Pa
P_d	dynamic pressure	Pa
P_s	static pressure	Pa
P_t	total pressure	Pa
Q	flow rate	$m^3\ s^{-1}$
ρ	flow density	$kg\ m^{-3}$
ρ_a	density of air	$kg\ m^{-3}$
ρ_w	density of water	$kg\ m^{-3}$
R	radius of pipe	m
r	radial distance	m
Re	Reynolds Number	–
u_{mean}	mean flow velocity	$m\ s^{-1}$
u	flow velocity	$m\ s^{-1}$
u_{max}	maximum flow velocity	$m\ s^{-1}$
$\dot{V}$	volume flow rate	$m^3\ s^{-1}$

References

[1] Maxwell J.C. A Treatise on Electricity and Magnetism, 3rd ed., vol. 2. Oxford: Clarendon, 1892, pp.68–73.

[2] Pennel W. T., Eckert E. R. G., Sparrow E. M. Laminarization of turbulent pipe flow by fluid injection. J. Fluid Mech. (1972), vol. 52, part 3, pp. 451-464

[3] Zubkov V.G. Numerical study of laminarization effects in turbulent boundary layers of accelerated flows. Zhurnal Prikladnoi Mekhaniki I Teknicheskoi Fiziki, No 2, pp. 71-78, March-April, 1985. 0021-8944/85/2602-0215809.50, Plenum Publishing Corporation.

[4] Riasi, A., Nourbakhsh M., Raisee M. (2009). Unsteady turbulent pipe flow due to water hammer using k–θ turbulence model. J. Hydr. Res. 47:4, 429-437. DOI:10.1080/00221686.2009.9522018

[5] Cardoso A.H., Graf W.H., Gust G. (1991) Steady gradually accelerating flow in a smooth open channel, J. Hydr. Res. 29:4, 525-543, DOI:10.1080/00221689109498972

[6] Cederwall K., Sellgren A. (1973). Polymeradditiv (Polymer Additive), Chalmers University of Technology, Gothenburg, Sweden. (In Swedish)

[7] Toms, B.A., (1948). Some observations on the flow of linear polymer solutions through straight tubes at large Reynolds numbers. Proc. 1st International Congress on Rheology, Vol. 2, North Holland Publishing Co., Amsterdam, Holland, p. 134.

[8] Afzelius B. (1976). The role of cilia in man. Elsevier/North-Holland Biomedical Press.

[9] Wu T.Y., Brokaw, C. J., Brennen, C. (1975). Swimming and Flying in Nature, Vol 2, p 939-952. Plenum Press, New York and London.

[10] Graham RN. (1934). The silent flight of owl. J of Royal Aeronautic Soc., pp. 837-843.

[11] Gruschka H.D., Borchers I. U., Coble J. G. (1971). Aerodynamic Noise produced by a Gliding Owl. Nature 233, 409 - 411 (08 October 1971); doi:10.1038/233409a0

[12] Thorpe W.H., Griffin D.R. (1962). The lack of ultrasonic components in flight noise of owl compared with other birds. Ibis 104:256-257.

[13] Lindblad J. (1973). I Ugglemarker (Owl Land), Bonniers, Stockholm, Sweden. 211 p. ISBN 91-0-037720-1 (In Swedish)

[14] Gray, J. (1936) Studies in animal locomotion VI. The propulsive powers of the dolphin. J. Exp. Biol. 13: 192–199.

[15] Norberg R.Å. (1975). Skull asymmetry, ear structure and function, and auditory localization in Tengmalm's owl, Aegolius funereus (L.), with aspects on the evolution of ear asymmetry among owls. Doctoral Thesis, Gothenburg University.

[16] Cousteau J.Y., Diolé P. (1975) Dolphins. Doubleday & Company Inc. Garden City, New York.

[17] Nordell B., Gawelin R. (2014). Laminarization of Highly Turbulent Air Flow. Measurements in Excel format. http://dx.doi.org/10.5281/zenodo.14784)

[18] Nilsson R., Nordell B., (1979). Experiments in reducing turbulence in pipe flow. (Försök med turbulensminskning). MSc. Thesis 1979:040E, Div. Water Res. Eng., Luleå Univ. of Techn., Sweden. (In Swedish) http://pure.ltu.se/portal/files/91305512/1979_040E.pdf

[19] NVG (1974). Nordic Ventilation Group, Guidelines for air flow rate determination. Swedish Council for Building Research, R51:1974. (In Swedish)

[20] Batchelor G.K. (1967). An Introduction to Fluid Dynamics. Cambridge University Press, U.K.

7

Research on cutting tool wear based on fractional Brownian motion

Liang Jian-kai, Song Wan-qing, Li Qing

College of Electronic and Electrical Engineering, Shanghai University of Engineering Science, Shanghai, P.R. China

Email address:

liangjiankai126@126.com (Liang Jian-kai)

Abstract: Cutting tool wear is a very complex process. Various factors have a direct or indirect effect on cutting tool wear, resulting in uncertainty, so it is difficult for experimental data and result to have good stability. However, Vibration analysis is a very important means for condition monitoring and fault diagnosis. This paper aims to study the methods of tool vibration signal processing, pattern recognition and trend prediction. Collected on tool vibration signal at different times, wavelet noise reduction is used to pretreat the vibration signals. Then, for the self-similar vibration signals, we propose the fractional Brownian motion (FBM) theory with long-range dependence (LRD). Combined with Wigner-Ville spectrum, characteristic parameter can be extracted, so the cutting tool wear state can be determined according to fractal dimension and average slope of the fitting curve of the logarithm power spectrum. Finally, we use FBM model to predict the trend of tool vibration signals. Experiments show that the methods have a good effect on tool wear state recognition and trend prediction.

Keywords: Tool Wear, Fractal Dimension, Wigner-Ville Spectrum, FBM Model, Trend Prediction

1. Introduction

In the machining process, the tool wear will affect the surface quality and dimensional accuracy of the artifacts. Therefore, monitoring and predicting tool wear more accurately are currently an urgent issue to be solved in the automatic machining process [1]. Studies have shown that tool condition monitoring and diagnosis face a lot of non-stationary signals. Researching the new practical methods are the need to promote the continuous development of tools fault diagnosis.

Bhattacharyya used cutting force signal to estimate tool wear in face milling. Stephenson and Ali performed studies on tool temperature effects on interrupted metal cutting and reported theoretical and experimental results [2-3]. Iwata and Moriwaki used an acoustic emission signal to monitor tool wear condition in cutting processes. Zhang et al. used a Hall Effect sensor to measure the current supplied to the spindle motor drive of a vertical NC miller together with the cutting forces [4-5]. According to trend prediction, prediction experts have put forward a prediction method, which adopts a particle swarm optimization extended memory method which combines support vector regression (SVR) and a prediction method which combines support vector machines (SVM) and

wavelet neural network optimization [6-7]. This improves the accuracy of prediction, but has a complex computing process.

In the paper, combined FBM theory with Wigner-Ville spectrum, the cutting tool wear state can be determined according to fractal dimension and average slope of the fitting curve of the logarithm power spectrum. Then, the paper also proposes the FBM model and calculates the model parameters to predict the future tool vibration signals.

2. The Method of Tool Wear State Recognition

2.1. Fractional Brownian Motion (FBM)

If $0 < H < 1$, the fractional Brownian motion (FBM) with Hurst parameter H is the continuous Gaussian process $\{B_H(t,w), t > 0\}$, $B_H(0,w) = b_0$ and so whose definition is given by [8]:

$$
B_H(t,w) - B_H(0,w)
$$

$$
= \frac{1}{\Gamma(1+\alpha)} \left\{ \int_{-\infty}^{0} [(t-s)^{\alpha} - (-s)^{\alpha}] dB(s,w) + \int_{0}^{t} (t-s)^{\alpha} dB(s,w) \right\} \quad (1)
$$

Where $\alpha = H - 1/2$, $B(s,w)$ is a weiner process and $\Gamma(1+\alpha) = \int_0^\infty x^\alpha e^{-x} dx$.

If $H = \dfrac{1}{2}$ then $B_H(t,w)$ coincides with the standard Brownian motion B(t, w).

The constant H determines the sign of the covariance of the future and past increments. This covariance is positive when $H > \dfrac{1}{2}$, zero when $H = \dfrac{1}{2}$ and negative when $H < \dfrac{1}{2}$.

2.2. Wigner-Ville Spectrum

FBM is a non-stationary process. For the non-stationary process, its power spectrum does not have a clear definition, so it cannot be obtained by the usual methods. But by Wigner-Ville spectral decomposition, we get the power spectrum.

If the average power of fractional Brownian function $B_H(t)$ is limited, then

$$\lim \frac{1}{2T} \int_{-t}^{t} |B_H(t)|^2 dt < +\infty \qquad (2)$$

So the power spectrum $B_H(t)$ is proportional to f^{-2H-1} [9], then

$$\lim \frac{1}{2T} \left| \int_{-t}^{t} B_H(t) e^{-j2\pi ft} dt \right|^2 = Cf^{-2H-1} \qquad (3)$$

For non-stationary random process X (t), if the correlation function R (t, s), then the Wigner- Ville spectrum of X (t) [10] is

$$W_x(t,w) = \int_{-\infty}^{+\infty} R_x \left[t + \frac{\tau}{2}, t - \frac{\tau}{2} \right] e^{-iwt} d\tau \qquad (4)$$

So Wigner-Ville spectrum is a time-dependent spectrum, Wigner-Ville spectrum of FBM $S_{B_H}(w)$ is

$$S_{B_H}(w) = \lim_{T \to \infty} S_{B_H}(w,T) = \left(1 / |w|^{2H+1} \right) \qquad (5)$$

If the function Z(x) subjects to FBM of variable x, then the power spectrum of Z (x) is bound to obey the formula 4 and get the power spectrum of Z(x):

$$S_Z(w) = \frac{1}{|w|^{2H+1}} \qquad (6)$$

Take the logarithm to formula 6:

$$\log(S_Z(w)) = -(2H+1)\log(w) \qquad (7)$$

If the slope of the line is k, you can get the fractal parameter H=-(k+1)/2.The relationship about fractal parameters H and fractal dimension D of FBM is: D = 2-H, so we can get D = (5-k) / 2.

By calculating the power spectrum of vibration signals can we plot the logarithmic plot of power spectrum, the slope of the fitted line can be obtained by the least squares fitting. By analyzing the values of the parameter D in different time periods, we can find the regularity and judge the state of tool wear.

2.3. The FBM Model

2.3.1. FBM Incremental Simulation

If time interval [0, T] is divided into N equal parts, then each length is $\Delta t = T / N$. For each time interval t_j $(j = 0,1,2\cdots,N)$, FBM increment is discretized by Maruyama extended symbols [11-12]:

$$\Delta B(t_j, H) = \Delta B(t_j + \Delta t, H) - B(t_j, H) = w(t)(\Delta t)^H \qquad (8)$$

If T = 1, N = 200, = 0.005, H = 0.65, then we get incremental simulation diagram of FBM and see FBM with steady increments.

According to $B(t_{j+1}, H) = B(t_j, H) + w(t)(\Delta t)^H$, we get the FBM simulation curves, shown in Fig1.

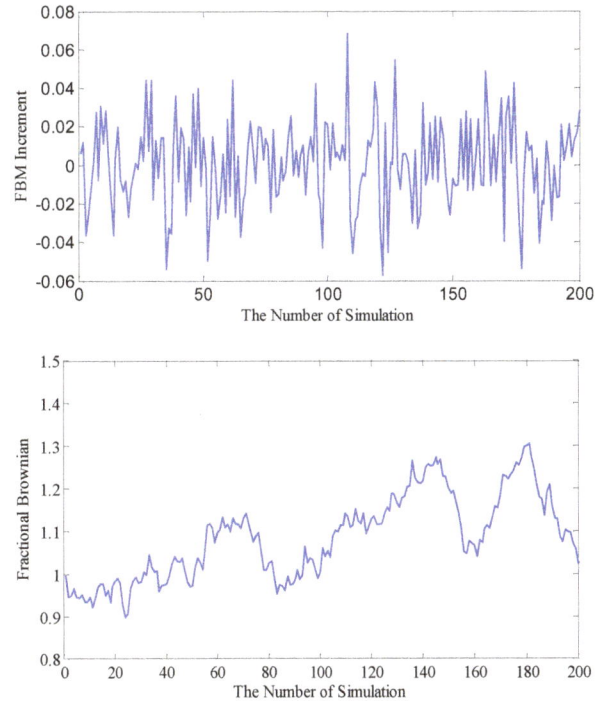

Fig. 1. FBM and its incremental simulation

2.3.2. FBM Stochastic Simulation

Simulate the vibration signals and its changes meet the following fractional stochastic differential equation.

$$dS = \mu S dt + \sigma S dB_H + \lambda S (dB_H)^2 \qquad (9)$$

Where μ is the drift rate, σ is the fluctuation rate of gain, λ is the interference term of gain.

The essential difference between this model and standard

Brown motion-driven model is that fractional Brown motion can be described the long-term memory of returns, and the increment is not independent of each other.

In this paper, the expansion of Maruyama symbol $dB_H = w(t)(dt)^H$ is used to simulate the FBM increment, the time period is divided into M equal-spaced intervals, the time interval is Δt, discrete stochastic differential equation [13]:

$$S_{t+1} = S_t + \mu S_t \Delta t + \sigma S_t w_1(t)(\Delta t)^H + \lambda S_t w_2(t)(\Delta t)^{2H} \tag{10}$$

Where $w_1(t)$ and $w_2(t)$ are independent which are standard normal distributions.

Take the initial signal S_0, FBM simulation of the time period is shown in Fig2.

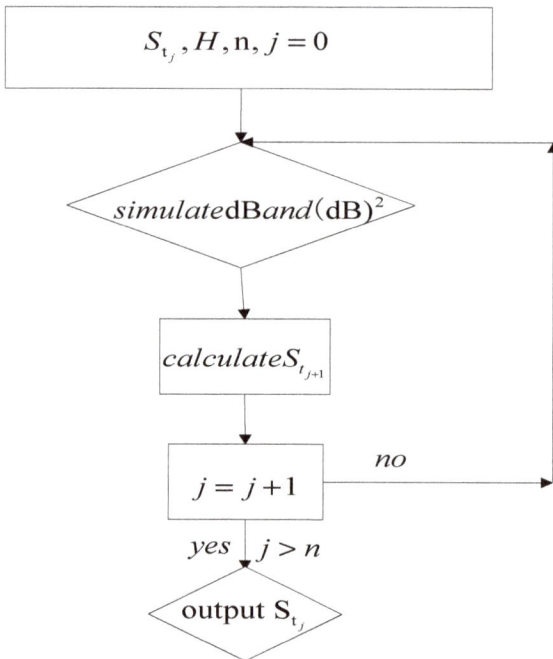

Fig. 2. Flow chart of signal simulation

2.3.3. Parameters Estimation

Assuming time interval of the gain data is Δt, the N +1 data of observation vector consist of $y = (y_0, y_{\Delta t}, \cdots y_{N\Delta t})$, time vector is $t = (0, \Delta t, \cdots, N\Delta t)$, fractional Brown motion vector is $B_H(t) = [B_H(0), B_H(\Delta t), \cdots, B_H(N\Delta t)]$, so the parameters μ and σ by the maximum likelihood estimation is as follows[14]:

$$\tilde{u} = \frac{t\Gamma_H^{-1} y'}{t\Gamma_H^{-1} t'} \tag{11}$$

$$\tilde{\sigma}^2 = \frac{1}{N} \frac{(y\Gamma_H y')(t\Gamma_H t') - (t\Gamma_H y')^2}{t\Gamma_H^{-1} t'} \tag{12}$$

Parameter λ can be obtained by fourth-order matrix describing the extreme phenomenon:

$$\tilde{\lambda} = \sqrt[4]{E(y - E(y))^4} \tag{13}$$

Where

$$y = \frac{S_{t+1} - S_t}{S_t}(t = 0,1,2\cdots, N) \tag{14}$$

$$\Gamma_H = \left[\text{cov}[B_H(i\Delta t), B_H(j\Delta t)]\right]_{i,j=0,1,2\cdots,N}$$
$$= \frac{1}{2}(\Delta t)^{2H} \cdot \left(i^{2H} + j^{2H} - |i-j|^{2H}\right)_{i,j=0,1,2\cdots,N} \tag{15}$$

In this paper, R/S method is calculated to Hurst parameter H of input sequence. Hurst coefficient $H \in (0,1)$ is a measure of a persistent random phenomenon. $H \in (0.5,1)$ can have a long- range dependence. The greater the value H is, the stronger the long- range dependence is. Hurst parameter on signal series is 0.6497, as shown in Fig3.

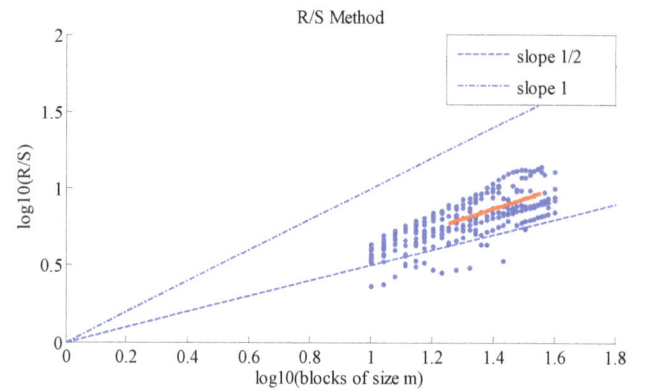

Fig. 3. Hurst index of vibration signals

According to the selected data as a calculation cycle, the parameters result is

$$\mu = 0.6011, \quad \sigma^2 = 0.0013, \quad \lambda = 0.0281$$

Through multiple simulations, we can get number of vibration signals approximate curves, then average for each time point of all possible approximation load curves, you can get on the approximate vibration signals values of each time point, that is the most likely path changes.

3. The Experiment Simulation

3.1. Tool Wear State Recognition

In this paper, we adopt db wavelets to denoise signals. Due to use different db wavelets, we can get different effects. And db4 wavelet relative to other wavelets has the shortest time window, and better time resolution, so we use db4 wavelet to denoise signals [15]. Because the selection and quantification of threshold is directly related to the quality of signal denoising, the paper selects the default threshold to denoise signals.

Collected 4096 tool vibration signals data for a period of time, it was divided four sections for analysis, the corresponding logarithmic power spectrum and fitted curve are shown in Fig4.

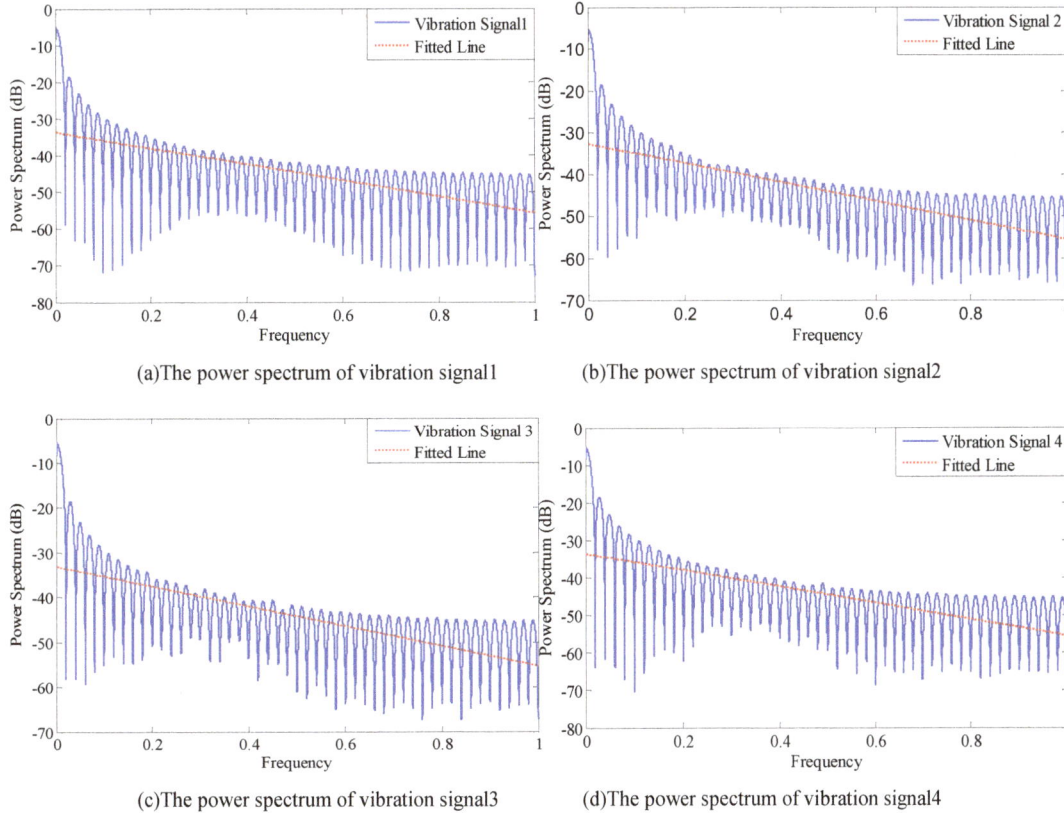

(a)The power spectrum of vibration signal1

(b)The power spectrum of vibration signal2

(c)The power spectrum of vibration signal3

(d)The power spectrum of vibration signal4

Fig. 4. *The power spectrum and fitted curve of vibration signals*

According to the Fig4, the average slope of the time period and fractal dimension D can be calculated by the fitting slopes, and we can in turn to calculate the fractal dimension D of other periods.

As can be seen from the Tab1, with the increase of the tool cutting time, the fractal dimension D of the Brownian motion of the vibration signal is also increasing. Thus, the larger fractal dimension D, the greater the tool wear.

Table 1. *The fractal dimension parameters*

Time Parameter	0 minute	20 minute	40 minute	60 minute	80 minute
D	13.54	13.69	14.44	14.96	15.7

3.2. Trend Forecast

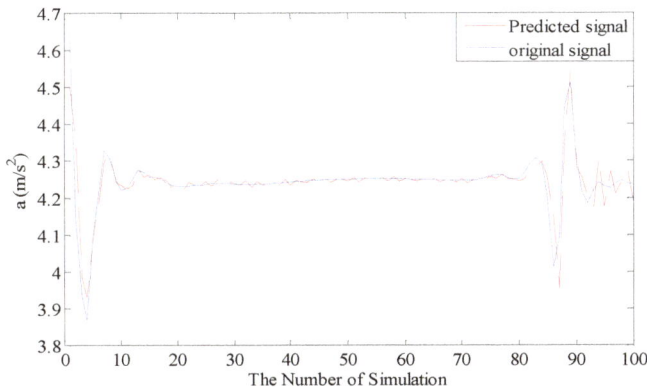

Fig. 5. *Original and predicted values of vibration signals*

According to the FBM model, we collect the tool vibration signals at a certain time and predict the future situation of vibration signals [16]. The next 100 vibration signals data are simulated and compared with the real values by using MATLAB, as it is shown in Fig5.

Figure 5 can clearly be seen that FBM model can relatively accurate predict the future values, the maximum relative error of FBM model is within 6%. From the terms of the maximum and minimum relative error, or the average relative error, the method has slightly satisfactory prediction results and provide a new method for predicting tool state.

4. Conclusions

For the characteristics of vibration signals of the tool wear state, it is studied on the application of FBM and Wigner-Ville methods to the tool wear state identification, three main conclusions are as follows:

- In terms of the instability, nonlinearity and long-range

dependence of vibration signals, the paper proposes the wavelet theory and FBM method combined with Wigner-Ville Spectrum;

- FBM can reflect the long-range dependence advantages of time series, combined FBM theory with Wigner-Ville spectrum, the cutting tool wear state can be determined according to fractal dimension and average slope of the fitting curve of the logarithm power spectrum;

- FBM model used in economy and network traffic is proposed to establish stochastic differential equation and simulation has a better prediction result and provides an effective method to study the long-term trend prediction.

Acknowledgements

This project is supported by Shanghai Natural Science Foundation of China (Grant No.14ZR1418500). The authors would like grateful to the editors and reviewers for their constructive comments.

References

[1]   D.E.Dimla, P.M.Lister, On-line metal cutting tool condition monitoring: force and vibration analyses, Int. J. Mach. Tools Manuf., 5(40), pp. 739–768, 2000.

[2]   P. Bhattacharyya, D. Sengupta, S. Mukhopadhyay, Cutting force-based real-time estimation of tool wear in face milling using a combination of signal processing techniques. Mechanical Systems and Signal Processing, 6(20), pp. 2665-2683, 2007.

[3]   D. A. Stephenson, A. Ali, Tool temperatures in interrupted metal cutting, Winter Annual Meeting of the ASME, pp. 261–281, 1990.

[4]   K.Iwata, T. Moriwaki, Application of acoustic emission measurement to in-process sensing of tool wear, Ann. CIRP 26 (1-2), pp. 19–23, 1977.

[5]   D. Zhang, S. Dai, Y. Han, D. Chen, On-line monitoring of tool breakage using spindle current in milling, In:1st Asia–Pacific and 2nd Japan–China International Conference Progress of Cutting and Grinding, Shanghai, China, pp. 270–276, 1994.

[6]   Hong W C, Chaotic particle swarm optimization algorithm in a support vector regression electric load forecasting model, Energy Conversion and Management, 50(1), pp.105-117, 2009.

[7]   Zhang Q, Lai K K, Niu D, Optimization Combination Forecast Method of SVM and WNN for Power Load Forecasting, IEEE Transactions on Computational Sciences and Optimization (CSO), pp.249-253, 2011.

[8]   Lundahl T, Ohley W J, Kay S M, et al, Fractional Brownian motion: A maximum likelihood estimator and its application to image texture, IEEE Transactions on Medical Imaging, 5(3), pp.152-161, 1986.

[9]   Jeon, Jae-Hyung, Fractional Brownian motion and motion governed by the fractional Langevin equation in confined geometries, Physical review, 2(81), 2010.

[10] S. Ghofrani, D.C. McLernon, Auto-Wigner–Ville distribution via non-adaptive and adaptive signal decomposition, Signal Processing, 8(89), pp.1540-1549, 2009.

[11] Longjin Lv, Ren Fu-Yao, The application of fractional derivatives in stochastic models driven by fractional Brownian motion, Physical a-statistical mechanics and its applications, 21(389), pp.4809-4818, 2010.

[12] Didier, Gustavo, Pipiras, Vladas, Integral representations and properties of operator fractional Brownian motions, Bernoulli, 1(17), pp.1-33, 2011.

[13] Wang li-li, Numerical Calculation and Empirical Analysis of American Options Pricing Based On Fractional Brownian Motion, Huazhong University of science and technology, 2012.

[14] Xiao we-lin, Research on the pricing method for warrants of long memory processes. South China University of science and technology, 2010.

[15] M. Kious, A. Ouahabi, Detection process approach of tool wear in high speed milling, Measurement, 10(43), pp.1439-1446, 2010.

[16] H. Saglam, A. Unuvar, Tool condition monitoring in milling on cutting forces by a neural network, Int. J. Prod. Res., 41 (7), pp. 1519–1532, 2003.

Effect of welding parameters on weld bead shape for welds done underwater

Joshua Emuejevoke Omajene, Jukka Martikainen, Paul Kah

LUT Mechanical Engineering, Lappeenranta University of Technology, Lappeenranta, Finland

Email address:

Joshua.omajene@yahoo.com (J. E. Omajene)

Abstract: The desire to model a control system so as to optimize the welding process parameters and the effect of the environment during underwater wet welding makes it necessary to study the effects of these parameters as it affects the weld bead geometry of welds achieved in underwater welding. The objective of this paper is to analyze how welding arc current, voltage, speed, and the effect of the water environment affect the weld bead geometry such as bead width, penetration, and reinforcement height. Comparing the differences of the effects of welding input parameters for air and wet welding as it affects the welding output quality parameter is the method employed in this research paper. The result of this study will give a better understanding of applying control mechanism in predicting the quality of a weld during underwater welding. A clearer insight of the weldability of structural steels for offshore applications as it relates to underwater welding, having a full knowledge of the nonlinear multivariable parameters is indicative of better control methods.

Keywords: Bead Geometry, Process Parameter, Water Depth, Water Temperature, Underwater Welding

1. Introduction

Underwater welding is used for repair welding of ships and other offshore engineering structures. The quality of welds achieved from underwater wet welding faces some quality challenges because of the rapid cooling of the weld by the water surrounding the weld zone. There is reduction in ductility and tensile strength of 50% and 20% respectively of the heat affected zone (HAZ) of welds in underwater welding as compared to air welding [1]. The bead geometry of an underwater weld is important in determining the mechanical properties of a weld joint [2].This paper gives a background for the design of an artificial neural network control of the welding process parameters as it affects the weld bead geometry. The optimization of the welding parameters (Fig. 1) which are nonlinear multivariable inputs will be discussed in the subsequent paper by the author. The water temperature and water depth are not welding process parameters but parameter of the water environment that affects the welding quality of underwater wet welding. The illustration of the weld bead geometry (Fig. 2) parameters, where W is the weld bead width (mm), R is the height of reinforcement (mm), and P is height of penetration (mm). WPSF (penetration shape factor) = W/P, WRFF (reinforcement form factor) = W/R. The strength of a weld is influenced by the composition of the metal, distortion and also the weld bead shape. The desired weld bead shape is dependent on the heat energy which is supplied by the arc to the base metal per unit length of weld, welding speed, joint preparation and the water environment in the case of wet welding [3, 4].

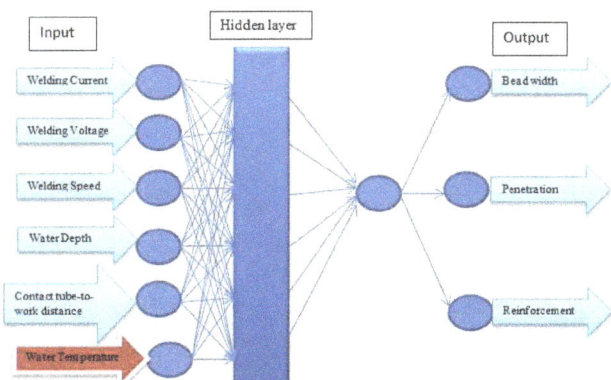

Fig. 1. Welding input vs output parameters.

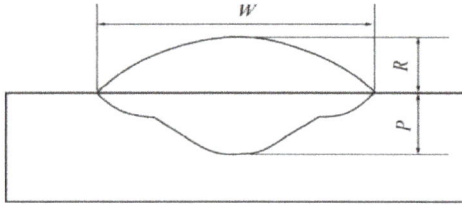

Fig. 2. Weld bead geometry of underwater wet FCAW [2].

2. Underwater Welding Processes

2.1. Shielded Metal Arc Welding (SMAW)

Shielded metal arc welding process is one of the most widely used welding processes. The process employs the heat of the arc to melt the base metal and the tip of the consumable covered electrode, and the flux covering the electrode melts during the welding. The melted flux forms gas and slag which act as a shield to the arc and molten weld pool (fig 3). The flux adds deoxidizes, scavengers and alloying elements to the weld metal. The welding current for SMAW can either be alternating current (AC) or direct current (DC) depending on the electrode used. Underwater SMAW uses a DCEN polarity. The use of DC is common in SMAW because of fewer arc outages, less spatter, easier arc starting, less sticking, and better control in out-of-position welds. Underwater welding electrodes are usually water proofed, and the flux creates bubbles during the welding process and displaces water from the welding arc and weld pool area. The welding speed and voltage have to be properly adjusted to create a stable bubble formation by the flux, because the gas bubble makes welding possible underwater. Underwater SMAW process uses two basic techniques known as the self-consuming techniques and the manipulative or weave technique. In the self-consuming technique, the electrode is dragged across the workpiece and pressure is applied by the welder. The manipulative technique, the arc is held as in surface welding and little or no pressure is required on the electrode. The manipulative technique requires more skills than the self-consuming technique [5, 6].

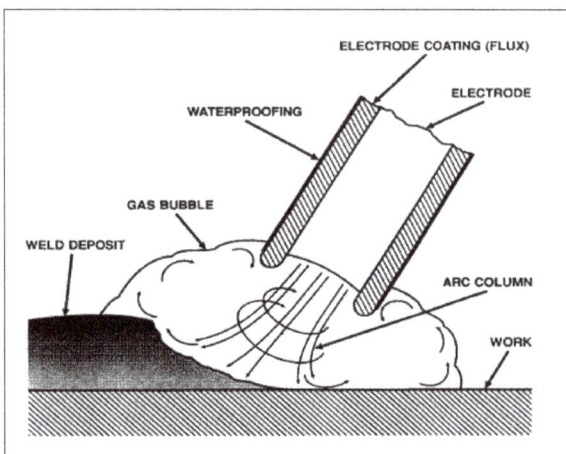

Fig. 3. Schematic illustration of SMAW process [5].

2.2. Flux Cored Arc Welding (FCAW)

FCAW uses heat generated by a DC electric arc to join the metal in the joint area. This welding process utilizes a continuously fed consumable welding wire that consists of a metal sheath which is composed of flux and alloying elements (Fig. 4). The flux formed is used to stabilize the welding arc, formation of slag, and produces shielding gas which protects the molten metal transfer and weld puddle as in SMAW process. The shielding in FCAW is achieved either by an additional gas shield supplied from an external source, or by the decomposition of the fluxing agent within the wire also known as self-shielding.The need for automation, out of position welding proficiency, high deposition rate and self-shielding capability has led to the development of underwater FCAW welding process [7]. FCAW technique using rutile type wire has a potential for high deposition rate and can be adapted to automated equipment [8]. The FCAW is suitable for underwater wet welding because of its self-shielding possibilities, ease of automation and out of position welding efficiency.

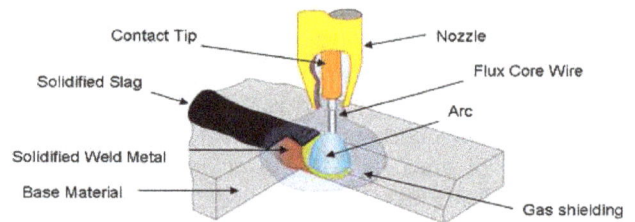

Fig. 4. Schematic illustration of FCAW process [9].

3. Welding Inputs

3.1. Welding Current

Welding current is an important parameter which affects the weld bead shape. The current controls deposition rate because it controls the melting rate of the electrode. The amount of melted base metal, the HAZ, depth of penetration is controlled by the welding current. An increase in welding current increases penetration and reinforcement. A low welding current will lead to unstable arc, low penetration and overlapping. Changing the polarity from direct current electrode posistive (DCEP) to direct current electrode negative (DCEN) affects the amount of heat generated at the electrode and workpiece which affects the metal deposition rate, weld bead geometry and mechanical properties of the weld metal. The positive electrode generates two third of the total heat, while the negative electrode generates one third of the total heat. The DCEN polarity produces higher deposition rate and reinforcement than the DCEP polarity in submerged arc welding. The level of hydrogen absorption in underwater wet welding which results in porosity can be minimized by using low current DCEP or a high current DCEN [3, 10, 11].

3.2. Welding Voltage

The length of the arc between the electrode and molten

weld metal determines the variation of the welding voltage. An increase in the arc length increases the voltage. The voltage determines the shape of the weld bead cross section and external appearance. Increasing the voltage at a constant current will result in a flatter, wider, and reduced penetration, which also leads to reduced porosity caused by rust on steels. Increase in arc voltage results to an increase in the size of droplets and thereby reduce the number of droplets. Increase in voltage enhances flux consumption, but a further increase in voltage will increase the possibility of breaking the arc and hinder normal welding process. Increase in arc voltage beyond the optimum value leads to an increase in loss of alloying elements which affects the metallurgical and mechanical properties of the weld metal. Arc voltage beyond the optimum value produces a wide bead shape that is susceptible to cracking, increase undercut and difficulty of slag removal. Lowering the arc voltage results in stiffer arc that improves penetration. Excessively lowering of the arc

voltage results in a narrow bead and difficulty of slag removal along the bead edges. A decrease in the welding voltage will decrease the diffusible hydrogen content during underwater welding. In many of the MILLER electric power sources, the constant current output are equipped with a feature called Arc Force, Dig or Arc Control, and Hot Start. In electric arc welding process, as the arc length increases, the voltage increases. The load voltage of a constant current welding machine is controlled by controlling the arc length. However these power sources does not work well for TIG welding process because it is better if the current does not change as the arc length changes in TIG welding. Power sources that have Arc Force (fig. 5B) allow the operator to change the shape of the volt/amp curve to meet the requirement of the operation being performed. Power sources with no Arc Force (fig. 5C) gives a more vertical shape of the volt/amp curve meaning that the current will not change much as the arc length is changed [10, 12-15].

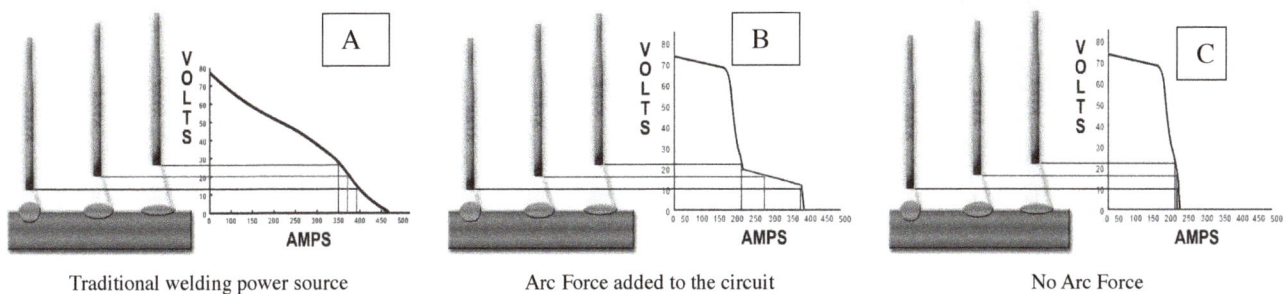

Fig. 5. Electric arc welding power sources [14].

3.3. Welding Speed

The linear rate at which the arc is moved along the weld joint influences the heat input per unit length of the weld. An increase in the welding speed will decrease the heat input and less filler metal is applied per unit length of the weld, which will result in less weld reinforcement and a smaller weld bead. The welding speed is the most influencing factor on weld penetration than other parameters except welding current. Excessive welding speed may result in porosity, undercutting, arc blow, uneven bead shape, cracking and increased slag inclusion in the weld metal. Higher welding speed results in less HAZ and finer grains. A relatively slow welding speed gives room for gases to escape from the molten metal, thereby reducing porosity. The bead width at any current is inversely proportional to the welding speed. A fast welding speed with low DCEN or high DCEP can minimize the level of porosity in wet welding [10, 11, 16].

3.4. Contact tube-to-Work Distance

Experimental evidence indicate that weld bead geometry is influenced by the change of the contact tube-to-work distance (CTWD) (Fig. 6) during welding. The CTWD influences the formation of the weld pool and the resulting weld shape by changing the arc length and welding current. the arc length which is related to the welding voltage and CTWD are

closely related to the weld bead geometry. The convection in the weld pool affects the weld pool geometry. An increase in arc length due to an increase in CTWD increases the bead width because of the widened arc area at the surface of the weld. Increase in arc length reduces the reinforcement height because the same volume of filler metal is used [17].

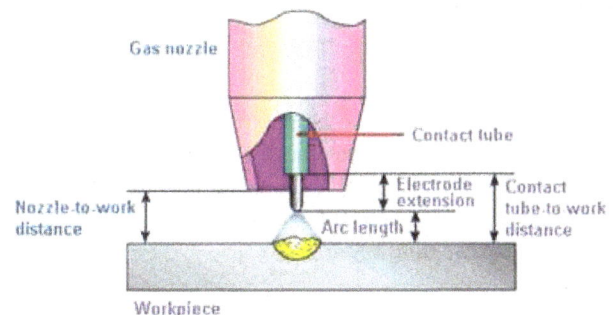

Fig. 6. Contact tube-to-work distance [18].

4. Environmental Parameters

4.1. Water Depth

Increased water depth constricts the welding arc resulting in an increased weld penetration on higher rate of filler metal transfer. The arc constriction results in an increased voltage and current as the water depth increases [19]. Weld metal oxygen, carbon, manganese, and silicon are fairly constant at water depth of 50 m to 100 m. oxygen content controls weld metal manganese and silicon content. At depth greater than 50 m, the diameter of the arc column decreases with increasing pressure (depth). Water reaction becomes increasingly important at depth greater than 50 m. The operating process parameter space decreases with increasing depth because high ionization potential for hydrogen makes it difficult to sustain the welding arc [11].

4.2. Water Temperature

The presence of water and the type of water surrounding the weld metal affects the welding process and the resulting temperatures. This effect is due to the greater convective heat transfer coefficient of water as compared to air which results in the rapid cooling in underwater welding [20]. At higher water temperature, for higher oxygen content, the diffusible hydrogen contents are likely to be less thereby, lowering the tendency of hydrogen assisted cracking. When welding at lower water temperature, there are few inclusions resulting in higher diffusible hydrogen level in the coarsed grain heat affected zone which leads to greater cracking tendency. The weldment sample (fig. 6) at 2.8^0C has cracking over two thirds of the weld height. The macrograph shows that the cracking progressed along the HAZ and a bit into the weld metal. At 10^0C the weldment has two cracks on the left hand edge, and the crack starts from the HAZ and unto the weld metal. The macrograph shows that the cracking is perpendicular to the direction of maximum tensile stress as a result of the cooling. The sample test conducted at the highest water temperature of 31^0C shows the highest volume fraction of inclusions of slag and oxide. This resulted in highest porosity level. Again, because of the higher oxygen content leading to lesser diffusible hydrogen content at higher temperature, thereby, resulting in a lower cracking tendency [19]. Underwater weld pool shape differs from air welds because of the heat transfer conditions caused by the cooling effect of the water and the presence of steam bubble [8].

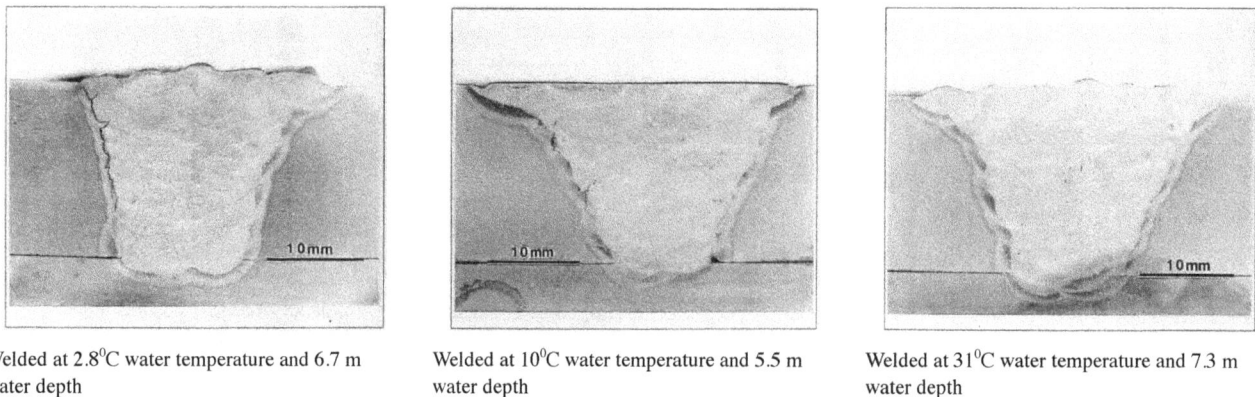

Welded at 2.8^0C water temperature and 6.7 m water depth

Welded at 10^0C water temperature and 5.5 m water depth

Welded at 31^0C water temperature and 7.3 m water depth

Fig. 7. Macrophotograph of different weld samples [19].

5. Welding Outputs

5.1. Bead Width

The bead width of a weld is the maximum width of weld metal deposited. This influences the flux consumption rate. The bead width is in direct proportion to the arc current, welding voltage and diameter of the electrode. It is inversely proportional to the welding speed [21, 22].

5.2. Penetration

Penetration is the maximum distance between the top of the base plate and depth of fusion. Penetration is influenced by welding current, welding speed, polarity, electrode stick out, basicity index and physical properties of the flux. Penetration is directly proportional to welding current and inversely proportional to welding speed and diameter of electrode. Increase in thermal conductivity of the weld metal decreases penetration. Deepest penetration is achieved with DCEP polarity than DCEN polarity. A stable arc increases penetration because the arc wander is minimized which allows more efficient heat transfer [19, 20].

5.3. Reinforcement

The reinforcement influences the strength of the weld and wire feed rate. Increasing the wire rate increases the reinforcement irrespective of welding current and polarity. Reinforcement is inversely proportional to welding voltage, welding speed and diameter of electrode. A bigger reinforcement is achieved with DCEN polarity than with DCEP polarity [21, 22].

6. Discussion

Weld joint is considered to be sound and economical if it has a maximum penetration, minimum bead width, reinforcement and dilution [23]. The relationship between speed, voltage, current, bead width, WRFF, and WPSF (Fig. 8) is explained for air welding in the figure below.

Howerever, underwater wet welding show some differences as compared to air welding. Water depth has a great influence on bead geometry when compared to other welding parameters when welding at a water depth less than 10 m.

Welding carried out at a water depth greater than 10 m and changing the welding speed highly influence the bead geometry than other welding parameters [2].

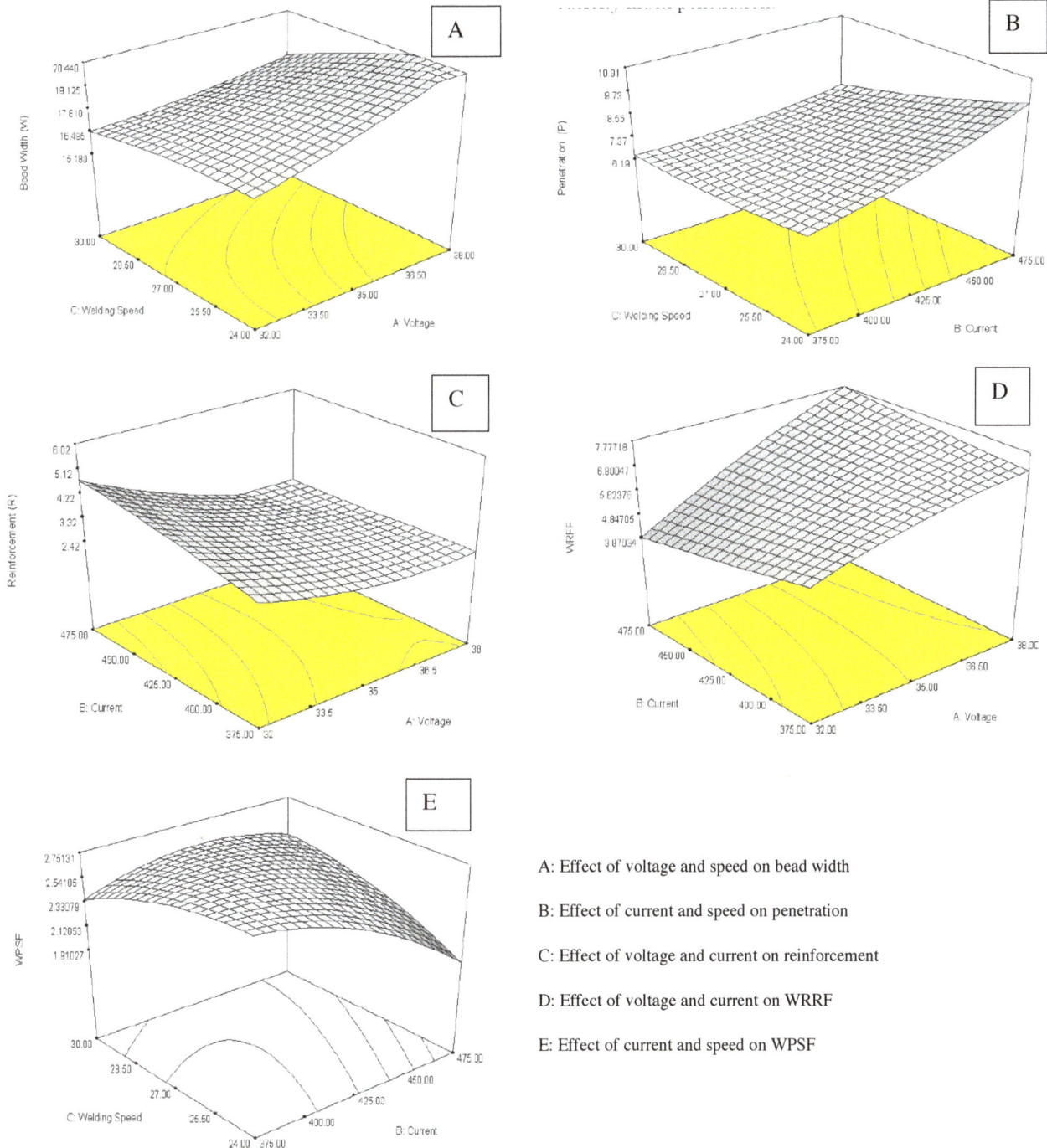

Fig. 8. Effect of various welding parameters on weld bead geometry for air welding [4].

A: Effect of voltage and speed on bead width

B: Effect of current and speed on penetration

C: Effect of voltage and current on reinforcement

D: Effect of voltage and current on WRRF

E: Effect of current and speed on WPSF

For all (Fig. 8A) values of speed an increase in voltage increases W. The bead width increases from 17.40 to 21.07 mm and from 16.49 to 17.75 mm with an increase in welding voltage from 32 to 38 volts, at welding speed of 24 and 30 m/hr respectively. This implies that voltage has a positive effect while welding speed has a negative effect on weld bead width [4].

The metal penetration (Fig.8B) increases with an increase in welding current for all values of welding speed. The metal penetration increases from 6.63 to 9.67 mm, and from 6.38 to 7.12 mm when welding current increased at welding speed of 24 and 30 m/hr respectively. Current has a positive effect on penetration while speed has a negative effect on penetration [4].

From (Fig. 8C), reinforcement decreases with increase in voltage with an increase in current from 375 to 475 amperes [4].

WRFF (Fig.8D) increases for all values of current when the voltage is increased from 32 to 38 volts. Voltage has a positive effect on WRFF while current has a negative effect.

Weld bead width increases with an increase in voltage but almost remain constant with change of current, while reinforcement decreases with increasing voltage and increase with increase in current [4]. The effect of current and speed on WPSF is shown in Fig. 8E.

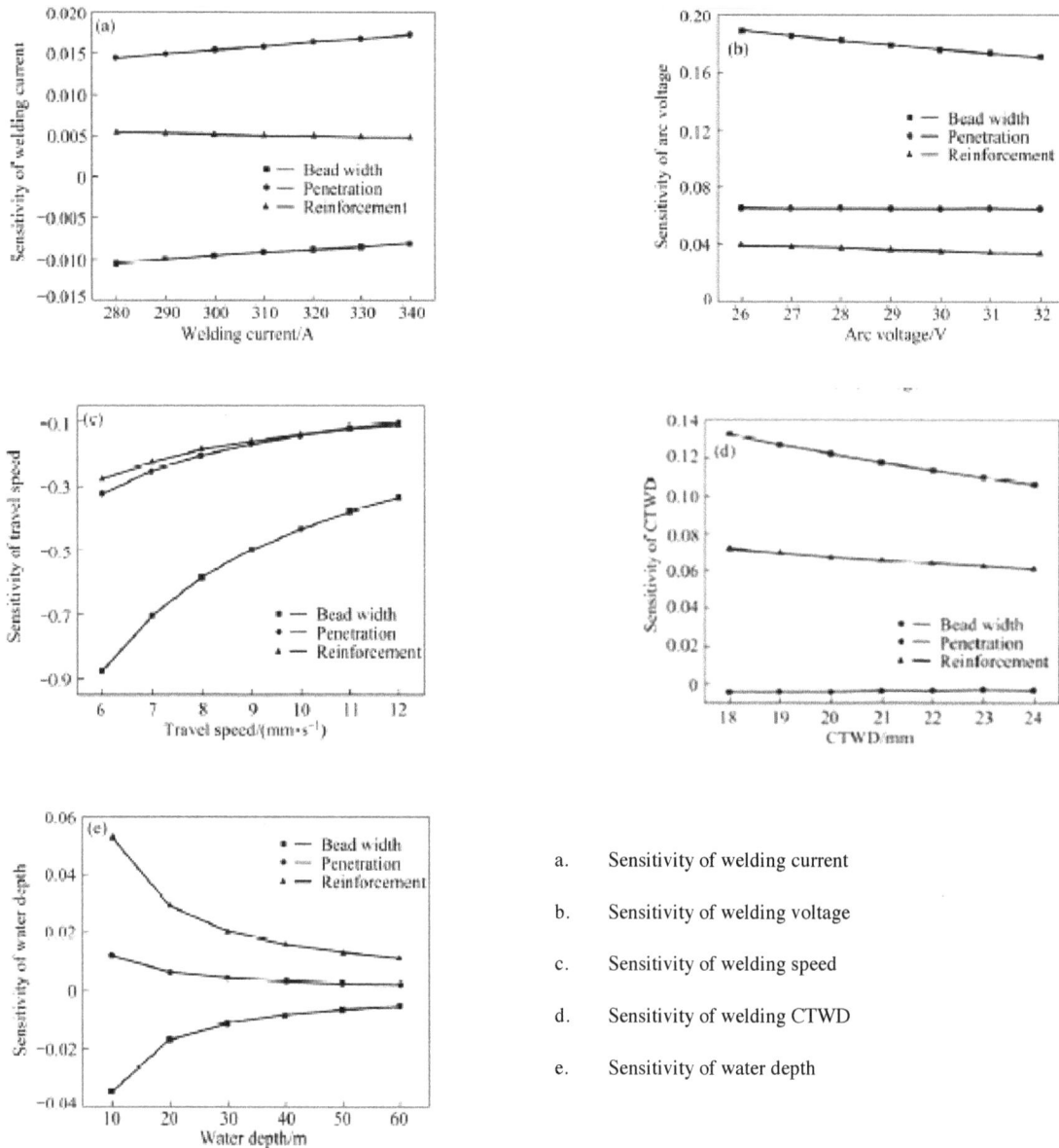

a.   Sensitivity of welding current

b.   Sensitivity of welding voltage

c.   Sensitivity of welding speed

d.   Sensitivity of welding CTWD

e.   Sensitivity of water depth

Fig. 9. Sensitivity of various welding parameters (welding current, speed, voltage) and water depth on weld bead geometry in underwater wet welding [2].

7. Conclusions

The major challenge facing underwater welding is the fast cooling rate of the base metal by the surrounding water. This leads to the formation of microstructures that are susceptible to cracking. The effect of water depth and the cooling rate of the weld metal from the water temperature influence the effect of welding current, welding speed, and voltage on the weld bead geometry during underwater welding. It is evident that a proper adjustment of welding process parameter to

minimize hydrogen pick up and alter the influence of the water environment can yield a sound weld in underwater welding as compared to air welding. The weld bead shape of a welded joint determines the mechanical properties of the joint. DCEN polarity is the most suitable power source for underwater wet welding. Arc stability increases penetration because of the reduction of arc wander and efficient heat transfer. Increase in water depth or pressure decreases bead width, but increases penetration, and reinforcement. Increase in voltage will increase bead width, penetration, and reinforcement. Welding current has greater influence on penetration than on bead width and reinforcement. Porosity in underwater welding can be minimized by the use of low current DCEP or high current DCEN. Voltage increase beyond the optimal value will result in the loss of alloying elements.

References

[1] R. T. Brown & K. Masubuchi, "Fundamental Research on Underwater Welding," *Welding research supplement*, pp. 178-188, 1975.

[2] Yong-hua SHI, Ze-pei ZHENG & Jin HUANG, "Sensitivity model for prediction of bead geometry in underwater wet flux cored arc welding," *Transaction of nonferrous metals society of China*, pp. 1977-1984, 2013.

[3] B. K. Srivastava, S. P. Tewari & J. Prakash, "A review on effect of arc welding parameters on mechanical behaviour of ferrous metals/alloys," *International Journal of Engineering Science and Technology*, vol. 2, no. 5, pp. 1425-1432, 2010.

[4] V. Kumar, "Modeling of weld bead geometry and shape relationships in submerged arc welding using developed fluxes," *Jordan journal of mechanical and industrial engineering*, vol. 5, no. 5, pp. 461-470, 2011.

[5] U. NAVY, "Underwater Cutting and Welding Manual," Naval Sea Systems Command, USA, 2002.

[6] J. D. Majumdar, "Underwater Welding - Present Status and Future Scope," *Journal of Naval Architecture and Marine Engineering*, vol. 3, pp. 39-47, 2006.

[7] WeiMin Zhang , GuoRong Wang, YongHua Shi & BiLiang Zhong, "LSSVM Model for Penetration Depth Detection in Underwater Arc Welding Process," *Journal of Information and Computing Science*, vol. 5, no. 4, pp. 271-278, 2010.

[8] M. Rowe & S. Liu, "Recent Development in Underwater Wet Welding," *Science and Technology of Welding and Joining*, vol. 6, no. 6, pp. 387-396, 2001.

[9] "https://www.hera.org.nz/Category?Action=View&Category_id=522," Hera Innovation in Metals, 2014. [Online]. Available: https://www.hera.org.nz. [Accessed 16 June 2014].

[10] V. Kumar, "Use of Response Surface Modeling in Prediction and control of flux consumption in submerged arc weld deposits," *Use of Response Surface Modeling in Prediction and*, vol. 2, pp. 1-5, 2011.

[11] S. Liu, D. L. Olson & S. Ibarra, "Underwater Welding," *ASM Handbook*, vol. 6, pp. 1010-1015, 1993.

[12] S. P. Tewari, A. Gupta & J. Prakash, "Effect of welding parameters on the weldability of material," *International Journal of Engineering Science and Technology*, vol. 2, no. 4, pp. 512-516, 2010.

[13] J. Singh, C. D. Singh, J. S. Khamba & F. P. Singh, "Influence of Welding Parameters on Flux Consumption in Submerged Arc Welding," *International Journal for Multi Disciplinary Engineering and Business Management*, vol. 2, no. 1, pp. 1-3, 2013.

[14] "http://www.techtrain123.com/publicdownloadsallfiles/Stick%20Welding.pdf," [Online]. Available: http://www.techtrain123.com. [Accessed 15 June 2014].

[15] Miller, "http://www.millerwelds.com/pdf/guidelines_smaw.pdf," July 2013. [Online]. Available: http://www.millerwelds.com. [Accessed 15 June 2014].

[16] L. J. YANG, R. S. Chandel & M. J. Bibby, "The Effects of Process Variables on the Weld Deposit Area of Submerged Arc Welds," *WELDING RESEARCH SUPPLEMENT*, pp. 11-18, 1993.

[17] J. Kim & S. Na, "A Study on the Effect of Contact Tube-to-Workpiece Distance on Weld Pool Shape in Gas Metal Arc Welding," *Welding Research*, pp. 141-152, 1995.

[18] "http://2dayforme.blogspot.fi/2012/11/wps-variables.html," [Online]. Available: http://2dayforme.blogspot.fi. [Accessed 29 June 2014].

[19] R. L. Johnson, "The effect of water temperature on underbead cracking of underwater wet weldments," Naval Postgraduate School, California, 1997.

[20] P. Ghadimi, H. Ghassemi, M. Ghassabzadeh & Z. Kiaei, "Three dimensional simulation of underwater welding and investigation of effective parameters," *Welding Journal* , pp. 239-249, 2013.

[21] B. Singh, Z. A. Khan & A. N. Siddiquee, "Review on effect of flux composition on its behaviour and bead geometry in submerged arc welding," *Journal of mechanical engineering research*, vol. 5, no. 7, pp. 123-127, 2013.

[22] V. K. Panwar & D. K. Choudhary, "Application of Two-Level Half Factorial Design Technique for Developing Mathematical Models of Bead Penetration and Bead Reinforcement in SAW Process.," *International Journal of Innovative Research in Science, Engineering and Technology*, vol. 2, no. 6, pp. 2011-2023, 2013.

[23] J. E. R. Dhas & M. Satheesh, "Sensitivity analysis of submerged arc welding parameters for low alloy steel weldment," *Indian Journal of Engineering and Materials Sciences*, vol. 20, pp. 425-434, 2013.

9

Volume of Material Removal on Distortion in Machining Thin Wall Thin Floor Components

Garimella Sridhar, Ramesh Babu Poosa

Department of Mechanical Engineering, University College of Engineering, Osmania University, Hyderabad, India

Email address:
garimella_s@yahoo.com (G. Sridhar), prbmechou@yahoo.com (R. B. Poosa)

Abstract: Thin wall thin floor monolithic components produced from prismatic blocks are machined on CNC machines by removing material up to 95 %. Components distort because of stresses induced due to severe heat generated and plastic deformation during cutting. Distortion of the components after machining aluminium alloys is major problem faced in aerospace and automobile industries. The volume of the material removed has direct bearing on machining time, which may lead to distortion. The objective of this study is to determine the effect of volume of material removal on the distortion of aluminium 2014 T651 alloy machined from prismatic block. Machining experiments were carried out with 5 different blank sizes to produce a representative component with same machining parameters. Distortion is measured on the face opposite to the machined surface and maximum deviations were used for comparisons between distortion and volume of material removal.

Keywords: Volume of Material Removal, Distortion, Blank Size

1. Introduction

With high strength to weight ratios and reduced assembly costs, sheet metal and multiple part assemblies are replaced by single piece monolith high strength aluminium alloy designs. With huge amount of material to be removed from large areas machining of these on CNC machines has become quite common and inevitable. Previous studies of authors on challenges in machining these components show that one of the biggest problems is distortion [1]. In a general manufacturing scenario all the monolithic thin structured components are produced on CNC machining centres from Pre-machined Prismatic blanks of planned sizes and thickness. The blanks are generally cut from rolled sheets of pre-planned thickness and are sized to the required dimensions on a conventional milling machine before machining them on CNC machining centres. As the volume of material removed is up to 95 %, the components distort due to stresses induced into the component because of temperature gradient and differential plastic deformations during cutting [2-4]. The magnitude, distribution and type of residual stresses induced during cutting is the main cause of distortion which is a function of machining parameters, tool geometry, cutting strategy and clamping methodology for a component of given material [5].

In recent years, lot of research was done to control the distortion of parts during machining. Hengbo Cui, Jong-Yun Jung and Dug-Hee Moon applied Taguchi method to know affect of deformation caused by heat during cutting of AL 7050/T7451 and found that cutting speed is the most influencing factor which causes deformation due to heat and the change of feed range has an insignificant effect on heat deformation [6]. Dong, Hui-yue, and Ying-lin KE carried out comparison of simulation and machining experiments on wing spar made of aluminium 7075 alloy using single tool- tooth milling simulation using Deform 3D, importing the force and temperature data into Abacus and simulating the machining experiments further by restart calculation and local re-meshing. The experiments showed good agreement with the simulation results and demonstrated that this method can be used to select optimal tool-path and machining sequences for minimizing distortion [7]. J-F. Lalonde, M.A. Gharghouri and J-F. Chatelain in their experiments and measurement of residual stresses by neutron diffraction method on controlled pre-processed blanks and standard blanks found that, the standard aluminium blanks distorted more because of asymmetrical bulk residual stresses in the blank because of previous operations before machining [8]. Younger, Mandy S., and Kenneth H. Eckelmeyer in their study concluded that distortion increases with increasing residual stress magnitude, increasing machining depth, and increasing machining asymmetry [9]. Denkena, B., and L. de León conducted machining experiments and showed in the results that machining operation and cutting edge geometry has a definite influence in the residual depth profile and residual stress distribution which effect distortion [10]. Marusich T.D., DA. Stephenson, S. Usui, and S. Lankalapalli proposed a methodology for distortion in thin-walled components due to both bulk and machining induced residual stresses which will aid in selecting parameters which can minimize distortion [3]. O. Belgasim and M. H. El-Axir in their study on aluminium magnesium alloy using response surface method, conducted experiments and concluded that residual stresses are sensitive to tool nose radius and feed rate. Cutting speed and depth of cut are significant parameters affecting maximum residual stresses which in turn affects distortion [11]. Similar results were achieved by author in experiments conducted using taguchi method to know the effect of distortion in machining aluminium alloy AA2014 T651. Depth of cut and width of cut were found to be significant contributors in machining distortion [12]. Keleshian, N, et al., in their work conducted machining experiments on aluminium alloy 7249 and observed that solution treatment temperature, quenching media, and various machining sequences affect distortion [13]. Chatelain, Jean-François, Jean-François Lalonde, and Antoine S. Tahan conducted experiments on a thin representative part to compare distortions with and without existing residual stresses in work pieces and concluded that initial residual stresses embedded within raw material has an effect on the final part deformation [5]. Songtao Wang, Zheng Minli, Fan Yihang, and Li Zhe conducted machining experiments on aluminium alloy 7475-T7351 and concluded that machining deformation of thin-walled components can be effectively controlled with the parameter combination of big radial cutting depth and small axial cutting depth in the condition of high spindle speed [14]. Huang, Xiaoming, et al., conducted high speed milling experiments on aluminium alloy 7050-T7541, measured residual stresses on the surface & subsurface of the work piece and showed that decrease of the cutting speed with increase of the feed rate lead to increase of compressive residual stresses [15]. Huang, Xiaoming, Jie Sun, and Jianfeng Li studied the effects of bulk residual stresses & machining induced residual stresses and observed that machining induced residual stresses is the primary cause of the distortion and has greater effect when the thickness is below 1.25 mm, also location of the part in the raw material has some effect on distortion [16].

Although, much research was done on distortion during machining of thin structured components, the impact of volume of material removal on distortion was not studied. Volume of material removed is total amount of material machined out from the prismatic blank to form component. As thickness of the blank increases the volume of the material to be removed and time of machining increases. This may lead to increase in influence of magnitude and distribution of residual stresses and distortion of the component after machining and removing from the fixture. So, in order to understand the effect of volume of material removal on distortion, machining experiments were carried on representative thin wall thin floor part from prismatic blocks of different thicknesses on CNC machining centre using aluminium alloy 2014 T651 under same machining conditions.

2. Experimental Setup and Procedure

In this study milling experiments were carried out on Vertical CNC machining centre on a representative thin wall thin floor part to know the effect of volume of material removal on distortion of the component.

2.1. Workpiece

Aluminium alloy 2014 T651 is used as work piece material. The mechanical properties of the material are shown in Table 1 and chemical properties of the material are shown in Table 2. The alloy is copper based aluminium alloy which is solution treated, artificially age hardened and stress relieved by stretching. This alloy is commonly used for airframes and mechanical packages for avionics. The blank sizes of length 105 mm and width 40 mm with different thicknesses i.e., 12 mm, 14 mm, 16 mm, 18 mm, and 20 mm are used for the experiment. The representative thin wall thin floor component for experimental work is shown in Figure 1.

Table 1. Mechanical Properties.

Property	Value
Yield strength	380 Mpa
Tensile strength	405 Mpa
Hardness Rockwell B	82
Density	2.80 g/cc
Poisson's Ratio	0.2

Table 2. Chemical composition.

Property	Value
Copper	3.8 to 4.8
Magnesium	0.2 to 0.8
Silicon	0.6 to 0.9
Iron	0.7 max
Manganese	0.2 to 1.2
Aluminium	Reminder

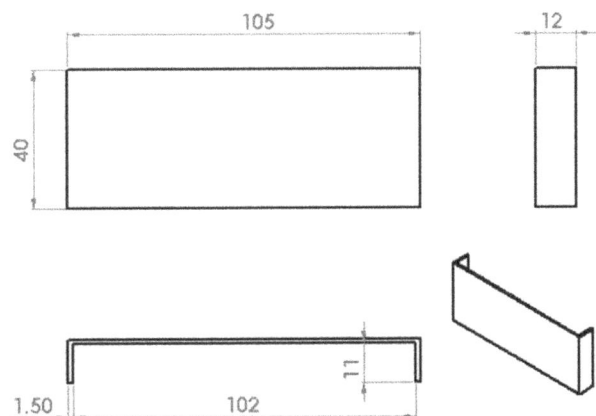

All Dimensions are in mm

Figure 1. Experimental Workpiece.

Figure 2. Solid Carbide slot Drill.

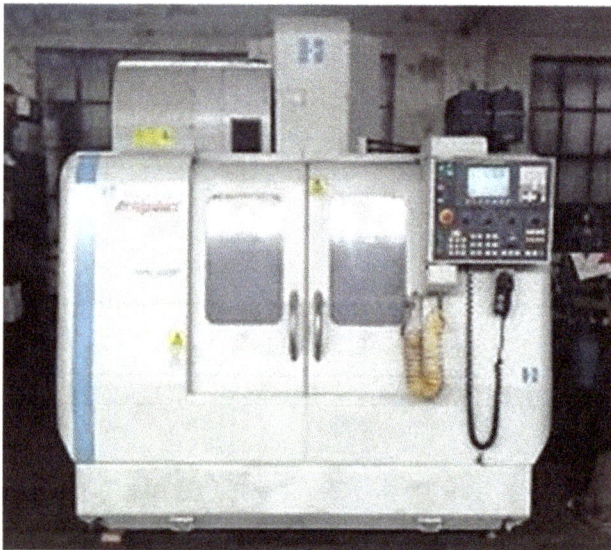

Figure 3. Hardinge Bridgeport VMC 600 P3.

Table 3. Specification of Machine.

Maximum RPM	8000 RPM
No. of Axes	3-Axes
X Axis Travel	600mm
Y Axis Travel	510mm
Z Axis Travel	510mm
Rapid Traverse	30 m/min
Feed	12 m/min
Power	13kW
Tool Station	20 Tools
Maximum Load	700kg

2.2. Tool

All the machining experiments were carried out using low helix two flute solid carbide Slot Drill Ø10 mm. Figure 2 shows the picture of the tool. New cutter is used for each

machining experiment to eliminate the affect of tool wear. The machining experiments were carried out on Hardinge Bridgeport VMC 600 P3 3-axis Vertical Machining Centre as shown in Figure 3. Table 3 shows the specifications of the machine tool. All the experiments were carried by holding the component from the bottom using specially made vacuum fixture as shown in Figure 4.

Figure 4. Vacuum Fixture along with work piece.

2.3. Measurement

Before Experiments, 18 points are marked on the opposite side of the face to be machined on all the work pieces as shown in Figure 5. The distortion is measured by taking the difference of Deviation before and after machining on the 18 marked places and maximum deviation is taken for comparison. Distortion measurements were carried using Metris LK Integra using CAMIO 4.4 software with Specifications: Size 800 mm X 700 mm X 600 mm, Accuracy 1.9+L/450 µ, Repeatability 2.2 µ and probe error 3.6 µ as shown in Figure 6. Twist in the components was also measured using Feeler gauges. The measurement of twist is shown in Figure 7. Comparisons between the experiments were done by taking the maximum deviation of the work piece after machining.

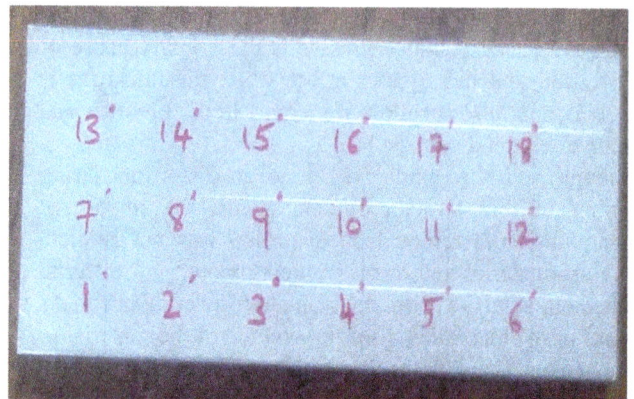

Figure 5. Marking of measuring point.

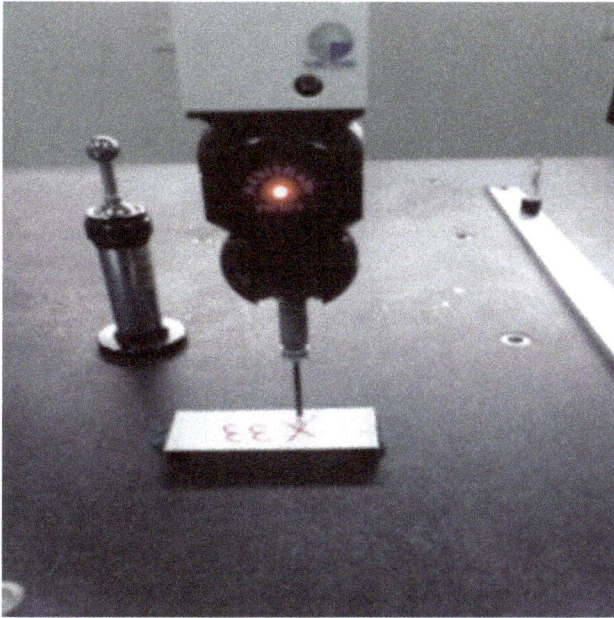

Figure 6. *CMM with work piece.*

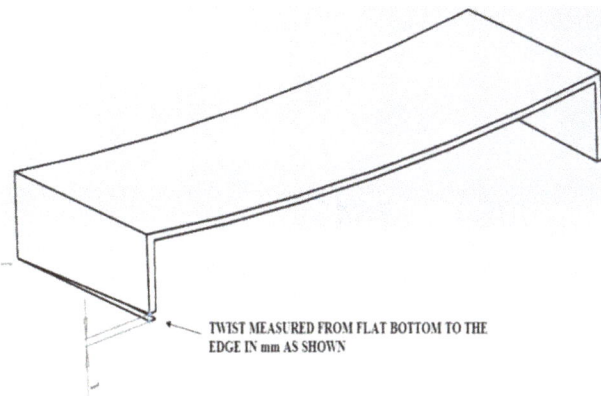

Figure 7. *Picture showing measurement of twist.*

Table 4. *Machining Parameters.*

Title	Value
Feed	0.1 mm/Tooth
Speed	120 m/min
Depth of cut	1mm
Width of cut	7mm
Coolant	Dry machining

2.4. Methodology

Aluminium rolled plates of 12 mm, 14 mm, 16 mm, 18 mm and 20 mm thick were cut into sizes 110 mm X 45 mm. The blanks were then sized to 105 mm X 40 mm on a vertical milling machine. After sizing, all the blanked were stress relieved, there by assuming very little or almost zero bulk residual stresses before machining them on CNC machining centre. The blanks after stress relief were machined by holding the part on specially made vacuum fixture from the bottom using low helix two flute solid carbide Slot Drill ø 10 mm. The machining conditions used for milling all the blanks is shown in Table 4. All the machining experiments

were carried out under same machining conditions.

Each blank during machining is taken out of fixture for every depth of cut which is 1mm and measurements were taken for distortion and twist. The cutting strategy adopted for the experiments was pocket Inside out as shown in Figure 8. Three sample work pieces were machined for each thickness of blank.

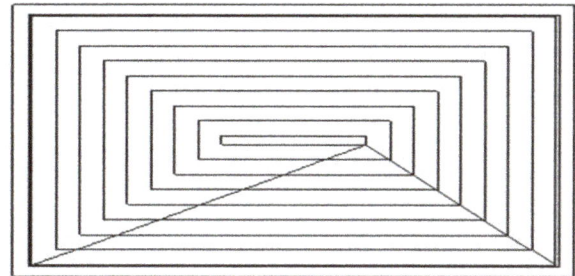

Figure 8. *Tool Path Strategy pocket inside out.*

3. Experimental Results and Discussions

Distortion and Twist measurements were taken for every 1mm depth of machining for all the blanks. The distortion and twist measurements for blank thickness of 20 mm, 18 mm, 16 mm, 14 mm, and 12 mm for every 1 mm depth of cut are taken for 3 samples each and maximum distortion and twist were recorded for comparison. Table 5 shows the values of distortion and twist at every 1 mm depth of cut for blank thickness 20mm. The maximum distortion and twist values for the blanks of all thicknesses at every 1 mm depth of machining are shown in Table 6. Comparative values of distortion & twist with respect to volume of material removed is shown in Table 7.

It can be observed from Table 6 that distortion and twist was observed for all the components machined. Significant distortion and twist was noticed only after the thickness of the components in the floor (bottom) is less than 3 mm after machining, for all the components. Figure 9 shows distortion and twist with respect to volume of material removal. It can be seen that, there no significant increase in distortion and twist with increase in volume of material removal.

The results clearly show that distortion of the components does not depend on the amount of material removed. As the amount of material machined increased from 44880 mm^3 to 700800 mm^3 there was no significant increase in distortion from initial distortion of 0.45 mm indicating that the distortion of the components may depend on machining parameters and tool parameters. The slight increase in twist for 20 mm thick blanks may be attributed to the variation of bulk residual stresses distribution existing in the blank prior to machining. It can be observed from Figure 10 and Figure 11 that the distortion and twist of the components was significant at thickness less than 3 mm for all the blanks indicating the effect of machining induced surface stresses causing distortion as reported in previous research [5, 15, 16].

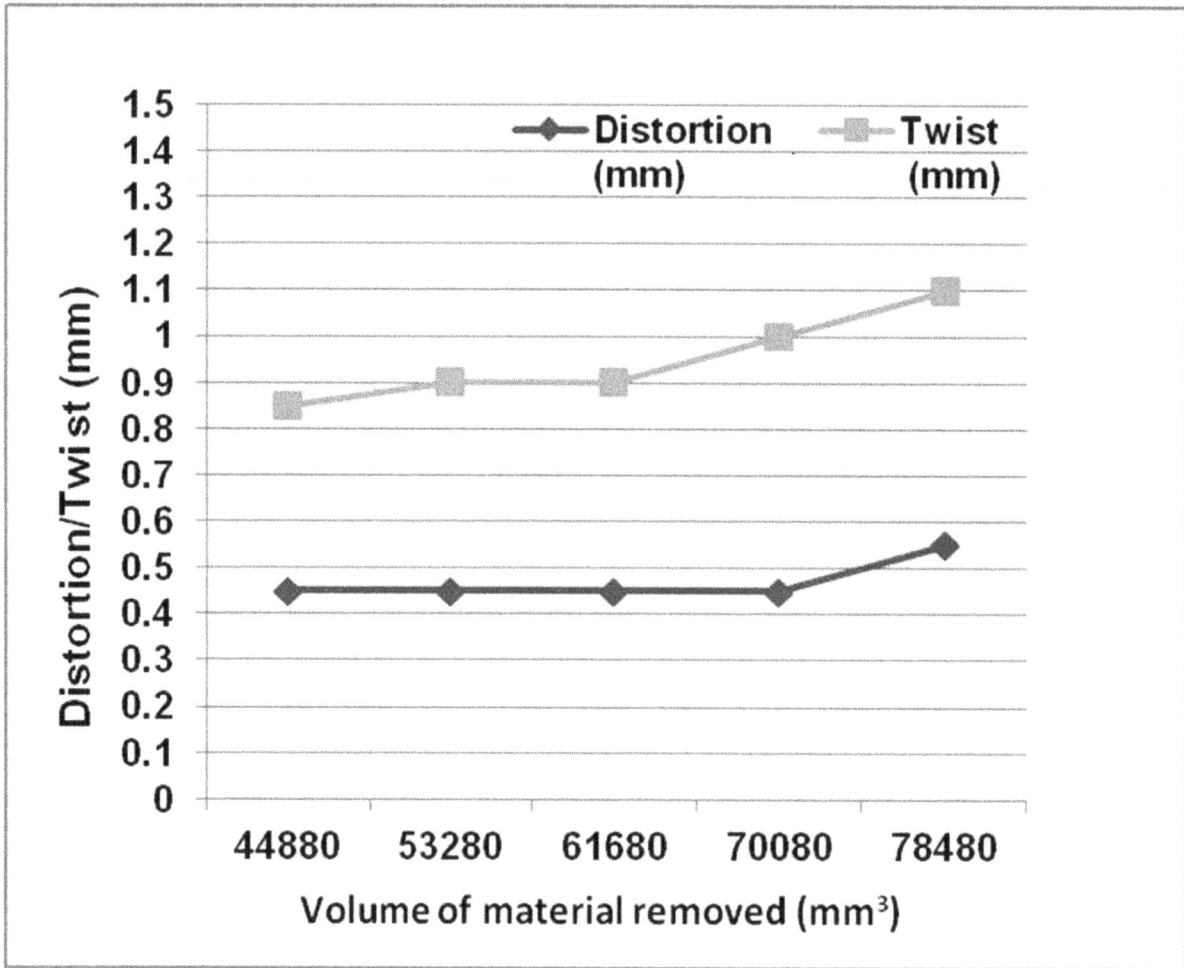

Figure 9. *Comparison of distortion and twist.*

Table 5. Distortion and Twist values for 20 mm Blank.

Blank (mm)	Distortion / Twist at Thickness	Sample 1 Dist. (mm)	Twist (mm)	Sample2 Dist. (mm)	Twist (mm)	Sample3 Dist. (mm)	Twist (mm)	Maximum Dist. (mm)	Twist (mm)
	20	0	0	0	0	0	0	0	0
	19	0	0	0	0	0	0	0	0
	18	0	0	0	0	0	0	0	0
	17	0	0	0	0	0	0	0	0
	16	0	0	0	0	0	0	0	0
	15S	0	0	0	0	0	0	0	0
	14	0	0	0	0	0	0	0	0
	13	0	0	0	0	0	0	0	0
	12	0	0	0	0	0	0	0	0
20	11	0	0	0	0	0	0	0	0
	10	0	0	0	0	0	0	0	0
	9	0	0	0	0	0	0	0	0
	8	0	0	0	0	0	0	0	0
	7	0	0	0	0	0	0	0	0
	6	0	0	0	0	0	0	0	0
	5	0	0	0	0	0	0	0	0
	4	0.05	0.1	0.08	0.1	0.08	0.12	0.08	0.12
	3	0.21	0.4	0.22	0.45	0.2	0.45	0.22	0.45
	2	0.35	0.7	0.34	0.75	0.35	0.75	0.35	0.75
	1	0.55	1.1	0.49	1.05	0.52	1.05	0.55	1.1

Table 6. *Maximum values of Distortion and Twist (mm) at various thicknesses.*

Blank (mm)	Thick-ness (mm)	20	19	18	17	16	15	14	13	12	11	10	9	8	7	6	5	4	3	2	1
20	DIST.	0	0	0	0	0	0	0	0	0	0	0	0	0	0	0	0	0.08	0.22	0.35	0.55
	TWIST	0	0	0	0	0	0	0	0	0	0	0	0	0	0	0	0	0.12	0.45	0.75	1.1
18	DIST.	-	-	0	0	0	0	0	0	0	0	0	0	0	0	0	0	0	0.05	0.33	0.45
	TWIST	-	-	0	0	0	0	0	0	0	0	0	0	0	0	0	0	0	0.25	0.55	1
16	DIST.	-	-	-	-	0	0	0	0	0	0	0	0	0	0	0	0	0	0.05	0.22	0.45
	TWIST	-	-	-	-	0	0	0	0	0	0	0	0	0	0	0	0	0	0.25	0.6	0.9
14	DIST.	-	-	-	-	-	-	0	0	0	0	0	0	0	0	0	0	0	0	0.25	0.45
	TWIST	-	-	-	-	-	-	0	0	0	0	0	0	0	0	0	0	0	0.15	0.6	0.9
12	DIST.	-	-	-	-	-	-	-	-	0	0	0	0	0	0	0	0	0	0.15	0.22	0.45
	TWIST	-	-	-	-	-	-	-	-	0	0	0	0	0	0	0	0	0	0.15	0.44	0.85

Table 7. *Maximum values of Distortion and Twist with respect to material removal volume.*

Blank (mm)	Part Volume (mm^3)	Blank Volume (mm^3)	Material removal Volume (mm^3)	% Material Removed	Distortion (mm)	Twist (mm)
12	5520	50400	44880	89.10	0.45	0.85
14	5520	58560	53040	90.60	0.45	0.9
16	5520	66720	61200	91.80	0.45	0.9
18	5520	74880	69360	92.70	0.45	1
20	5520	83040	77520	93.40	0.55	1.1

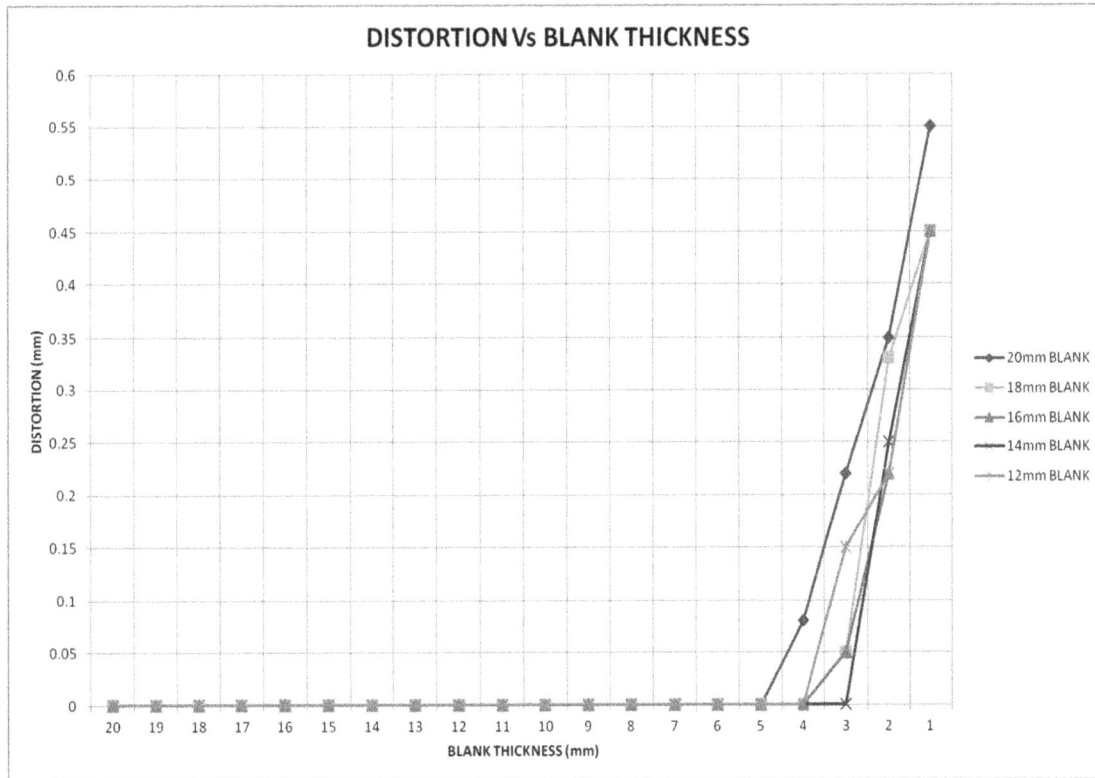

Figure 10. *Values of Distortion at different thicknesses of blanks.*

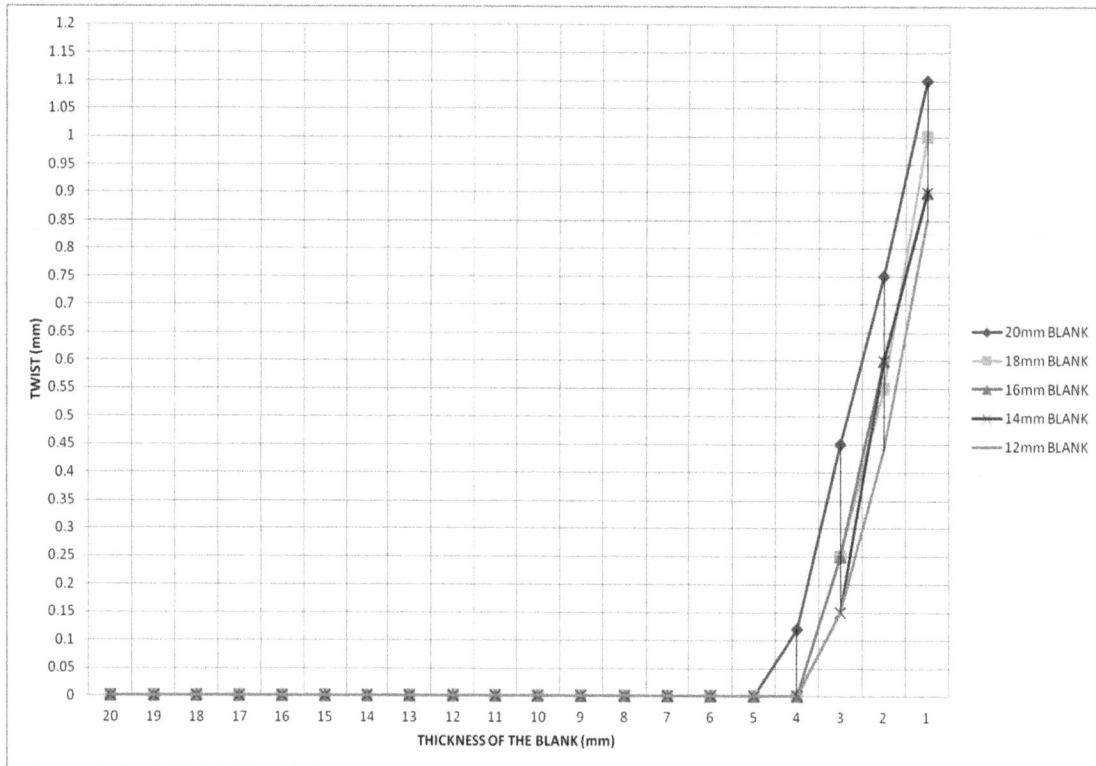

Figure 11. Values of Twist at different thicknesses of blanks.

4. Conclusions

A comparison of distortion and twist vis-a-vis material removal volume was done. Machining experiments were carried with constant machining process parameters on stress relieved aluminium alloy 2014 T651 with different thickness of blanks. It was found that the amount of material removal has no significant affect on distortion. Distortion may be due to process parameters. Further, it was observed that the distortion was significant only below 3mm thickness because of surface induced stresses due to machining. Further study should be focused on the effect of machining and tool parameters on the distortion of the parts machined.

Acknowledgements

The authors are thankful to the Head of Department, Osmania University for his constant encouragement and Support for conducting experiments. The authors are also thankful to the reviewers for their valuable inputs in improving quality of the manuscript.

References

[1] Sridhar, Garimella, and P. Ramesh Babu. "Understanding the challenges in machining thin walled thin floored Avionics components, " International Journal of Applied Science and Engineering Research 2.1: 93-100, 2013.

[2] Stephenson, D. A., Agapiou, J. S., Metal Cutting Theory and Practice, Second Edition, CRC, Boca Raton, FL, 568-9, 2006.

[3] Marusich T. D., D. A. Stephenson, S. Usui, and S. Lankalapalli, "Modeling Capabilities for Part Distortion Management for Machined Components." Third Wave Systems (2009).

[4] Totten, George E., and D. Scott MacKenzie, eds. Handbook of Aluminum: Vol. 1: Physical Metallurgy and Processes. Vol. 1. CRC Press, 2003.

[5] Chatelain, Jean-François, Jean-François Lalonde, and Antoine S. Tahan, "Effect of Residual Stresses Embedded within Workpieces on the Distortion of Parts after Machining," International Journal of Mechanics 6: 43-5, 2012.

[6] Cui, Hengbo, Jong-Yun Jung, and Dug-Hee Moon, "The Selection of Machining Parameters to Minimize Deformation caused by Heat," Proceedings of the Fall Conference of Society of Korea Industrial and Systems Engineering, Korea. 2005.

[7] DONG, Hui-yue, and Ying-lin KE, "Study on machining deformation of aircraft monolithic component by FEM and experiment," Chinese Journal of Aeronautics 19.3: 247-254, 2006.

[8] Lalonde, J. F., M. A. Gharghouri, and J. F. Chatelain, "Effect of Residual Stresses on the Distortion of Components after Machining.", NRC-CNBC Annual Report 2007.

[9] Younger, Mandy S., and Kenneth H. Eckelmeyer. "Overcoming residual stresses and machining distortion in the production of aluminium alloy satellite boxes," Sandia Report SAND2007-6811, Sandia National Laboratories, 2007.

[10] Denkena, B., and L. de León, "Machining induced residual stress in wrought aluminium parts, " Proceedings of 2nd International Conference on Distortion Engineering. 2008.

[11] Belgasim, O., and M. H. El-Axir, "Modeling of residual stresses induced in machining aluminum magnesium alloy (Al–3Mg)," Proceedings of the world congress on engineering, Vol. 2. 2010.

[12] Garimella Sridhar and P. Ramesh Babu, "Cutting Parameter Optimization for Minimizing Machining Distortion of Thin Wall Thin Floor Avionic Components using Taguchi Technique", International Journal of Mechanical Engineering & Technology (IJMET).Volume:4,Issue:4,Pages:71-78, 2013.

[13] Keleshian, N., et al., "On the distortion and warpage of 7249 aluminum alloy after quenching and machining," Journal of materials engineering and performance 20.7: 1230-1234, 2011.

[14] Songtao Wang, Zheng Minli, Fan Yihang, and Li Zhe, "Cutting Parameters Optimization in Machining Thin-Walled Characteristics of Aircraft Engine Architecture based on Machining Deformation," Advances in Information Sciences & Service Sciences 4.10, 2012.

[15] Huang, Xiaoming, et al., "An Experimental Investigation of Residual Stresses in High-Speed End Milling 7050-T7451 Aluminum Alloy," Advances in Mechanical Engineering, 2013.

[16] Huang, Xiaoming, Jie Sun, and Jianfeng Li, "Effect of Initial Residual Stress and Machining-Induced Residual Stress on the Deformation of Aluminium Alloy Plate," Strojniški vestnik-Journal of Mechanical Engineering 61.2: 131-137, 2015.

[17] Ma, K., R. Goetz, and S. K. Srivatsa. "Modeling of residual stress and machining distortion in aerospace components." ASM Handbook 22, 2010.

10

Stress Analysis of Gun Barrel Subjected to Dynamic Pressure

H. Babaei[*], M. Malakzadeh, H. Asgari

Department of Mechanical Engineering, Engineering Faculty, University of Guilan, Rasht, Iran

Email address:

ghbabaei@guilan.ac.ir (H. Babaei), hghbabaei@Gmail.com (H. Babaei)

Abstract: In the optimal design of a modern gun barrel, there are some aspects to be considered. One of the main factor is internal ballistic which consist of pressure-time, pressure-distance, velocity-time and distance-time curves. In this paper, a simple analytical solution for the plastic stress of an internally pressurized open-ended thick-walled cylinder made of hardening steel which is the closest model to gun barrel is obtained in perfectly plastic and plane stress condition by using energy method and the yield criterion of Von Mises and adding rifle grooves and choosing stress components as basic unknowns and ballistic pressure equation as known. Then results of analytical solution are compared to a numerical model and verified a very well and reliable accuracy. So the resultant can be used easily in calculation of radial expansion velocity and compressive pressure.

Keywords: Gun Barrel, Stress, Dynamic Pressure

1. Introduction

Many papers are published around the matter of gun design and especially about gun tube design from the beginning of improved guns design generation until now, but reaching to variation sources around this subject is impossible because of preventing of publishing of this technology and security and military problems of some countries. Therefore many engineers and designers who work on this field assume a reasonable simplification that considers gun barrel as an open-ended thick-walled cylinder in which an explosive causes stress and deformation on the wall by creation an immediate pressure. According to this assumption investigation must be done under plane stress condition ($\sigma_z = 0$). Few satisfactory theoretical solution based on Von Mises yield criterion that are reliable and convenient for engineering use have been obtained for an open-ended thick-walled cylinder. A closed-form solution for the stress components was given by Nadai using an auxiliary-variable method and the deformation theory of Hencky [1]. A set of analytical expressions for the elasto-plastic stress and displacement components were obtained by Davidson et al. on the basis of an empirical relationship resulting from tests [2]. Two mathematically consistent analytical solutions to the strains and displacements were

obtained according to the deformation theory of Hencky and to the flow theory of Prandtl-Reuss, respectively by P.C.T. Chen using a modified Nadai's auxiliary variable method [3]. D.R. Bland et al. derived equations of stress distribution in thick-walled cylinder under internal and external pressure using numerical solution in 1956 [4]. They studied terms of equivalent stress and strain for the two conditions of open-ended ($\sigma_z = 0$) and close-ended ($\varepsilon_z = 0$) for this analysis but the way of calculating of stresses in the open-ended condition which is the appropriate model for gun barrel had not been mentioned. V. A. Adintsov et al. achieved an equation for radial and hoop stresses and radial expansion velocity by analytical investigating on perfectly plastic thick-walled cylinder behavior under internal explosive pressure using energy method [5]. One of the weak points of this study is analyzing the material as a perfectly plastic one but not considering the strain rate. An analytical pattern and some numerical results for an internal-pressurized open-ended thick-walled cylinder made of linear-hardening material was given by Xu Hong and Chen Shuning on the basis of the Prandtl-Reuss flow theory [6]. A theoretical and numerical analysis for a similar problem was performed by

LiuYong using the elasto-plastic mixed-boundary-element method [7]. Gao Xin-lin et al. obtained a closed form analytical solution for stress, strain and displacement components for an internal-pressurized elasto-plastic open-ended thick-walled cylinder in 1991 [8]. In this paper it is shown that the solution is a general one with on the one hand Nadai's known solution for stress components and on the other P.C.T. Chen's solution for strain components of an open-ended thick-walled cylinder made of elastic-perfectly-plastic material as its two specific cases. In this analysis it is assumed an auxiliary function for radial stress in terms of current radius after performing pressure for obtaining stress components at plane stress condition in plastic zone which is very complicated and Time-consuming. X.-L Gao et al. exhibited a technique for elastic-plastic internally pressurized cylinder analyzing in that both Tresca and Von Mises criterion is used for three conditions of plane stress, plane strain and close-ended in order to calculate the stresses and strains in 2003 [9]. Also the deformation energy of cylinder wall is obtained after the pressure is performed. Lack of considering the term of wall expansion acceleration is one of it's weaknesses. Li Mao-lin et al. investigated plastic limit load of viscoplastic thick-walled cylinder and spherical shell subjected to internal pressure analytically using a strain gradient plasticity theory in the paper which is presented in 2008 [10]. Results show that the size effect is more evident with increasing strain or strain rate sensitivity index, but the weak point of this approach is that the viscoplastic model analysis doesn't include material behavior in all part of the wall. Bagheri et al. presented a paper with the purpose of deriving a mathematical model for expansion isotropic thick-walled aluminum cylinder containing TNT in which, JWL equation of state is considered for explosive products [11]. As a result the equations of radial and hoop stress and radial expansion velocity is obtained. However, it must be noted that, each of these known analytical solutions is either over-simplified in the material model but time-consuming in calculation [12,13] or too complicated in the expressions proposed but incomplete in content [14,15].

In this analysis after-explosion phenomena in a thick walled steel cylinder, such as radial pressure on the wall of cylinder, the radial drift, the rate of expansion of the radial drift and radial and circumferential stresses, are considered. This study starts with a simple analytical model using energy-based methods for mentioned parameters. Subsequently, by choosing a cylinder which has specific material and specific geometric dimensions as well as an explosive substance, the mentioned analysis is carried out. One of the advantages of this study is simplicity in calculating the rate of the radial expansion and hence, changes in the radius of the cylinder and its stresses and also the acceleration of the wall are measurable.

2. Theoretical Analyses

2.1. Basic Equations of Thick-Walled Cylinders

Considering Figure 1, the following equations can be obtained [16]

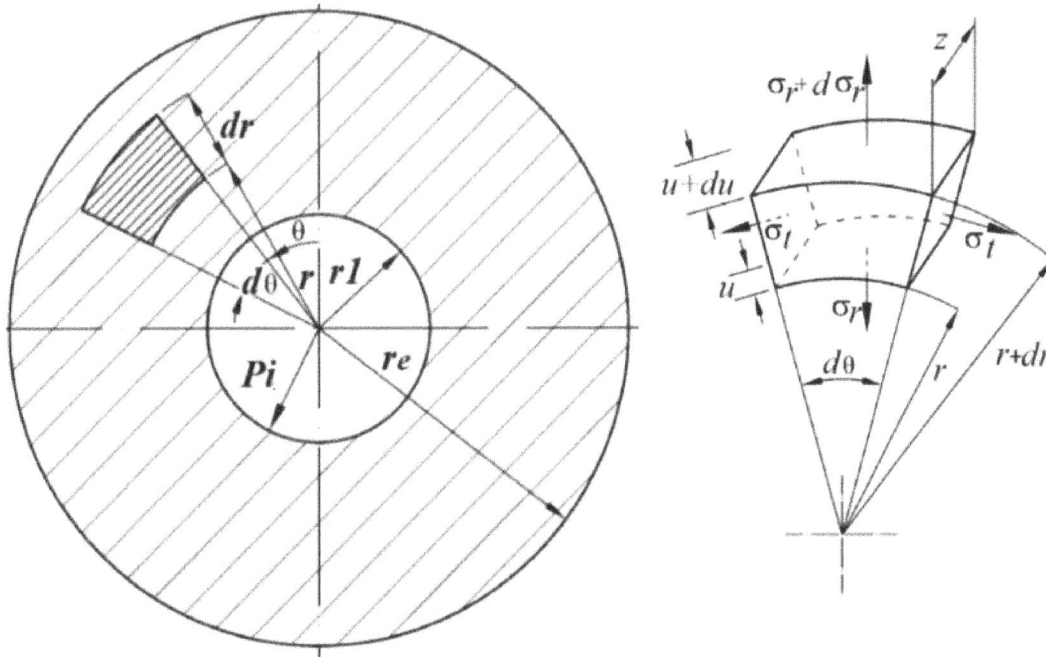

Figure 1. An element of a cylindrical body and its related components.

$$\frac{d\sigma_r}{dr} + \frac{\sigma_r - \sigma_\theta}{r} = 0 \qquad (1)$$

Where σ_r and σ_θ are radial and hoop stresses respectively. The equation above shows the equation of the thick-walled cylinder in static state [17] and considering the open-ended condition ($\sigma_z = 0$), and in a dynamic state it turns to the following equation considering the movement of the element shown in figure 1. It should be mentioned that the term related to density in acceleration is added to the equation [18].

$$\frac{d\sigma_r}{dr} + \frac{\sigma_r - \sigma_\theta}{r} = \rho\frac{dv_r}{dt} = \rho\dot{v}_r \qquad (2)$$

Where v_r and $\dot{v}_r$ are radial expansion velocity and acceleration respectively, and ρ is the density of cylinder material.

2.2. Conservation Law

Conservation law in physics expresses that the whole energy of an isolated system is consistent with time [19]. Energy neither produces nor wastes, but transforms from a state to another. In this study the chemical energy of explosives turns to kinetic energy. this law is explained in the form of the equation below.

$$E + W + E_f + W_r = E_0$$

where, E_0 and E represent initial internal energy and current energy of the substances of explosion, respectively. W is the kinetic energy of the wall of the cylinder, E_f is the work of plastic deformation and W_r is the work of friction force of the rifle. If we represent the equation above per unit of length, will have the following equation.

$$\bar{E} + \bar{W} + \bar{E}_f + \bar{W}_r = 1 \qquad (3)$$

The terms of the equation above are determined during the trend of analysis in this study.

2.3. Adiabatic Expansion Law

In this research the expansion of the wall due to explosion is considered adiabatic and obeys the equation below [20].

$$pV^\gamma = const \qquad (4)$$

Where p and V is pressure and special volume respectively and γ is adiabatic expansion coefficient which is specified for any material.

2.4. Gunnery Internal Ballistic Equations

Internal ballistic equations consist of formulas and diagrams related to time-pressure, distance-pressure and distance-time. It should be mentioned that, because the selected weapon in this study is M24 which has a 500 mm barrel and the firing time is 1 ms, we consider the pressure curve yielded from Russel Model within these ranges and subsequently, we curve fit this diagram in software. The results are as follows.

The equation of pressure based on time:

$$P = a_1 e^{-\left(\frac{t-b_1}{c_1}\right)^2} + a_2 e^{-\left(\frac{t-b_2}{c_2}\right)^2} + a_3 e^{-\left(\frac{t-b_3}{c_3}\right)^2} \qquad (5)$$

where P represents internal ballistic pressure and t represents the time interval between triggering and firing bullet through the barrel.

$a_1 = 2.287 \times 10^8, b_1 = 0.000358, c_1 = 0.000147,$

$a_2 = 7.827 \times 10^7, b_2 = 0.0004937, c_2 = 0.000350,$

$a_3 = 1.999 \times 10^8, b_3 = 0.0004937, c_3 = 0.0002388$

Hence, figure 2 shows the diagram of equation (5)

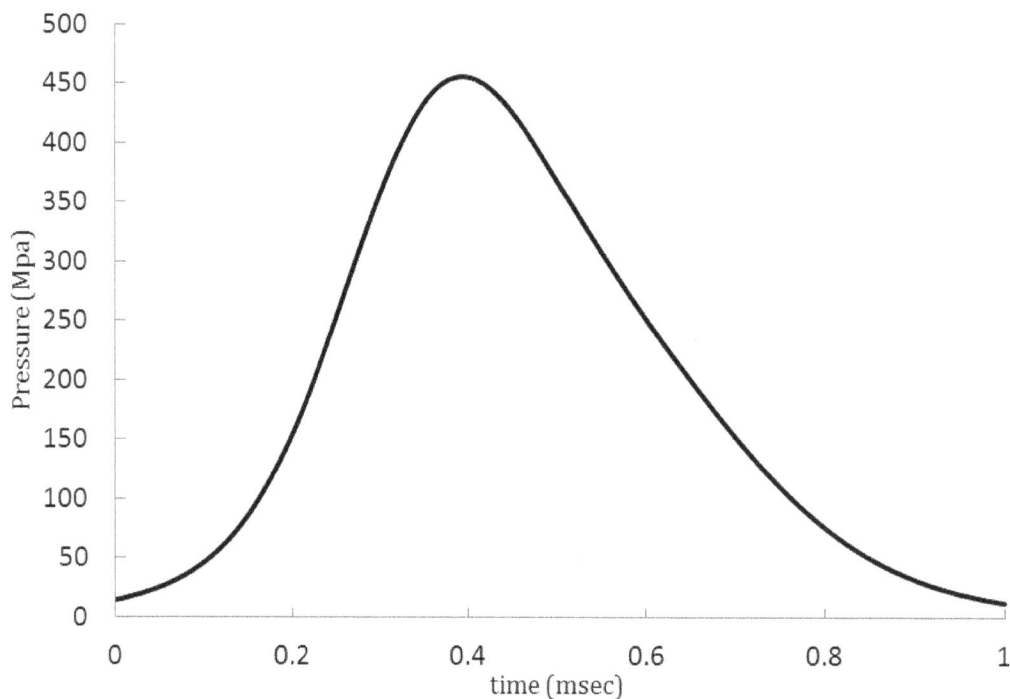

Figure 2. the internal ballistic pressure versus triggering time.

The equation of distance (the amount of space passed by the bullet into the barrel) based on time is presented as follows.

$$x = me^{n \times t} + qe^{s \times t} - m - q \qquad (6)$$

In which x is the length of the barrel

$$m = -0.01431 \, , n = 9424, q = 23.44, s = 3384$$

The yielded diagram from the equation above is demonstrated in the Figure 3.

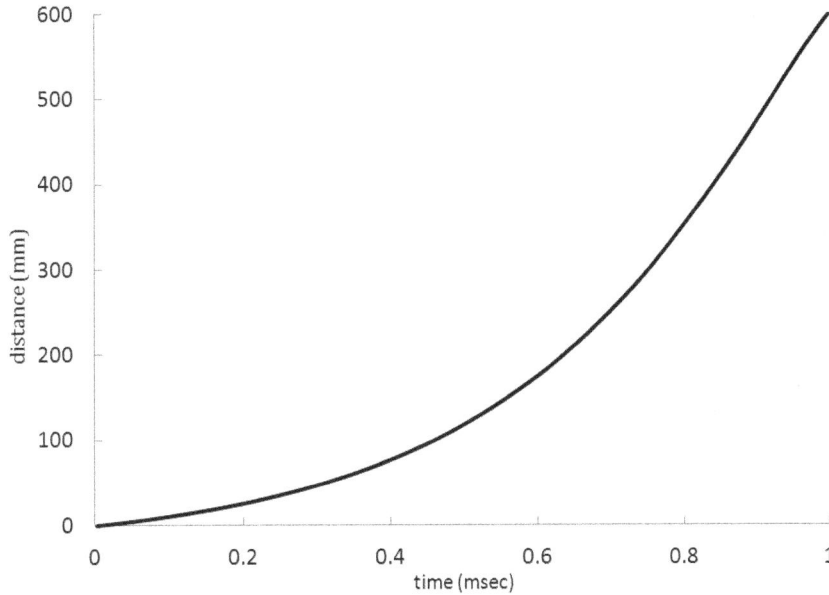

Figure 3. *the distance passed by the bullet into the barrel against time.*

As a result, the diagram of pressure versus distance is demonstrated as Figure 4.

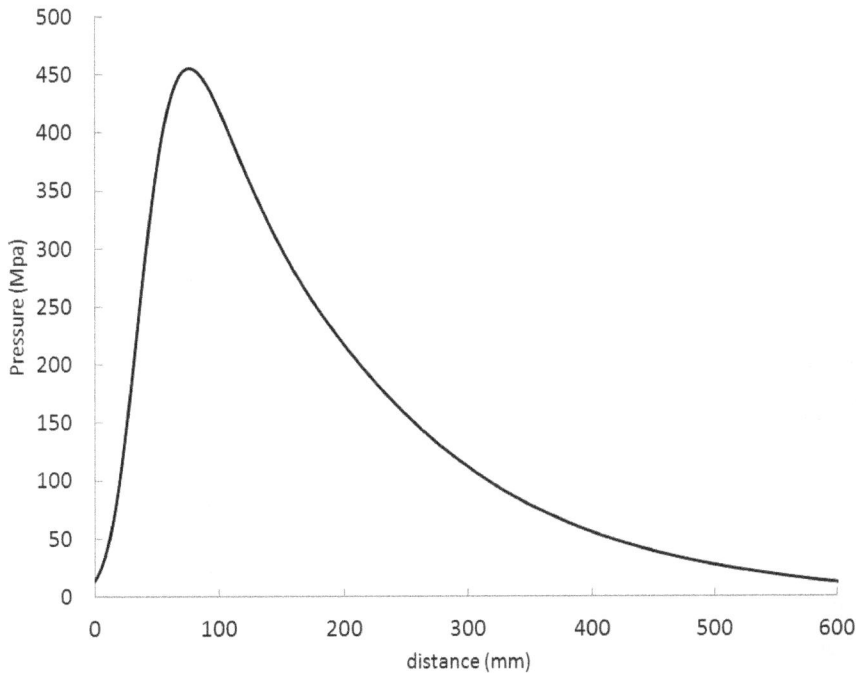

Figure 4. *inside ballistic pressure versus distance.*

3. Research Method

3.1. Assumptions

• In this analysis the cylinder containing explosives is assumed perfectly plastic, homogeny and incompressible.

• The used criterion is Von Mises. As a result of recent studies and comparing with the analysis a more appropriate criterion is introduced which is Von Mises criterion. Due to utilizing plastic work, the simulation and the analysis are independent of stiffness strains and thermal variations.

- The volume of the barrel is assumed constant before and after explosion and there is no body force affecting the barrel but the pressure of explosion.
- In this analysis the shock wave loading and swinging effects on the diameter of the barrel are ignored.
- The barrel analysis is carried out under the conditions of plane stresses ($\sigma_z = 0$).
- It is assumed that axial stresses of the barrel caused by the friction force of the projectile do not affect the analysis results.
- In the equation of conservation law, the term related to the energy of explosion is neglected.

3.2. Analysis

The stresses σ_r and σ_θ are principal. The internal and external initial cylinder radii before the deformation due to explosion are denoted by a_0 and b_0 respectively and radii after deformation by a and b, and r is the current radii. We assume the deformation of the tube wall is perfectly plastic, so there is no changing in volume, therefore from equality of current and initial volume we have:

$$r^2 - a^2 = r_0{}^2 - a_0{}^2 \tag{7}$$

By substituting b in (5) and differentiating, v_a (internal radius expansion velocity) and v_r (current radius expansion velocity) is obtained, so [21]:

$$v_r = \frac{a}{r} v_a \tag{8}$$

By differentiating form (6) and substituting in (2) we have

$$\frac{d\sigma_r}{dr} + \frac{\sigma_r - \sigma_\theta}{r} = \rho \left(\frac{v_a^2 + a\dot{v}_a}{r} - \frac{a v_a^2}{r^3} \right) \tag{9}$$

The yield criterion adopted here is the Von Mises criterion according to recent studies which say it is more accurate than others like Tresca, therefore

$$\sigma_\theta - \sigma_r = \alpha \sigma_y \tag{10}$$

Where $\alpha = \frac{2}{\sqrt{3}}$ for Von Mises criterion and σ_y is yield stress. By integrating (7) and applying the boundary condition $\sigma_r = -p$ in $r = a$:

$$\sigma_r = \left(\alpha \sigma_y + \rho v_a^2 + \rho a \dot{v}_a \right) ln \frac{r}{a} + \frac{1}{2} \rho v_a^2 \left(\frac{a^2}{r^2} - 1 \right) - p(x, t) \tag{11}$$

Where $p(x, t)$ is the pressure dependent to distance and time that is considered as p from now on. We have $\sigma_r = 0$ in $r = b$, the equation of internal radii acceleration is as follow:

$$\dot{v}_a = \frac{p}{\rho a ln \frac{b}{a}} - \frac{\alpha \sigma_y}{\rho a} - v_a^2 \left(\frac{1}{a} + \frac{a^2 - b^2}{2ab^2 ln \frac{b}{a}} \right) \tag{12}$$

Now we want to calculate the amount of v_a and $\dot{v}_a$. Accordingly, in order to preventing of using complicated calculation of differential equation, we present a simple solution using energy method (conservation energy law) and thermodynamic fundamental. Hence we calculate each term of equation (3).

3.2.1. Calculation of Detonation Products Internal Energy Per Unit of Length ($\overline{E}$)

p_0 is instantaneous pressure and obtained from

$$p_0 = \frac{\rho_0 D^2}{8} \tag{13}$$

ρ_0 is the density of the explosive and D is detonation velocity which are specified for any explosive and obtained empirically and accessible in engineering handbooks.

Internal energy for an ideal gas is in form $E = \frac{pV}{\gamma - 1}$ in which $V = \pi a^2$ is volume per unit of length, so we have

$$\overline{E} = \frac{E}{E_0} = \frac{p}{p_0} \left(\frac{a}{a_2} \right)^2 \tag{14}$$

3.2.2. Calculation of Wall Kinetic Energy Per Unit of Length ($\overline{W}$)

The equation of wall kinetic energy is in form $W = \int_a^b \frac{1}{2} v_r^2 dm$, substituting (8) and $dm = 2\pi \rho r dr$ in the

equation, it turns to $W = \pi \rho a^2 v_a^2 ln \left(\frac{b}{a} \right)$, in which ρ is barrel material density, so

$$\overline{W} = \frac{W}{E_0} = \frac{(\gamma - 1)\rho v_a^2 \left(\frac{a_0}{a} \right)^2 ln \left(\frac{b}{a} \right)}{p_0} \tag{15}$$

3.2.3. Calculation of Work of Plastic Deformation Per Unit of Length ($\overline{E}_f$)

Work of plastic deformation equation is $E_f = \sqrt{3} \alpha \sigma_y \pi \int_a^b \varepsilon_i r dr$, where ε_i is equivalent strain, We assume $E' = \int_a^b \varepsilon_i r dr$, for solving the integral we need to obtain an equation for the equivalent strain in plane stress condition [4].

$$\varepsilon_i^p = \frac{2}{\sqrt{3}} \left(\frac{C}{r^2} - \frac{1 - v^2}{E} \left(\alpha \sigma_y + \rho r \dot{v}_r \right) \right) \tag{16}$$

Where $C = \frac{1 - v^2}{E} \left(\alpha \sigma_y r_c^2 \right)$, in which r_c is the boundary of elastic and plastic domain and v is poisson coefficient. Since the workplace of this research is perfectly plastic, we have $r_c = b$, as a result the equation of equivalent is

$$\varepsilon_i = \frac{2}{\sqrt{3}} \frac{1 - v^2}{E} \left(\frac{\alpha \sigma_y b^2}{r^2} - \left(\alpha \sigma_y + \rho r \dot{v}_r \right) \right) \tag{17}$$

In order to obtain the amount of $\dot{v}_r$ we differentiate from (8), therefore

$$\dot{v}_r = \frac{v_a^2 + a \dot{v}_a}{r} - \frac{a v_a^2}{r^3} \tag{18}$$

Substituting (12) and (18) into (17)

$$\varepsilon_i = \frac{2}{\sqrt{3}} \frac{1-v^2}{E} \left(\alpha\sigma_y \left(\frac{b^2}{r^2} - 1 \right) + \rho \left(\frac{a^2 v_a^2}{r^2} - \frac{p}{\rho ln\frac{b}{a}} + \frac{\alpha\sigma_y}{\rho} + \frac{(a^2-b^2)v_a^2}{2b^2 ln\frac{b}{a}} \right) \right) \tag{19}$$

As a result

$$E' = \int_a^b \varepsilon_i r dr = \frac{2}{\sqrt{3}} \frac{1-v^2}{E} \left[\alpha\sigma_y \left(b^2 ln\frac{b}{a} \right) + v_a^2 \rho \left(a^2 ln\frac{b}{a} + \frac{(a^2-b^2)(b^2-a^2)}{4b^2 ln\frac{b}{a}} \right) - \frac{p(b^2-a^2)}{2ln\frac{b}{a}} \right]$$

Finally work of plastic deformation equation is in form

$$\bar{E}_f = \frac{E_f}{E_0} = \frac{\sqrt{3}\alpha\sigma_y(\gamma-1)}{p_0 a_0^2} E' \tag{20}$$

3.2.4. Calculation of Work of Rifiling Friction Force Per Unit of Length ($\overline{W}_r$)

According to number of grooves the equation $F = PA$ can be in form $F(x,t) = P(x,t)\left(\frac{\pi}{4}D^2 + nhb_g\right)$, in which $P(x,t)$ is gas pressure in the place x and time t, $F(x,t)$ is gas force on the chamber basis in the place x and time t, D is the caliber, n is the number of grooves, h is the grooves depth, and b_g is land width.

On the other hand work of friction force is

$$W_r = \mu_0 \left(\frac{2i}{d} \right)^2 tan\alpha \int_0^L F(x,t)dx$$

Where μ_0 is average friction coefficient, $\frac{2i}{d}$ is of physical properties of bullet that is 0.74 on average, α is groove angle and equal to 8 degree according to compare variation of guns. So $tan\alpha = 0.15$. we choose the barrel length 500 mm. finally rifle work of friction force per unit length is

$$\overline{W}_r = \frac{W_r}{E_0} = \frac{73.26\mu_0 F(x,t)(\gamma-1)}{\pi p_0 a_0^2} \tag{21}$$

Substituting equations

$$\frac{p}{p_0}\left(\frac{a}{a_2}\right)^2 + \frac{(\gamma-1)\rho\left(\frac{a_0}{a}\right)^2 ln\left(\frac{b}{a}\right)}{p_0} v_a^2 + \frac{2\alpha\sigma_{y(\gamma-1)}(1-v^2)}{E p_0 a_0^2} \left[\alpha\sigma_y b^2 ln\frac{b}{a} + \left(\rho a^2 ln\frac{b}{a} + \frac{\rho(a^2-b^2)(b^2-a^2)}{4b^2 ln\frac{b}{a}} \right) v_a^2 - \frac{p(b^2-a^2)}{2ln\frac{b}{a}} \right]$$
$$+ \frac{73.26\mu_0 F(x,t)(\gamma-1)}{\pi p_0 a_0^2} = 1 \tag{22}$$

For simplification of recent equation, we assume $A = \frac{p}{p_0}\left(\frac{a}{a_2}\right)^2$, $B = \frac{(\gamma-1)\rho\left(\frac{a_0}{a}\right)^2 ln\left(\frac{b}{a}\right)}{p_0}$, $C = \frac{2\alpha\sigma_{y(\gamma-1)}(1-v^2)}{E p_0 a_0^2}$, $D = \alpha\sigma_y b^2 ln\frac{b}{a}$, $F = \rho a^2 ln\frac{b}{a} + \frac{\rho(a^2-b^2)(b^2-a^2)}{4b^2 ln\frac{b}{a}}$, $G = \frac{p(b^2-a^2)}{2ln\frac{b}{a}}$, and $H = \frac{73.26\mu_0 F(x,t)(\gamma-1)}{\pi p_0 a_0^2}$

So v_a is obtained as

$$v_a = \sqrt{\frac{1}{B+CF}(1 - A - CD + CG - H)} \tag{23}$$

Now by substituting v_a into (12), (11) and (10) we obtain $\dot{v}_a$, σ_r and σ_θ respectively.

3.3. Numerical Simulation

In this research ABAQUS 6.12 and dynamic explicit solver is used to simulate and solve the problem.

CATIA software is used to model the M24 barrel. The barrel have four grooves with the pitch of a round per twelve inches in accordance with Figure 5.

This model imported to ABAQUS to numerical analysis. Other assumptions and geometrical and physical properties of material and explosive properties is shown in table 1 and table2 respectively.

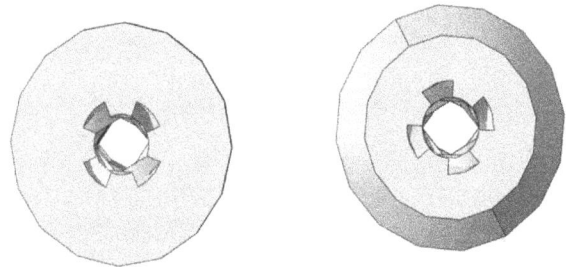

Figure 5. behind and front of barrel with 4 grooved.

Table 1. stainless steel 416R tube properties of M24 model gun used in analysis and simulation.

Dimension and size	
internal radius (mm)	3.81
External radius (mm)	14.6
Final expansion radius (mm)	14
Barrel length (mm)	600
Groove width (mm)	3
Groove depth (mm)	2
Number of rifles	4
Physical properties	
Density (kg/m^3)	7800
Yield stress (Mpa)	330
Poisson coefficient	0.3

Table 2. *explosive properties [22].*

Density (kg/m^3)	1717
Detonation velocity (m/s)	7980
Gas coefficient γ	2.7

The equation of pressure-distance imported in analytical field section to investigate pressure effect. The initial part of barrel is bounded to create boundary condition.

According to the analytical solution, period of 0.001 sec is chosen to numerical analysis in step section in ABAQUS. Mesh section is used for modeling and reticulation of problem geometry. The model have free reticulation and the elements are quadratic. Form of meshing used in barrel is shown as Figure 6.

Solving the problem is done by job section. Other setting related to problem solution in parallel processing condition, output numbers accuracy, etc can be done in this section. Stress contour among barrel length is shown in Figure 7.

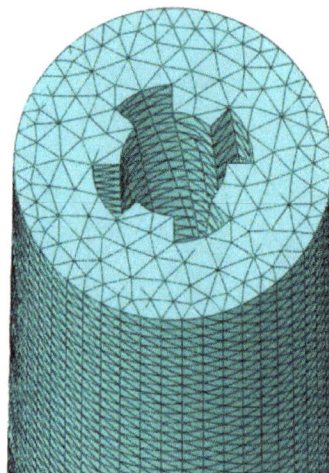

Figure 6. *form of meshed barrel.*

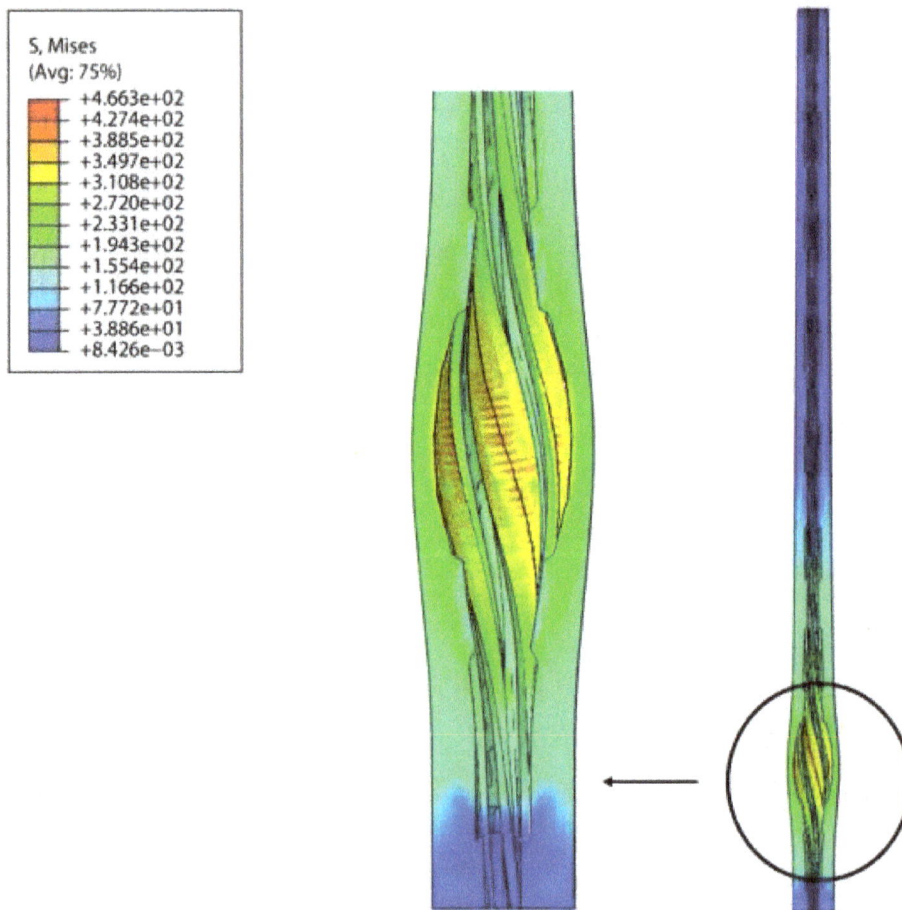

Figure 7. *Stress contour among barrel length.*

Graph of stress distribution on grooves among of barrel length is presented in Figure 8.

barrle length (mm)

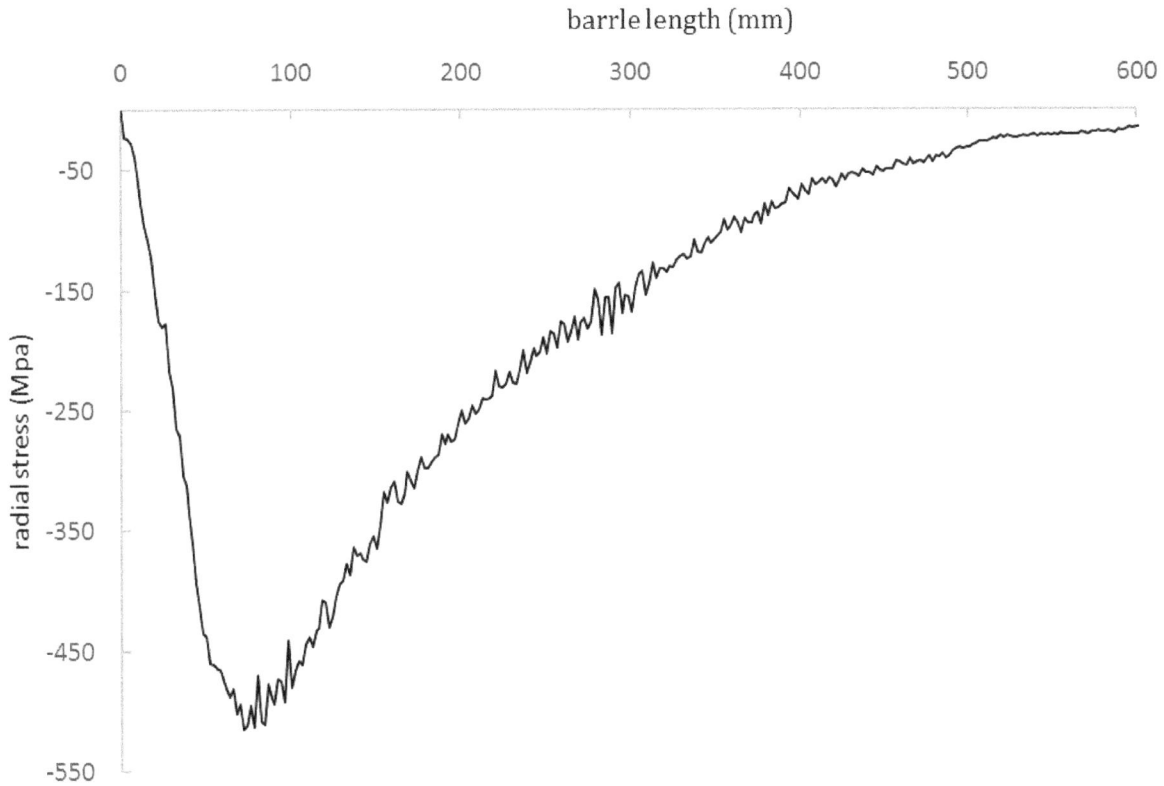

Figure 8. Graph of stress distribution per unit of length.

4. Conclusion

Diagram of radial expansion velocity variation per time is shown in Figureure 9 as one of the important results of this research. Maximum tube expansion velocity is 722.04 m/s in 0.39 sec for 4 rifiled barrel and 709.02 m/s in same time for 3 rifiled barrel on internal surface. The tube has maximum deformation and minimum thickness in this time. So the more number of grooves, the more expansion velocity.

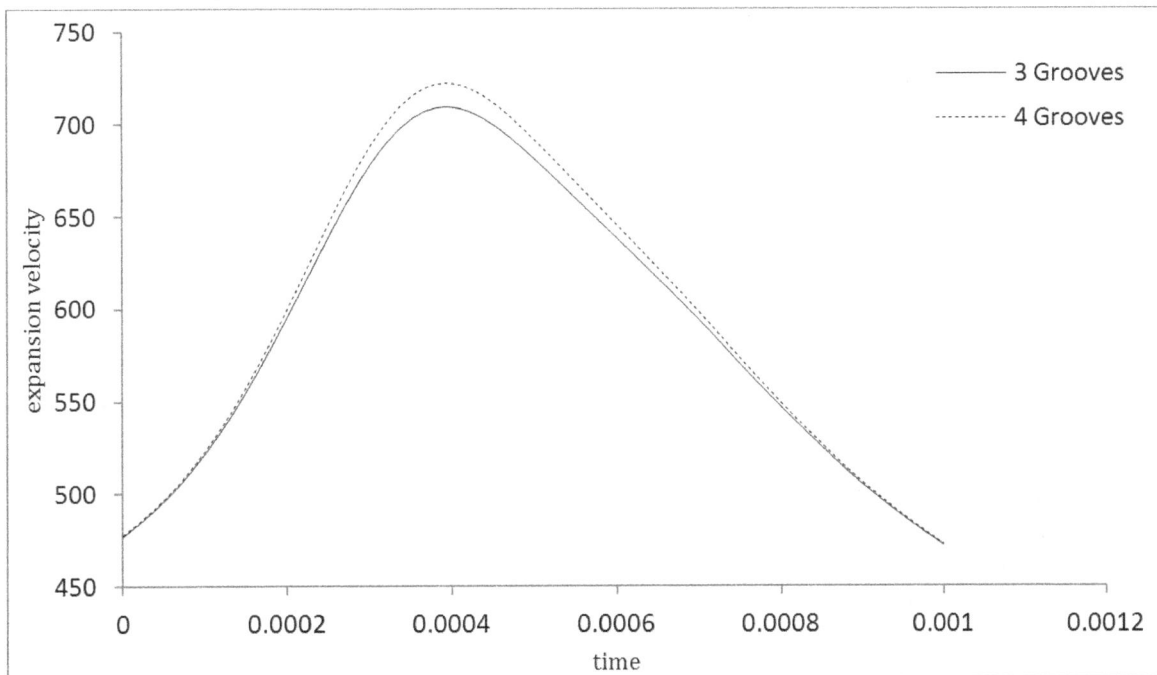

Figure 9. graph of radial expansion velocity variation per time.

Figureure 10 shows comparison of stress diagram per barrel length between two condition of analytical and numerical solution on internal surface. Maximum compressive pressure is about 455 Mpa due to analytical solution and about 465 Mpa due to numerical one. As a result the equations obtained from theoretical analysis are in well accordance on numerical solution for calculating the amounts of compressive pressure in order to acceptable difference between the two graphs.

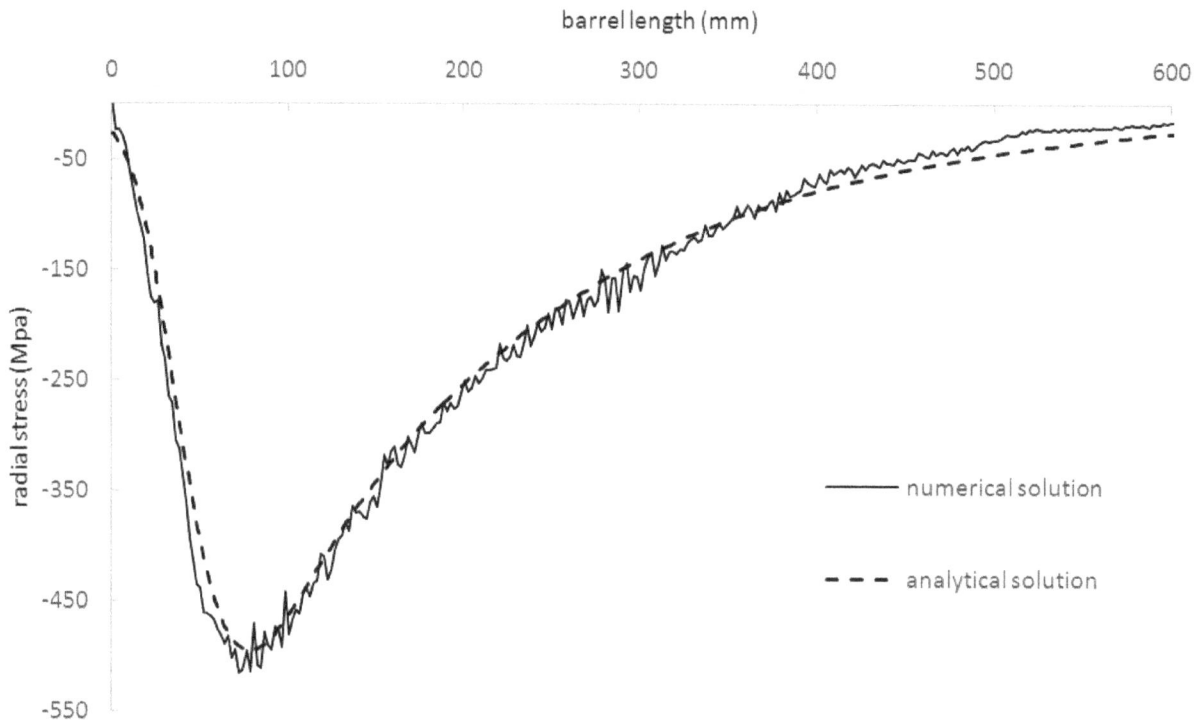

Figure 10. comparison between analytical and numerical solution for stress.

- In this paper, equilibrium equations of cylinder is considered as the closest model to gun barrel and a simple equation for radial expansion velocity is obtained considering boundary conditions. Expansion velocity can be calculated in every times entering boundary condition. The method used in this research is unique according to comparison analytical and numerical output, therefore it can be used in real situation, lab environment and testing experimental related to gunnery in order to reducing costs and time-consuming.
- Basically choosing the sort of equation of state governing the explosive after explosion and material model is not affecting on the analytical technique but it is necessary to be chosen in such a way to conform to the terms and actual results. Hence, internal ballistic pressure equation is considered in general form, but it should be mentioned the equation has been extracted with very high precision software. But from the convergence of analysis and simulation results can be deduced the equation of state describes explosive products expansion behavior very well with the related coefficients. So the resultant can be used in calculation easily.
- Radial velocity in the analysis is verified according to comparison with simulation therefore other extractive parameters accuracy like radial and hoop stresses depended to radial velocity is sufficient.

- Generally it can be stated that analysis technique introduced in this paper is accurate and the application of material models in the process of expansion recall is complete according to the simulation results and a small amount of errors. So this model can be used to predict the results before testing experimental.

References

[1]   Nadai, A., Theory of Flow and Fracture of Solid, Vol. 1, McGraw-Hill, New York, 1950, pp. 472-81.

[2]   Davidson, T. E., et al., Overstrain of high-strength open-ended cylinders of intermediate diameter ratio, In Proceedings of the 1st International Congress on Experimental Mechanics, Pergamon Press, New York, 1963, pp. 335-52.

[3]   Chen, P. C. T., A comparison of flow and deformation theories in a radially stressed annular plate. Trans. ASME, J. Appl. Mech., 40 (1973) 283-7

[4]   Bland, D. R. "Elastoplastic thick-walled tubes of work-hardening material subject to internal and external pressures and to temperature gradients."*Journal of the Mechanics and Physics of Solids* 4.4 (1956): 209-229.

[5]   Odintsov, V. A., and V. V. Selivanov. "Behavior of a rigidly plastic cylindrical shell exposed to internal pressure." *Journal of Applied Mechanics and Technical Physics* 16.3 (1975): 457-460.

[6] Xu Hong &Chen Shuning, An analysis of an autofrettaged cylinder considering strain-hardening effect, Bauchinger effect and temperature dependence of material properties (Part 1 Analysis for open-ended cylinder), In Proceedings of the 6th International Conference on Pressure Vessel Technology, Vol. 1. Pergamon Press, Oxford, 1988, pp. 423-430.

[7] Liu, Yong. "Solution of an open-ended autofrettaged thick-walled cylinder by mixed BEM." *J. Zhejang Institute of Technology* 2 (1991): 55-60.

[8] Gao, Xin-lin. "An exact elasto-plastic solution for an open-ended thick-walled cylinder of a strain-hardening material." *International journal of pressure vessels and piping* 52.1 (1992): 129-144.

[9] Gao, X-L. "Elasto-plastic analysis of an internally pressurized thick-walled cylinder using a strain gradient plasticity theory." *International journal of solids and structures* 40.23 (2003): 6445-6455.

[10] Li, Mao-lin, and Ming-fu Fu. "Limit analysis of viscoplastic thick-walled cylinder and spherical shell under internal pressure using a strain gradient plasticity theory." *Applied Mathematics and Mechanics* 29 (2008): 1553-1559.

[11] Bagheri, Seyed Masood, "Introduce a new Model for Expansion Behavior of Thick-Walled Cylinder under Internal Explosive Loading with Numerical Analysis" *Modares mechanical engineering*, 15.3 (2015): 251-259

[12] Lu, W. Y. & Hsu, Y. C. "Elastic-plastic analysis of a flat ring subjected to internal pressure" *Acta Mech.*, 27 (1977) 155-72

[13] Lu An-qi, "Elastic-plastic analysis of stresses around a circular hole in an infinite sheet subjected to equal biaxial tension" *Acta Mech. Solida Sinica*, 5 (1984) 449-53

[14] Xu Hong & Chen Shuning, "An analysis of an autofrettaged cylinder considering strain-hardening effect, Bauchinger effect and temperature dependence of material properties (Part 1 Analysis for open-ended cylinder)" *In Proceedings of the 6th International Conference on Pressure Vessel Technology, Vol. 1. Pergamon Press, Oxford*, 1988, pp. 423-430

[15] Liu Yong, "Solution of an open-ended autofrettaged thick-walled cylinder by mixed BEM" *J. Zhejang Institute of Technology*, (2) (1991) 55-60

[16] Vullo, V. "Circular Cylinders and Pressure Vessels Stress Analysis and Design"; *Springer International Publishing*: Switzerland, 2014.

[17] Paffumi, E.; Taylor, N. "Structural Response of a Large Pressure Vessel to Dynamic Loading"; *JRC Sci.Tech. Rep.*, 2008.

[18] Yong, L.; Xin, W. L.; Wei, X. L.; Meng, Z. F. "Residual Stress Analysis of a Thick-Walled Cylinder in Dynamic Loading"; Int. J. Pres. Ves. Pip. (IJPVP), Vol. 60, 17-20, 1994

[19] Hojman, Sergio A. "A new conservation law constructed without using either Lagrangians or Hamiltonians." *Journal of Physics A: Mathematical and General* 25.7 (1992): L291.

[20] Elliott, J. Richard, and Carl T. Lira. *Introductory chemical engineering thermodynamics*. Upper Saddle River, NJ: Prentice Hall PTR, 1999.

[21] Fenner,Roger T. *Engineering elasticity*. Halsted Press, 1986.

[22] Dobratz, Brigitta M. *Properties of chemical explosives and explosive simulants*. No. UCRL--51319; UCRL--51319 (REV. 1). comp. and ed.; California Univ., Livermore (USA). Lawrence Livermore Lab., 1972.

11

Characterization of amorphous ribbon by means vibrating sample magnetometry as an interesting tool to investigate a possible detector of vector field

Arturo Mendoza Castrejón[1, *]**, Herlinda Montiel Sánchez**[2]**, Guillermo Alvarez Lucio**[3]**, Damasio Morales Cruz**[1]

[1]School of Mechanical and Electrical Engineering -IPN, 07738, U. P. Adolfo López Mateos, D. F., México
[2]Technosciences Department, Center for Applied Science and Technology Development -UNAM, 04510, C.U., D. F., México
[3]School of Physics and Mathematics -IPN, 07738, U. P. Adolfo López Mateos, D. F., México

Email address:

zaratustra_also@hotmail.com (A. M. Castrejón)

Abstract: Characterization of amorphous ribbon is made by using Vibrating Sample Magnetometry VSM technique with different geometric arrangements: P10, P190, P20 and P290. The purpose is to determine the evolution of the saturation magnetization M_S, retentivity M_R and magnetic anisotropy K_1 as a function of annealing time treatment and also as a function of the geometric arrangement. The rate of change of magnetization $\Delta M/\Delta H$ is determined for orientation P190 and orientation P290. These values of rate of change for the ribbon with no annealing treatment are: 0.122 emu/cm^3 and 0.11 emu/cm^3, respectively. The highest values of anisotropy are for orientation P190 and for orientation P290, these values are: $K_1 = 2,365,100$ erg/cm^3 and $K_1 = 2,405,520$ erg/cm^3, respectively. Thus we establish that the amorphous ribbon is a strong candidate for technological applications in the area of the magnetic industry, because they can be designed vector field detectors in three directions: longitudinal, transverse (to the ribbon axis) and normal to ribbon plane.

Keywords: Retentivity, Magnetic Anisotropy, Saturation Magnetization

1. Introduction

Due to rapid technological progress, we have seen an increasing interest in materials that can respond quickly to excitation fields; in particular we refer to ferromagnetic materials which respond to DC magnetic fields. We can find many techniques for determining the magnetic response of materials, such as magneto-optical method, vibrating-coil magnetometer and magnetic force, between some others. However, in this paper the vibrating-sample magnetometer is used because it minimizes any error source, additionally the technique is simple, inexpensive and most importantly, it allows high accuracy in the measurement of the magnetic moment.

A field in a volume element generates an energy gradient by means a force; this force can be detected by the moving of the charge carrier or by the torque on the magnetic dipoles. Both these effects cause changes in the magnetic structure, which is formed by longitudinal and transverse domains (longitudinal and transverse anisotropy) [1,2]. These effects lead to changes in material properties and these properties are explained because the nanocrystals length D is smaller than the exchange length L_{ex} [3-5], see (1),

$$\langle K \rangle = \frac{v_{cr}^2 D^6}{A^3} K_1^4 \qquad (1)$$

<K> is the average energy density anisotropy, v_{cr} is the volume fraction, D is the grain size, A is a constant depending on the exchange length L_{ex} y K_1 is the anisotropy energy. Thus, a very important factor to understand the micro-structural properties is to know the magnetic anisotropy and how to control it. For this purpose we use the Vibrating Sample Magnetometry VSM which was first described by Foner [6] and it is based on the change of flow in a coil when the sample vibrates perpendicular to magnetic field, this vibration causes a change of a scalar potential in the form $\phi_1 \exp(i\omega t)$, where

$\phi_1 = -a(\partial \phi / \partial z)$, with sufficiently small amplitude a, ϕ is the scalar potential of a fixed dipole, this technique gives information about the magnetization M [7]. Since this technique is very versatile and highly sensitive, in this work will made a characterization to establish both the magnetic properties and the geometrical conditions for the development of a measurement methodology and thereby propose criteria for the possible development of a vector field sensor.

2. Experimental Procedure

The alloy $Fe_{73.5}B_9Si_{13.5}Mo_3Cu_1$ was made at Materials Research Institute, México, using a conventional melt spinning method in a protective argon atmosphere, a casting speed of 40 m/s was employed. The resulting ribbon was 3 mm wide and 25 μm thick. Vibrating-sample magnetometer (LDJ 9600 model) is used, in sweep magnetic field cycle H_{DC} = ± 1000 Oe.

Four configurations were used: 1) P10, the plane of the amorphous ribbon AR is in the yz plane, the y axis is parallel to the longitudinal axis LA of ribbon ($y \| LA$), the z axis is parallel to the transverse axis TA of ribbon ($z \| TA$); 2) P190, the ribbon plane is in the xz plane, $x \| LA$, $z \| TA$; 3) P20, the ribbon plane is in the yz plane, $z \| LA$, $y \| TA$; and 4) P290, the ribbon plane is in the xz plane, $z \| LA$, $x \| TA$. The H_{DC} is always in the y axis direction ($H_{DC} \| y$). The furnace annealing was performed up to 400 C in a hydrogen flow atmosphere at different annealing times 10, 40 60 120 and 180 min. Magnetic properties were measured on sample with 6 mm long, retentivity M_R, saturation magnetization M_S, anisotropy energy K_1 were determined from hysteresis loop.

3. Results and Discussions

In a traditional VSM, when the orientation of the sample to the field is changed, the orientation of the sample relative to the coils is changed. As an important result, the response and the sensitivity of the sample will be different at every geometrical configuration (P10, P190, P20 and P290). This is especially true if the sample is not rotation symmetric, it is because we have a sample with longitudinal y transversal anisotropies. Since the sample has a positive magnetostriction and contains 73.5% iron, a longitudinal anisotropy due to the spin-orbit interaction is generated. Although the presence of copper is lower, only 1%, this also generates a transverse anisotropy. Even with rotation symmetric samples there will always be some angular variation due to rotation eccentricities. We show how depending on the orientation of the amorphous ribbon with respect to an applied external field, it has a characteristic response of the longitudinal and transverse domains, this answer can be see because there are important changes in: the anisotropy energy K_1, the retentivity M_R and the saturation magnetization M_S. High sensitivity of the material relative to field is shown in Fig. 1, the dotted line box represents the region of magnetic field created by the electromagnet poles, the field is always in the direction of the y axis. We can observe two important facts: the first, a clear

difference in the hysteresis loops between {P10, P20} and {P190, P290}, this is because in {P10, P20} the magnetic moment dynamic is in ribbon plane, but not for {P190, P290} where the dynamic of the moments is out of plane; the second, a difference between {P10, P20} and {P190, P290}, it is explained in terms of the effect of the field on the transverse and longitudinal anisotropy, respectively. These important facts establish the basic conditions to design of vector field sensors. But let analyze the properties of the amorphous ribbon, which determine the suitability of a material for a given application.

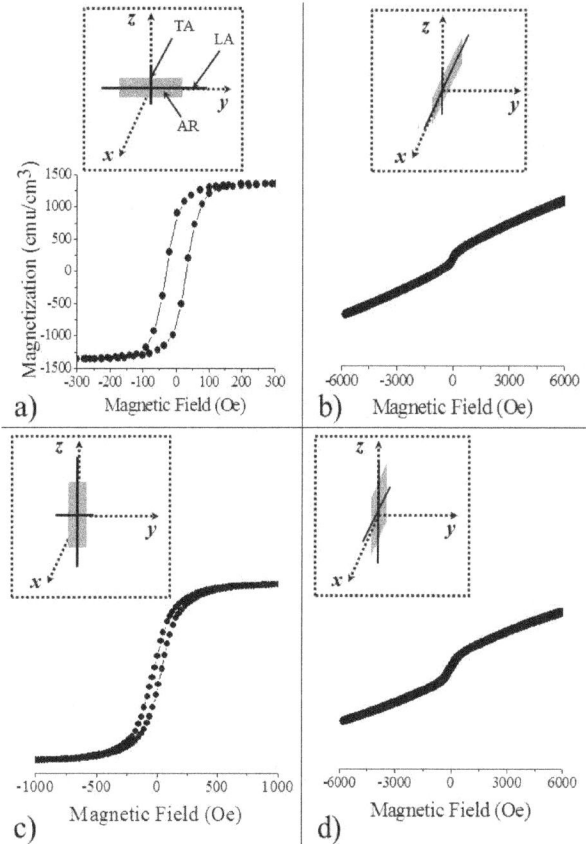

Figure 1. Hysteresis loop in configurations a) P10, b) P190, c) P20 and d) P290. The box with dotted line represents the region of magnetic field; the field is generated by the electromagnet poles.

Sample temperatures were estimated by monitoring the temperature dependence of chosen the physical properties (retentivity M_R, saturation magnetization M_S and coercive field H_C) and changes of this properties according the orientation of ribbon plane to field. In Fig. 2, a typical hysteresis loop is shown. M/M_S as a function of field present: M_R, ($H_{DC} = 0$), M_S saturation state (the magnetization vector is only in one direction) and H_C (where M = 0). Three important magnetization processes we can distinguish: the domain bulging, the domain walls displacement and the spin rotation. This hysteresis loop corresponds to amorphous ribbon in as cast AC (sample with no thermal treatment) state at orientation P10, by increasing the magnetic field; the transverse domains respond quickly to fields above 200 Oe, the saturation state is achieved. This is because the ribbon contains 1% Cu and thus a minimal energy is needed to achieve monodomain state. A comparison was made between orientation P10 and orientation P20 for the as cast ribbon, both hysteresis loops are shown in Fig. 3. We note that in the case of orientation P20, the monodomain state is reached until the fields above 1000 Oe, this is explained in terms of the amount of Fe (73.5 %); it means that the system needs more power to move more moments, these are located along the longitudinal axis of ribbon, *i.e.* more energy is needed to overcome the spin-orbit coupling. Thus, depending on the geometric arrangement of the ribbon, we have a characteristic response. At P10, M_R=815 emu/cm^3, M_S = 1353 emu/cm^3; at P20, M_R = 216 emu/cm^3, M_S = 1300 emu/cm^3. At P10 the most notable difference is in the retentivity, it could be explained as follows, to change the field direction, this field requires moving a large number of moments (which are along the longitudinal axis) to bring it to a state of lower energy (H_{DC} = 0); not for orientation P20, since in this case is reduced the number of moments that are in the transverse axis.

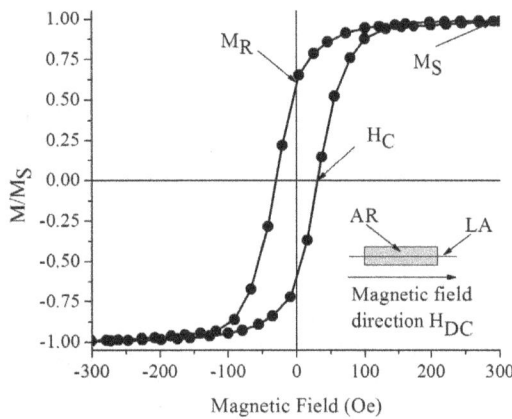

Figure 2. *Hysteresis loop in orientation P10 for as cast state.*

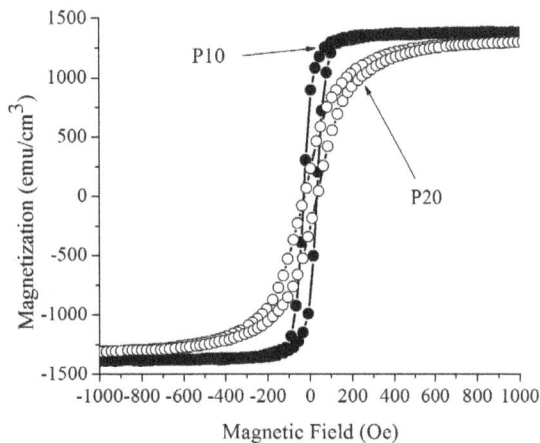

Figure 3. *Comparison between P10 and P20 in as cast state.*

In a simple physical model, if M_S in any one domain makes an angle θ with the positive field direction, the magnetization is $M_S\cos\theta$, the retentivity M_R in whole ribbon is given by

$$M_R = \int_0^{\pi/2} M_S \cos\theta \sin\theta d\theta = M_S / 2 \qquad (2)$$

From (2), we define M_R/M_S the retentivity ratio [8]. According with this model, we might qualitatively estimate the energy involve in the dynamic of domain walls. The M_R/M_S ratio at P10: 0.705, 0.704, 0.66 and 0.55 for 10, 40, 60 and 120 min, respectively; at P20: 0.161, 0.141, 0.119 and 0.153 for 10, 40, 60 and 120 min, respectively. These experimental results together with the theoretical model are in correspondence with the physical explanation given above. As the direction of easy axis of the ribbon is the direction of spontaneous magnetization in the demagnetized state, we can correlate the retentivity ratio with the anisotropy energy. In extreme cases, if the ratio is close to 1, indicates that the energy required to moving the vector of magnetization from its easy axis would be near to zero, if the ratio is near zero, the energy required to move the vector would be almost infinite.

From this way with the retentivity ratio is possible to determine the evolution of anisotropy depending not only the heat treatment time, but also on the orientation of amorphous alloy relative to the field. According to Cullity's model [9], with a very good approximation, to one-dimensional materials, the anisotropy calculation K_1 is performed by using (3),

$$H = 2\frac{K_1}{M_S} \qquad (3)$$

H is the magnetic field where the material is in saturation state. Anisotropy values K_1 in orientation P10 are: 2.5×10^5, 19×10^5, 1.81×10^5 and 1.86×10^5 erg/cm^3 for 10, 40, 60 y 120 min, respectively; in orientation P20 these values are: 5.6×10^5, 4.42×10^5, 4.46×10^5 and 4.89×10^5 erg/cm^3 for 10, 40, 60 y 120 min, respectively. In Fig. 4, the hysteresis loop for the ribbon with no thermal treatment at positive fields is shown. The induced anisotropy by the thermal treatment in the amorphous ribbon was evaluated qualitatively from the area formed by curve of hysteresis loop and the horizontal line located where the magnetization saturation is achieved.

The induced anisotropy is an important design parameter. In many works has already showed that step-induced anisotropy can be used in the development of magnetoresistive sensors based on Hall effect and spin-dependent tunneling [10,11]. We use the anisotropy energy K_1 as a design parameter. In P10, a shaded area is indicated and it is proportional to the anisotropy energy. In the inset of the figure, a shaded area is indicated for the case P20. This behavior is very important due to it sets the initial conditions of the anisotropy energy and we can see the change with the annealing time. Most pronounced change in properties of the amorphous ribbon and the energy involved in the dynamics of the magnetic moments is where the orientation of the magnetization vector is out plane {P190, P290}. Even though the saturation is not achieved, we use the Eq. 3 with M_S = 6000 Oe.

The Fig. 5 shows a shaded area for orientation P190, this area is proportional to the stored energy by the magnetic moments when they are oriented to the applied field direction. The inset shows the behavior of the ribbon in orientation P290. Clearly we note that the shaded area has changed, so K_1 is a very sensitive parameter of position. In both curves is not possible to determine the retentivity and the saturation magnetization, we can explain it as follow, the applied field is normal to ribbon plane, so the field do not distinguish the magnetic structure which is formed by domains and domain walls.

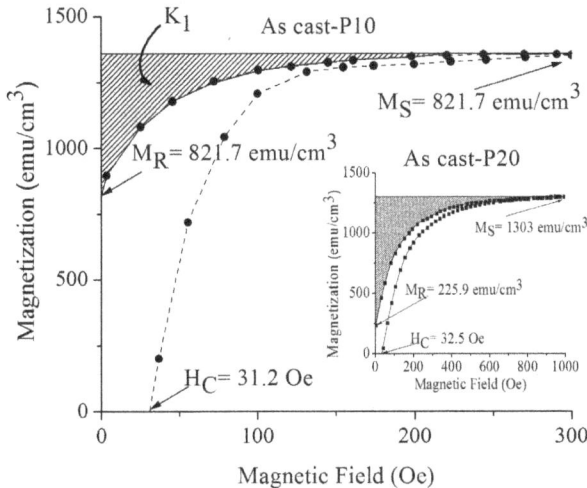

Figure 4. *Magnetization as a function of field for as cast at P10 and P20 (in the inset), K_1 is related to induced anisotropy.*

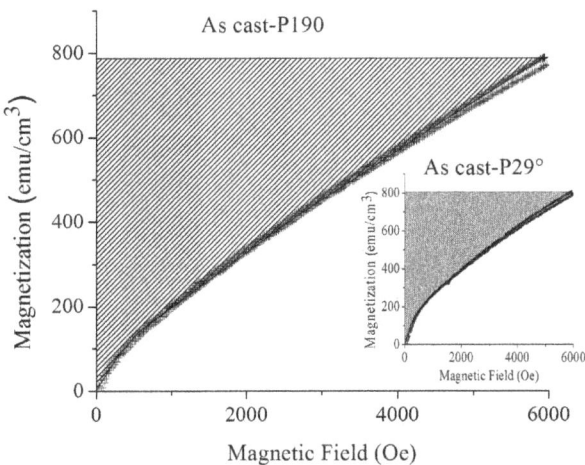

Figure 5. *Anisotropy energy in orientation P190 and orientation P290 (inset).The shaded area is proportional to induced anisotropy due to the thermal treatment.*

The magnetic properties in P190 and P290 are shown in Table 1. The column 2 and the column 3 corresponds to P190 (with a maximum value M_S = 798 emu/cm³, K_1 = 2.37×10⁶ erg/cm³), the column 4 and the column 5 corresponds to P290 (with a maximum value M_S = 802 emu/cm³, K_1 = 2.41×10⁶ erg/cm³).

Table 1. *The magnetic properties in P190 and P290. As cast (AC).*

Time (min)	Magnetization Saturation, M_S (emu/cm³)	Anisotropy Energy, K_1 (×10⁶ erg/cm³)	Magnetization Saturation, M_S (emu/cm³)	Anisotropy Energy, K_1 (×10⁶ erg/cm³)
AC	798	2.37	802	2.41
10	714	2.14	670	2.0
40	515	1.55	522	1.57
60	568	1.70	550	1.65
120	455	1.40	374	0.6

The effect of the annealing time over the magnetic properties in the alloy is shown in Fig. 6. Fig. 6a shows the dependencies of M_R/M_S and M_R on annealing time at P10. It can be seen that M_R/M_S changes very regular at annealing. M_R increases with increasing the time of annealing treatment until a maximum value of 870 emu/cm³ (for as cast), and then its consequent decrease until a minimum value of 516 emu/cm³ (for 120 min). A similar behavior we found in the retentivity ratio. M_R/M_S increases with increasing the annealing time until a maximum value of 0.705 (for 10 min), and then its decrease until a minimum value of 0.55 (120 min).

Fig. 6b, shows the dependencies of H_C and M_S on annealing time. The H_C initially decreases until a minimum 31 Oe, corresponding to amorphous ribbon with 40 min of thermal treatment, after 40 min H_C increases slowly to 60 min with 34 Oe. Then H_C decreases slowly to 120 min with 29 Oe. This behavior is due to magnetic softening originated by structural relaxation. It can be explained as the nucleation of a nanocrystal precursor matrix. This nanocrystals has a large anisotropy that the amorphous phases and are poorly couple to it [12,13], indicating that they are acting as effective pinning centers for the propagating domains [14,15]. M_S decreases slowly until a minimum value of 455 emu/cm³. This behavior has been reported in other works [16], and it has been explained as a softening and hardening state, respectively. In order to guarantee the magnetic softness of the amorphous ribbon with ultrafine FeMo structure, it is important to inhibit the formation of Fe-borides which is managed by the molybdenum addition in combination with the boron content [17].

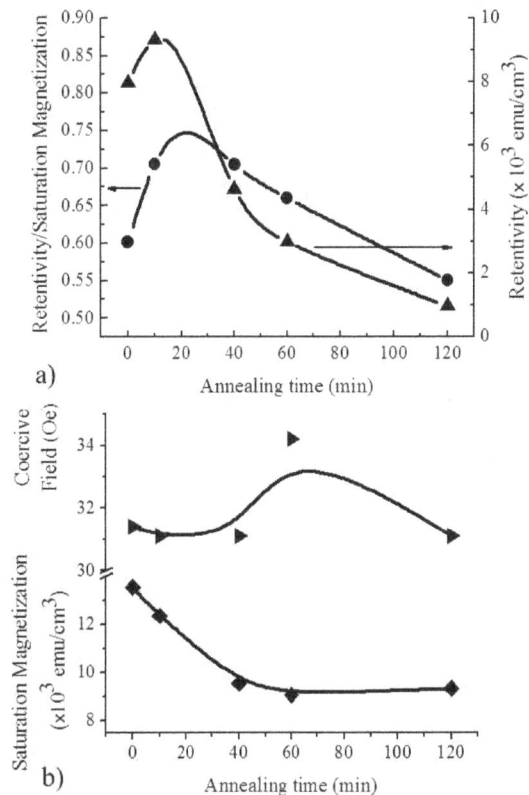

Figure 6. *a) Retentivity, retentivity ratio, b) coercive field and saturation magnetization as a function of annealing time.*

The response of the ribbon for orientation P190 is shown in Fig. 7. This behavior is very interesting, since the magnetization increases linearly, the rate of change of the magnetization is defined as $m = \Delta M / \Delta H$, we find the values of 0.122 (700/5776.9), 0.107 (563/5254), 0.069 (388/5582), 0.063 (346/5507) and 0.075 (406/5411) emu / Oe cm^3 corresponding to as cast, 10, 40, 60 and 120 min, respectively. For the as cast, we find the largest range of magnetic field detection. The minimum is for 10 min. According the magnetization change with the field, the maximum is for the as cast and the minimum for 60 min. In orientation P290, the values of the rate of change of magnetization are 0.11 (568/5157), 0.84 (437/5183), 0.089 (458/5137), 0.08 (334/4173) and 0.11 (322/2913) emu / Oe cm^3 corresponding to as cast, 10, 40, 60 and 180 min, respectively. For 10 min, we found the largest range of magnetic field detection. The minimum is for 180 min. According the magnetization change with the field, the maximum is for the as cast and the minimum for 180 min.

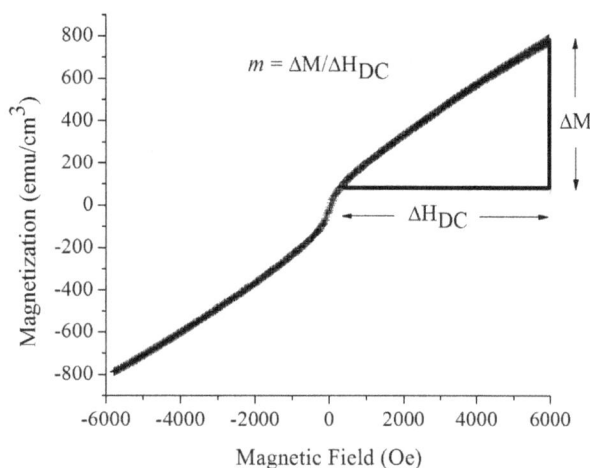

Figure 7. Rate of change of magnetization (slope of linear region in the hysteresis loop.

The variation of sensibility as a function of annealing time is shown in table 2. This is the major parameter to determine the response of the amorphous ribbon in front of DC magnetic field. There is a direct relationship between rate of change of magnetization and the saturation magnetization. A material with a high value of M_S is a very good candidate to design magnetic field sensors. The column 2 corresponds to orientation P190 and the column 3 corresponds to orientation P290. In both cases the maximum value is for the material with no thermal treatment (as cast) and the minimum value is for 60 min.

Table 2. The magnetization ratio to P190 and P290.

	$\Delta M / \Delta H$ (emu/Oe cm^3)	$\Delta M / \Delta H$ (emu/Oe cm^3)
As Cast	0.122	0.110
10 min	0.107	0.084
40 min	0.069	0.089
60 min	0.063	0.08
120 min	0.075	-----
180 min	-----	0.11

The evolution of magnetization as a function of heat treatment time to P190 is shown in Fig. 8. In all curves, we can see a clear change in the slope and consequently in the sensibility. This change gives information about the response of the amorphous ribbon with the temperature. An important requirement is that the shape of the hysteresis loop can be varied according to the necessities of technological applications.

An atomic pair ordering due to a thermal treatment induces an easy direction parallel to the applied field. It has been widely discussed elsewhere that this behavior is due to the evolution of the amorphous and nanocrystalline phases, the selected temperature is below the crystallization (500 and 630°C) [18]. All these factors affect to dynamic of the magnetic moments. The change of amorphous and nanocrystalline phases determines the hysteresis form. Hysteresis loop of the treatment samples are quite different when comparing to as cast. The different results obtained through the thermal treatment method show that the field induced anisotropies behave in a very different way with respect to the induced anisotropies by orientation P10 and orientation P20.

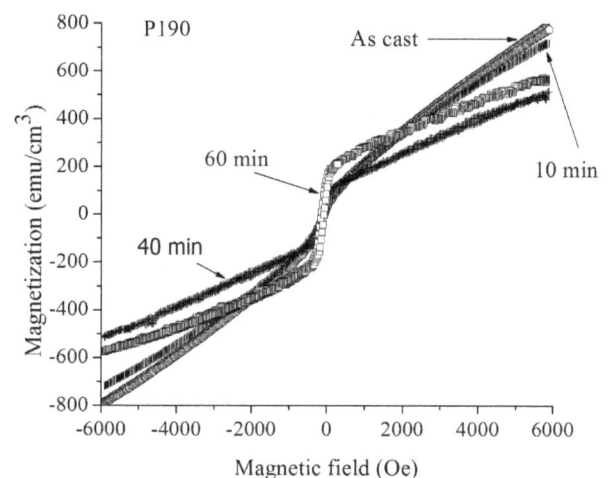

Figure 8. Variation of slope in the hysteresis loop with different annealing times.

For future works, we can determine the magnetic properties and the linear response of the magnetization at temperatures above the first and second nanocrystallization phase. Joule treatment is feasible, the purpose is to determine the stability of the material against an alternating current and thereby determine the critical point at which the magnetic properties such as anisotropy, saturation magnetization, and retentivity affected. In the design of any sensor is vital to know the thermal stability. This amorphous ribbon is wonderful, since according to their various responses can also be used as a temperature sensor.

The knowledge of the change of the anisotropy of the amorphous ribbon, not only with the sample orientation with respect to the field, but with the heat treatment time, a comprehensive methodology is set for declare that this material is a serious candidate for industrial applications.

4. Conclusions

The amorphous ribbon was studied in four geometric arrangements: P10, P190, P20 and P290. Changes of the retentivity, saturation magnetization, anisotropy energy, coercive field and $\Delta M/\Delta H$ in $Fe_{73.5}B_9Si_{13.5}Mo_3Cu_1$ amorphous alloy have been reported. The response of M_R, M_S, M_R/M_S and K_1 was analyzed, not only in terms of the geometrical arrangement, but also in terms of time of heat treatment. The best values are presented for orientation P10. In the case of the orientation P10, at as cast ribbon, the values are: $M_S = 1353$ emu/cm^3, $M_R = 815$ emu/cm^3 and $M_R/M_S = 0.6016$; at P20: $M_S = 1300$ emu/cm^3 $M_R = 216$ emu/cm^3, and $M_R/M_S = 0.166$, respectively. However, for orientation P20 a larger anisotropy value is achieved $K_1 = 650,000$ erg/cm^3 than to the case of orientation P10 ($K_1 = 270.600$ erg/cm^3). This difference is generated because at P10 the magnetization vector is in the ribbon plane, and the dynamic vector is regulated by the magnetic structure (domains and domain walls). In P20, the dynamic magnetization vector is outside the ribbon plane, so the domain dynamic is not regulated by the magnetic structure, in this case exchange interaction takes less importance.

The highest values of anisotropy are presented for P190 and P290 arrangements, these values for as cast are: $K_1 = 2,365,100$ erg/cm^3 and $K_1 = 2,405,520$ erg/cm^3, respectively. Regarding the rate of change $\Delta M/\Delta H$ (defined only for the study of the amorphous ribbon in P190 and P290), we have the following values. At P190: 0.122, 0.107, 0.069, 0.063 and 0.075 emu/Oe cm^3 corresponding to as cast, 10, 40, 60 and 120 min, respectively; At P290: 0.11, 0.084, 0.089, 0.08 and 0.11 emu/Oe cm^3 corresponding to as cast, 10, 40, 60 and 180 min, respectively. The best linear increase is for the ribbon without heat treatment.

According to our results, the amorphous alloy is a strong candidate as a vector field sensor in three directions: longitudinal, transverse and normal to the plane (saturation magnetization, retentivity and external magnetic field). In the subsequent, detailed study considering small steps of heat treatment time, in order to determine the critical point where the magnetic properties begin to change, so we can control the thermal stability of the amorphous ribbon.

References

[1]   W. H. Meiklejohn and C. P. Bean, "New Magnetic Anisotropy," Physical Review, vol. 105, N. 3, pp. 904–913, February 1957.

[2]   Andrei V. Palii, Boris S. Tsukerblat, Eugenio Coronado, Juan M. Clemente-Juan, Juan J. Borrás-Almenar, "Orbitally dependent kinetic Exchange in cobalt (II) pairs:origin of the magnetic anisotropy," Polyhedron, vol. 22, pp. 2537-2544, February 2003.

[3]   Giselher Herzer, "Nanocrystalline soft magnetic materials," Journal of Magnetism and Magnetic Materials, vol. 157/158, pp. 133-136, 1996.

[4]   Tadeusz Kulik, Antonio Hernando,"Magnetic properties of $Fe_{76.5-x}Cu_1Nb_xSi_{13.5}B_9$ alloys nanocrystallized from amorphous state," Journal of Magnetism and Magnetic Materials, vol. 160, pp. 269-270, 1996.

[5]   V. Franco, C. F. Conde, A. Conde, "Changes in magnetic anisotropy distribution during structural evolution of $Fe_{76}Si_{10.5}B_{9.5}Cu_1Nb_3$," Journal of Magnetism and Magnetic Materials, vol. 185, pp. 353-359, 1998.

[6]   Simon Foner, "Versatile and Sensitive Vibrating-Sample Magnetometer," The Review of Scientific Instruments, vol. 30, pp. 548-557, July 1959.

[7]   David Jiles, Introduction to Magnetism and Magnetic Materials, 1st ed., Chapman and Hall, 1991, pp. 49–52.

[8]   Shoshin Chikazumi, Physics of Magnetism, 1st ed., John Wiley and Sons, 1964, pp. 554.

[9]   B. D. Cullity, Introduction to Magnetic Materials, Addison-Wesley Publishing Company, Inc., 1972, pp. 225-229.

[10] R. S. Popovic, J. A. Flanagan, P. A. Besse, "The future of magnetic sensors," Sensors and Actuators A, vol. 56, pp. 39-55, 1996.

[11] G. Herzer, M. Vázquez, M. Knobel, A. Zhukov, T. Reininger, H. A. Davies, R. Grössinger, J. L. Sánchez Ll, "Round table discussion: Present and future applications of nanocrystalline magnetic materials," Journal of Magnetism and Magnetic Materials, vol. 294, pp. 252-266, 2005.

[12] C. Miguel, A. Zhukov, J. J. del Val. J. Gonzáles, "Coercivity and induced magnetic anisotropy by stress and/or field annealing in Fe- and Co- based (Finemet-type) amorphous alloys," Journal of Magnetism and Magnetic Materials, vol. 294, pp. 245-251, 2005.

[13] N. Murillo, J. González, "Effect of the annealing conditions and grain size on the soft magnetic character of FeCu(Nb/Ta)SiB nanocrystalline alloys," Journal of Magnetism and Magnetic Materials, vol. 218, pp. 53-59, March 2000.

[14] R. Schäfer, S. Roth, C. Stiller J. Eckert, U. Klement and L. Schultz, " Domain Studies on Mechanically Alloyed Fe-Zr-B-Cu- Nanocrystalline Powder," IEEE Transactions on Magnetics, vol. 32, No. 5, pp. 4383-4385, 1996.

[15] R. Valenzuela and J. T. S. Irvine, "Domain Wall dynamics and short-range order in ferromagnetic amorphous ribbons," Journal of Non-Crystalline Solids, vol. 156-158, pp. 315-318, 1993.

[16] Arturo Mendoza Castrejón, Herlinda Montiel Sánchez, Guillermo Alvarez Lucio, Rafael Zamorano Ulloa, "Nanocrystallization in $Fe_{73.5}Si_{13.5}B_9Mo_3Cu_1$ amorphous ribbon and its magnetic properties," Materials Science Forum, vol. 691, pp. 77-82, 2011.

[17] P. Kwapuliński, A. Chrobak, G. Haneczok, Z. Stokłosa, J. Rasek, "Structural relaxation and magnetic properties of $Fe_{86-x}Nb_xB_{14}$ amorphous alloys," Journal of Magnetism and Magnetic Materials, vol. 304, pp. e654-e656, March 2006.

[18] E. Illeková, D. Janičkovič, M. Miglierini, I. Škorvánek, P. Švec, "Influence of Fe/B ratio on thermodynamic properties of amorphous Fe-Mo-Cu-B," Journal of Magnetism and Magnetic Materials, vol. 304, pp. e636-e638, March 2006.

12

Analysis of Chinese Welding Industries Today and in the Future (Focus on Cost, Productivity, and Quality)

Xiaochen Yang, Paul Kah, Jukka Martikainen

Laboratory of Welding Technology, Lappeenranta University of Technology, Lappeenranta, Finland

Email address:

xiaochen.yang@lut.fi (Xiaochen Yang), paul.kah@lut.fi (P. Kah), jukka.martikainen@lut.fi (J. Martikainen)

Abstract: This paper make a detailed survey and research about the real situation about Chinese welding industry, and the future cooperation suggestion between Finland and China in this area. This topic was approached by searching for the latest and most accurate information through interviews, site visits, websites, books and promotional materials. The site visits are the most important. The destination cities in China cover a large range from northern to the southwestern areas in China. More than 15 professional welding factories or factories which have a welding workshop were visited, and based on discussions and an investigation in those factories; detailed and reliable data was collected. By using and analyzing that information, a general profile of modern Chinese welding factories was generated.

Keywords: Welding Quality, Finland-China Co-operation, Welding Cost, Welding Productivity, Welding, TWM

1. Introduction

Finnish and European welding factories are currently finding it difficult to hire professional or expert welders because the lack of labor force. Many companies choose to buy welding products from China, and therefore, more and more Chinese welding factories become subcontractors to Finnish contractors. As a result of this trend, Finnish companies wish to obtain the latest and most accurate information of the status of the Chinese welding industry and factories in order to support their business decisions. The background of this paper is the HitNet-Global project of TEKES. One aim of this project is to understand the real situation of the Chinese welding industry exactly. The information about Chinese welding industry will be collected from Chinese welding institute, Chinese welding factories and some of the Chinese universities. The information includes the welding productivity, the welding cost, and the quality control situation in China.

After the survey period, many problems were found in the welding processes in Chinese welding factories. How to balance the low cost and unsatisfactory quality of Chinese products nowadays becomes the top significant issue does not only exist in the welding industry. A more better and reliable regulatory approach or a more proper qualification

system of Chinese welding factories should be studied in future as quickly as possible[12, 15].

In conclusion, a mutual understanding of the culture, lifestyle and customs between the East and West also play an important role in future cooperation.

2. Overview of Welding Industries in China

The welding industry is one of the most important branches in the Chinese manufacturing industry. According to a report on the Beijing Essen Welding and Cutting Fair, the exhibitors' number has increased 12.36%, while the professional visitors' number has increased by 45.82% and the exhibition space by 33.23% during 2009 and 2010 [1] [2]. The increasing trend indicates that more people were interested in the welding industry, and more foreign companies regard China as their significant partner in cooperation and their present and future market.

This study on the Chinese welding industry is divided into two areas: research on welding equipment and new welding technologies, and the application of welding methods. Research on the first subject area is carried out in professional institutes, universities, and large enterprises. The examination of the second subject area covers nearly all of the areas of China which is include steel structure manufacturing, ship building industry, construction machine fabrication, high pressure vessel assembly, and chemical metallurgical industry.

There are all types of welding processing workshops in China: small and medium-sized workshops, and large-scale welding factories. The private small and middle-sized workshops occupy a large market share in China. Many large factories are closing their welding workshop and moving their welding work to smaller companies, thus focusing only on their assembly plant. This kind of cooperation pattern is becoming the prevailing trend not only between foreign companies and Chinese subcontractors, but also between state-owned enterprises and small private factories within China.

In the field of international commercial cooperation, Finnish companies also perform a great deal of commercial activities in China, but choose different cooperation patterns. Some companies choose Chinese welding factories as their subcontractors, and the final assembly work of their product is finished in Finland. Others choose to build their own or joint venture factories in China in which all work from welding to machining and assembly is carried out. However, the targets of all of the Finnish companies are same: higher productivity, lower costs, and reliable quality.

A Finnish company J, which is specialized in the design, manufacture, and service of hydraulic piling equipment, has chosen the first cooperation mode and cooperates with a specific Finnish agency called Ea (see Figure 1) [3].

The agency is in charge of the contact with Chinese industry, the quality control, machine part transport, and contracts with Chinese suppliers.

Fig. 1. *The cooperation model between Finnish and Chinese welding industries.*

The other collaboration model – building factories in China – requires Finnish companies to understand the Chinese welding industry more accurately and in detail. However, most of the small enterprises in Finland or Europe do not have much experience of collaboration with Chinese industry.

3. Case Studies

Three aspects of Chinese welding factories will be examined: the productivity, the economy and cost, and the quality control system. Productivity is determined based on the following factors: the number of workers (welders), the customers, workers' qualification, the plant situation, equipment and machines, auxiliary materials and gases, the environment and climate conditions, and other factors.

Economy and cost are analyzed based on the wages of staff (workers and management teams), the cost of materials (main material and auxiliary materials), the cost of processing, the cost of logistics, the margin and turnover, and other factors. Quality and quality control are determined by the processing methods, the welding procedure qualification record (WPQR) and welding procedure specification (WPS), human and cultural factors, the assessment and training of welders, inspection methods, the maintenance and repair of equipment, storage conditions, and environmental and climate conditions. The final objective is to find out the most suitable Chinese subcontractors.

In this section, all of the information above on six Chinese welding factories or workshops in different areas and industries will be listed. The information is based on the site visits by the author to each factory.

3.1. Company A

The main product of this factory is hydraulic support for coal mines. This factory is located in Beijing, North China [4].

Fig. 2. *Working environment of Company A welding factory (photo by author).*

Number of workers: 6 welding workshops, almost 1000 welders.

Workers' qualifications: 1/4 of the welders are senior technicians, and 3/4 of them are intermediate technicians. Every welder in this factory has a welding technician certificate.

Plants: Welding plant, machining workshop, and cutting workshop.

Manufacturing tools: CO2 MAG and MMA welding.

Materials: Main material: steel board, steel type: Q550, Q460, Q690 (based on GB221-79, the China national standards).

Welding wire: flux-cored welding wire: type 77, 78, 702; solid welding wires: JL-80M.

Environment: No special temperature and humidity control method.

Logistics costs: Highway and railway transportation, shipping.

Processing costs: Water, plant transport, propane gas, oxygen gas, and electricity.

Other costs: Equipment depreciation, warranty service,

and rework cost for defective products.

Turnover: 3 billion RMB yuan in the year 2010.

Processing steps: Raw material cutting $\rightarrow$ cold machining (some parts) $\rightarrow$ heat treatment $\rightarrow$ riveting and welding (spot welding and assembly) $\rightarrow$ aging process (some parts) $\rightarrow$ seam filling welding $\rightarrow$ cold machining (some parts) $\rightarrow$ assembly.

Cultural factors: Zhangjiakou city (Hebei province), Huludao city (Liaoning province), and Shijiazhuang city (Hebei province), North China.

Worker training: Security assessment and skill assessment.

Inspection methods: Personnel from the quality assurance office. The following aspects are inspected: appearance of the welding seam, size of the part, welding slag, gas stoma, oil pollution, and quality of the welding seam.

Machine maintenance: Everyday basic cleaning and weekly professional maintenance.

Material storage and treatment: Special warehouse and pretreatment in order to remove rust. The storage of steel board inside will reduces the corrosion.

Environment and climate: The dry and cold winter requires preheating of the materials. Warm method (heating system in the workshop) for welders.

3.2. Company B

The main business of this factory is machining for steel structures. This factory is located in Shijiazhuang city, North China [5].

Fig. 3. Working environment of Company B (photo by author).

Number of workers: 50 workers, 10 of them are welders.

Workers' qualifications: Two of the welders are senior technicians, five are technicians. All other workers are senior technicians.

Plants: Two workshops; the total area for welding is 2650 m2. The courtyard area for storage and truck parking is 1200 m2.

Manufacturing tools: MIG (Argon as shielded inert gas) welding machine×1; MAG (CO_2 as shielded active gas) welding machine×2 (Chengdu Gaoxin KH-350); AC MMA welding machine×5 (Hebei welding machine factory BX3 500J and BX1 500).

Materials: Stainless steel, aluminum, carbon steel, and copper.

Environment: All the welding works are certified by the Chinese National Environment Performance Assessment. Noise, waste gas and water emissions all meet the standards. The temperature and humidity do not influence welding works in the North China area.

Workers' salary: 2500 to 3000 RMB yuan per month. The workers' wages account for the 1/3 of the total costs of the factory.

Materials costs: According to the market prices. The cost of materials is approximately 1/3 of the total costs. The price of 5 mm thick steel board was approximately 5000 RMB yuan per ton in the year 2011.

Logistics costs: Logistics, processing and other costs are roughly 1/3 of the total costs, and the logistics costs amount to approximately 10% of the total costs. Truck, labor and petrol costs, highway fees and management fee to the government.

Processing costs: Water (4 to 6 RMB yuan per ton, general industrial water price for the year 2011 in North China), argon, oxygen, acetylene, carbon dioxide, hydrogen, electricity (1 RMB yuan per KWh, general industrial electricity price for the year 2011 in North China) and all types of welding rods.

Other costs: Depreciation cost for manufacturing machines and cost of equipment loss. For taxes, the middle-sized welding and machining factories in North China have a 10% margin. The tax costs are approximately 17%, but 12-13% will be paid by the customers. The taxes include 1/4 of the profit, a channel environment charge, the income tax of workers, a disabled worker tax, and an environment management tax.

Turnover: 5 to 6 million RMB yuan per year.

Processing steps: Engineering drawing reading and review $\rightarrow$ processing design $\rightarrow$ manufacturing plan $\rightarrow$ materials preparation $\rightarrow$ processing steps (heating treatment) $\rightarrow$ manufacturing (machining and welding) $\rightarrow$ inspection (every processing steps) $\rightarrow$ packaging $\rightarrow$ products ready.

Cultural factors: Shijiazhuang city, Hebei province (North China).

Worker training: Training in safety and professional skills.

Inspection methods: The welding size, flatness of welding seams, and cracks are the inspected items. The full non-destructive test (NDT) is conducted only upon the customers' request.

Machine maintenance: Machine maintenance on Saturdays and major machine maintenance annually.

Material storage and treatment: Steel board and materials: indoor environment in the workshop. Welding wire and welding rods: special warehouse. The oxygen: isolated warehouse.

Environment and protection: Overalls, special masks, welding glasses, and welder gloves. Fan to control the temperature in the summer.

3.3. Company C

This is one of the largest and most professional manufacturers of steel-manufacturing equipment in North

China. It is located in the city of Tangshan [6].

Fig. 4. Working environment of the Company C (photo by author).

Number of workers: Three welding plants, 20 senior welding technicians, 80 welders.

Workers' qualifications: Senior welding technicians with national boiler and pressure vessel welder certification.

Plants: 1.4 million mm2, of which 82000 mm2 consists of manufacturing workshops.

Manufacturing tools: MIG (argon as shielded inert gas) welding; MAG (CO_2 as shielded active gas) welding machine (NBC-500 and NB7-500, Tangshan Greatwall welding machine factory); AC MMA welding machine; SAW machine×1.

Materials: K235, Q345, Cr-Mo steel, heat-resistant steel, low alloy steel.

Environment: Pre-heating works for thick steel boards in the winter.

Workers' salary: 1200 RMB yuan per month, including the five insurances and one fund. The technician welders' wage is 2000 RMB yuan per month.

Materials costs: 60% - 70% of the total costs.

Logistics costs: 10% of the total costs. 90% of the transportation methods are highway transport.

Processing costs: Welding wire, welding rod, carbon dioxide, argon, oxygen, diesel oil, water, and electricity. 5% to 10% of the total costs.

Other costs: The worker's wage and depreciation cost for manufacturing machines and cost of equipment loss is 10% of the total costs. The margin is approximately 10%, of which 23% is the sales tax. The value-added tax is 17%.

Turnover: 0.21 billion RMB yuan for the year 2010. The average turnover for the past five years is 0.2 billion RMB yuan.

Processing steps: Material re-examination $\rightarrow$ material cutting (CNC flame cutting) $\rightarrow$ coiling and drilling $\rightarrow$ assembly (spot welding) $\rightarrow$ seam welding $\rightarrow$ NDT $\rightarrow$ heat treatment $\rightarrow$ painting $\rightarrow$ pressure testing (for boilers and containers) $\rightarrow$ products delivery.

Cultural factors: Tangshan or Qian'an city, Hebei province (North China).

Worker training: Safety training and certificate check.

Inspection methods: Material delivery period: ultrasonic,

chemical component, tensile strength and yield strength tests. Welding seam: size checking, ultrasonic, radiation, magnetic particle, and penetration detection.

Machine maintenance: Everyday machine maintenance by the maintenance department.

Material storage and treatment: Steel board: outside storage area. Welding wire and welding rods: inside warehouse.

Worker protection: Mask, gloves, flame retardant overalls, and workers' boots.

3.4. Company D

Three individual factories: Company D1, Company D2, and Company D3. Xuzhou, Jiangsu, eastern and coastal areas, construction machinery industry [7].

Number of workers: D1: 400 workers; 200 of them are welders. D2: The labor productivity is 300 000 RMB yuan per one worker per one year. The industry net value added is currently 20%.

Workers' qualifications: 95% of the workers are post-secondary technical school students, 70% of them are college students. In the key positions, 5% - 10% of the workers are undergraduate students.

Plants: D1: The entire area of the factory is 2666 acres; the area of the steel structure workshop is 200 meters×150 meters.

Manufacturing tools: D1: MAG (CO_2 as shielded active gas) welding, MAG inverter welding machine (PANASONIC), automation liner welding machine×2, welding robot×1.

D2: MAG (CO_2 as shielded active gas) welding, MAG inverter welding machine (PANASONIC: KRII 500 or LINCOLN welding machine) × 20. Welding wire: MAG WELDING CHW-50C6 (Shanghai Atlantic Welding Consumables Co., Ltd.). RE: Gas plasma cutting, contour cutting, MAG (CO_2 as shielded active gas) welding, laser welding, and robotic welding.

Materials: D1: Steel board, steel type: 70 steel. RE: Steel board, steel type: medium carbon steel, and Q235.

Environment: The entire plant basement is covered with a whole sealed steel board in order to prevent the leakage of petroleum. Thermostat and humidity controlled plants.

Workers' salary: D2: the average wage is 60 000 RMB yuan per year (workers and middle managers). D1: the average wage is 4500–4700 RMB yuan and 1500 RMB yuan housing accumulation fund per month for welders.

Materials costs: 40% of the total costs, the gross profit margin of the mechanical industry are currently approximately 20%.

Logistics costs: 1% of the total costs, road transport and railway transport.

Processing costs: Labor costs amount to 20% of the processing costs. The total processing costs are roughly 10% of the total costs.

Other costs: The selling expenses are 3% of the sales price. Office supplies, a shuttle bus for workers, business travel, hospitality, canteen costs, maintenance of other assets,

R&D expenses, testing costs, and rework cost; the percentage in the total costs is very small.

Turnover: D2: 2.2 billion RMB yuan last year. D1: 0.98 billion RMB yuan for 2009, 2 billion RMB yuan for 2010, and the prospective turnover for 2011 is 4.5 billion RMB yuan.

Processing steps: D1 and D3: incoming inspection (Chemical elements: carbon, manganese, silicon, sulfur and other alloying elements. Physical inspection: yield strength, tensile strength, and bending strength. Metallographic analysis: hardness) $\longrightarrow$ raw material cutting $\longrightarrow$ pretreatment (heat treatment) $\longrightarrow$ cold machining (some parts) $\longrightarrow$ heat treatment $\longrightarrow$ riveting and welding (spot welding and assembly) $\longrightarrow$ aging process (some parts) $\longrightarrow$ seam filling welding $\longrightarrow$ cold machining (some parts) $\longrightarrow$ assembly. D2: incoming inspection (Chemical elements: carbon, manganese, silicon, sulfur and other alloying elements. Physical inspection: yield strength, tensile strength, and bending strength. Metallographic analysis: hardness) $\longrightarrow$ pretreatment $\longrightarrow$ cutting $\longrightarrow$ riveting and welding $\longrightarrow$ painting.

Culture factors: Xuzhou city or other cities in the Jiangsu province (eastern and coastal areas).

Worker training: The regular staff has classes and training every year, taught by experts from institutes, professors form universities, and engineers from other companies.

Inspection methods: Ultrasonic testing, magnetic particle testing, and turbine testing. The inspection is carried out by personnel (200-person team, 10 of them quality engineers) from the quality assurance office.

Machine maintenance: Everyday basic cleaning and maintenance, and monthly professional maintenance.

Material storage: Satisfaction for the national standard requirement.

Worker protection: Protected by special protective gear for welding work.

3.5. Company E

This is a producer of globally high-end offshore vessels, located in Ningbo city, Zhejiang, in the Eastern and coastal area [8].

Fig. 5. *Working environment of Company E (photo by author).*

Number of workers: Two welding workshops, 1300-1400 welders from outsourcing labor force companies. 70–80 regular staff welders.

Workers' qualifications: 1000 welders are intermediate technicians. 300 are apprentice welders.

Plants: Welding workshops for ship body sections and small parts.

Manufacturing tools: Overall, 90% of the welding works were finished by MAG (CO_2 as shielded active gas) welding using a MAG inverter welding machine (OTC: XD 500S; PANASONIC: KR II 350; Zhoushan Donghai electrical welder manufacturing co., ltd.: MR3151), and 10% of the welding works consist of submerged arc welding (Zhoushan Donghai electrical welder manufacturing co., ltd.: MZ-1250-2) and MMA welding (Zhoushan Donghai electrical welder manufacturing co. ltd.: ZXE 1-500). TIG welding and fillet welding are using for patches, repair, tube welding or fine welding.

Materials: Carbon steel board and stainless steel (according to the A to E levels). Welding wires: 3Y level welding wire, Antai QJ 501 and Jintai 712C.

Environment: If it rains, welding works is stopped. In strong wind, MAG welding is stopped.

Workers' salary: The intermediate technician wages are 3000 RMB yuan per month.

Materials costs: According to the market price.

Logistics costs: Highway transport costs and ship transport.

Processing costs: Water, plant transport, argon, oxygen, carbon dioxide, acetylene, machine maintenance costs, workshop building repair costs and electricity.

Other costs: Inventory costs are zero because all the warehouse or storage cost is outsourced to storage companies.

Turnover: 4.2 billion RMB yuan for 2010. Before the financial crisis, the turnover was 10 billion RMB yuan per year. In 2009, the turnover was approximately 5 billion RMB yuan.

Processing steps: Raw material cutting $\longrightarrow$ small part welding or assembly (inside the workshops) $\longrightarrow$ welding or assembly of middle-sized parts (partly inside the workshops, partly outside) $\longrightarrow$ final assembly on the berth.

Cultural factors: Nantong city, Jiangsu province (Eastern and coastal areas).

Workers' training: Security assessment and skill assessment; 50% of the welders in the factory have welder certification.

Inspection methods: Incoming material inspection, pre-welding inspection, welding process inspection, non-destructive test, and final inspection by the ship register.

Machine maintenance: Everyday machine cleaning and weekly machine maintenance by the specialized personal.

Material storage: Steel board and welding wire storage in a special warehouse.

Worker protection: All protective equipment such as masks or earplugs should be used.

3.6. Company F

This is a producer of marine diesel engines, located in Shandong province, in the eastern and coastal areas [9].

Fig. 6. *Working environment of Company F (photo by author).*

Number of workers: 123 welders. The welder performance qualification is based on EN 287 for welding workers and EN 1418 for welding operators.

Workers' qualifications: Qualified by DNV (Det Norske Veritas), GL (Germanischer Lloyd), BV (Bureau Veritas), or ABS (American Bureau of Shipping). Only Chinese welding certification is not accepted for work in company F.

Plants: One welding workshop, one machining workshop, and one assembly workshop.

Manufacturing tools: Submerged arc welding, MAG (CO_2 as shielded active gas) welding, MMA welding, MIG (argon as shielded inert gas) welding, stud welding, and TIG welding.

Materials: S235 steel board.

Environment: Well-constructed workshops with constant temperature and humidity conditions, and dust exhaust machines.

Workers' salary: 3000 RMB yuan per month. Overtime pay on workdays is 1.5 times the usual pay and on holidays and weekends 2 times the usual.

Materials costs: According to the market price in China.

Logistics costs: Specific logistics company to manage transportation matters.

Processing costs: The cost of welding seams is carefully calculated and divided into processing costs, consumption costs, NDT costs, and post-treatment cost. The processing costs in company F include shield gas costs (CO_2, or 82% argon + 12% CO_2), propane, welding wire, welding rod, welding flux, water and electricity. The total processing costs are 20% of the total costs.

Other costs: UT test: 50 RMB yuan per hour, VT test: 30 RMB yuan per hour, and MT test: 40 RMB yuan per hour.

Turnover: 0.6 billion RMB yuan for the entire factory in 2010. The margin for the welding workshop is 20% to 30%.

Processing steps: Pre-treatment of materials →material cutting → part assembly and welding → painting → machining →assembly for the whole product →testing →product delivery. The pWPS is based on EN 15609-1. The WPQR is according to ISO 15614-1. The WPS is according to EN 15609-1.

Cultural factors: Shandong province (eastern and coastal areas).

Worker training: There is a welding training school in the welding workshop. Its purpose is to introduce new workers to the daily welding routines, focusing especially on the welding methods used in the factory.

Inspection methods: UT, VT, or MT tests are conducted throughout the welding process. All of the testing activities follow the standards EN 473, EN 12062, EN 1714, EN 571-1, EN 970, and EN 1290. Each welding cell is inspected by more than one inspection staff member in order to ensure the welding seam quality.

Machine maintenance: Maintenance and calibration follow the standards EN ISO 9001/9002, EN ISO 3834-2, EN ISO 17662, DVS techn. Bulletin 3009, and DVS techn. Bulletin 0714. [10, 11]

Material storage and treatment: The material storage and treatment rules cover every welding material. The handling and storage of covered welding electrodes, solid and cored welding wire, and welding flux have a significant influence on the weld quality.

Worker protection: Overalls, special masks, welding glasses, and welder gloves.

4. Summary of Chinese Welding Industries

The advantages of North China areas are: 1) two large harbors, Tianjin and Dalian, for easy export, 2) the worker qualifications in this area are easy to ensure, 3) the factories near Beijing have a higher business management level, 4) the workers' wage and material costs are at the medium level, 5) the workers in North China are known for their loyalty, 6) the North China does not have a humid season. The disadvantages of North China areas are: 1) the total welding management (TWM) is not completely implemented, 2) the ND test is inadequate in the production process, 3) the pWPS, WPQR, and WPS are usually overlooked, 4) the winter in North China is cold which will influence the welding quality if no sufficient heating works.

The advantages of eastern and coastal areas in China are: 1) many large harbors, such as Qingdao, Shanghai, Lianyungang and Ningbo, for easy export, 2) the highest level of business management, 3) worker qualifications in this area are easy to ensure, 4) a good history of cooperation with foreign companies, 5) the workers are known for their diligence, 6) the eastern and coastal Chinese areas do not have a cold winter, 7) the level of quality control and inspection is high. The disadvantages of eastern and coastal areas in China are: 1) workers' wages are the highest, 2) the eastern and coastal areas have a heavily humid season.

5. The Finnish-Chinese Cooperation Situation and Improvement Suggestions

Quality control and production management are two significant problems in Chinese small-scale welding factories, some medium-sized welding factories, and even some of the large and state-owned factories. To solve the two problems, advanced and standard production management theory, a suitable enterprise management system, and more welding production standards should be employed. Overall welding quality management, the Plan-Do-Check-Act (PDCA) cycle, the 5S methodology, the Six Sigma business management strategy, and more formal welding standards can help the factories to avoid management or quality problems efficiently [13, 14].

Cooperation with foreign companies can improve the enterprise management model, improve the reputation of the enterprise, enhance its credit, and increase its profitability. In the future, strengthening the cooperation between Chinese factories and Finnish or European companies will be an important issue that the Chinese factories should consider.

Suggestions for Finnish companies are: a) expanding their market in China, b) promoting their enterprise management system in China, and c) establishing a welding factory certification system in China. The implementation of these suggestions will reduce the obstacles of Finnish companies which aspire to find a Chinese subcontractor. Moreover, Chinese welding factories will feel more at ease in signing a contract with a Finnish contractor.

6. Conclusion

The production scale and conditions decide whether a factory can receive a foreign contract or not. The study shows that welding factories located in North China and eastern and coastal areas of China are more competitive than those in other areas. Large-scale or state-owned factories currently have enough domestic and foreign contracts, resulting in a lack of aspiration towards cooperation with new foreign companies. Choosing small and medium-sized Chinese factories as partners in cooperation is the suitable choice for Finnish or European companies today.

References

[1]   Report book of the 14th Beijing Essen Welding & Cutting Fair, p. 3.

[2]   Report book of the 15th Beijing Essen Welding & Cutting fair, p. 3.

[3]   Juha Vierros, Procurement Director of Global Sourcing, interviewed on 10 February 2011, Company company J.

[4]   Yanshi Bi, Vice Factory Rector, interviewed on 10 March 2011, Company A., Beijing, China.

[5]   Wang, Factory Rector, interviewed on 19 April 2011, Company B, Shijiazhuang, China.

[6]   Yanfeng Zhang, Deputy Chief Engineer, interviewed on 6 May 2011, Company C, Qian'an, China.

[7]   Zhao, General Manager, interviewed on 17 March 2011, Company D, Xuzhou, China.

[8]   Zhang, Chief Welding Engineer, interviewed on 24 March 2011, Company E, Ningbo, China.

[9]   Frank Wang, Welding Production Manager, interviewed on 10 May 2011, Company F, Qingdao, China.

[10] SFS-EN ISO 3834-1: en. 2006. Quality requirements for fusion welding of metallic materials. Part 1: criteria for the selection of the appropriate level of quality requirements. Brussels: European Committee for Standardization (CEN). 7 p.

[11] SFS-EN ISO 3834-2: en. 2006. Quality requirements for fusion welding of metallic materials. Part 2: comprehensive quality requirements. Brussels: European Committee for Standardization (CEN). 10 p.

[12] Xiaochen Yang, Paul Kah, Jukka Martikainen. Development of a Welding Product Quality Control and Management System Model for China. Mechanika-2015 conference, Kaunas, 2015.

[13] Xiaotong Yang, Jianjie Cui, Yongqi Zhang, Application Status and Improved Method of 5S Management in Pipeline Construction, Journal of Designation and Construction, Vol 33, No. 5, May 2014.

[14] Luciane de Oliveira Cunha, João Murta Alves, Application of Lean Manufacturing and Quality Management in Aeronautical Industry, International Review of Mechanical Engineering, Vol 8, No. 3, 2014.

[15] Jenni Toivinen, Paul Kah, Jukka Martikainen, Quality Requirements and Conformity of Welded Products in the Manufacturing Chain in Welding Network, International Journal of Mechanical Engineering and Applications, Vol 3, Issue 6, 2015.

13

Quality Requirements and Conformity of Welded Products in the Manufacturing Chain in Welding Network

Jenni Toivanen, Paul Kah, Jukka Martikainen

Laboratory of Welding Technology, Lappeenranta University of Technology, Lappeenranta, Finland

Email address:
paul.kah@lut.fi (P. Kah), jenni.toivanen@lut.fi (J. Toivanen)

Abstract: The objective of this study is to examine the manufacturing and conformity of welded products and the significance of co-operation of different functions to welding quality. This study focuses on costs arising from nonconformity from the manufacturing perspective. It briefly discusses unnecessary costs, claim costs and warranty costs in the production chain. It furthermore takes an overview of challenges in welding manufacturing in the engineering field with empirical research in the industry and shows that failures and defects are identifiable and known in companies but very rarely the root cause of imperfections is investigated. The requirements from manufacturing go unrecognized at the many levels of organisation. One of the main obstacles to improving welding functions is the lack of co-operation and knowledge of the demands on welding. This can cause continuous nonconformity in products and in welding manufacturing. The observations have been collected from welding networks in engineering workshops where GMAW welding is a commonly used process. The results provide a framework for future research to define the importance of actions of different functions to the quality and costs of manufacturing.

Keywords: Welding Manufacturing, Welding Network, Product Conformity, Welding Quality, ISO 3834, Welding Production

1. Introduction

The product life cycle starts with different requirements and needs that are followed in manufacturing over the course of development and design phases [1]. Manufacturing is linked with many other stages, like design, purchasing and quality, and what becomes emphasised in welding. The quality of a product can have many different dimensions, for example, with regard to performance, conformity, service [2] and design [3].

It is generally accepted that different standards and requirements coordinate the level of quality in manufacturing. However, if these demands are not understood and met in the many stages of the manufacturing chain, it can cause unnecessary costs. This study concentrates on explaining the effects of quality of conformity, quality of performance and quality of profitability on the manufacturing chain in the welding network. Quality of performance comprises the relationship between design engineering and manufacturing.

The study is based on empirical research in a project focused on the development of welding networks. The functional framework of the welding manufacturing network is presented and discussed from overall quality and demand aspects. The study takes an overview of the challenges of welding manufacturing. It briefly discusses the unnecessary costs, claim costs and warranty costs in the production chain. The paper reviews the linkages between design, purchasing, manufacturing and quality. The quality requirements of welding by different functions and standards are also discussed. The observations at the empirical part of the study are collected from welding networks in engineering workshops where GMAW welding is a commonly used process, and the range of defects and costs studied relate to the process. This review creates a framework for future research on the profitability of the welding network from the viewpoint of manufacturing.

2. Relation of Functions in Welding

Welding is a special manufacturing process [4] because it

is difficult to be verified and because of the many factors that affect the welding. Welding is nevertheless the most common joining process in the metal industry [5] and has an influence on several important aspects, for example, product reliability and human safety [6]. Operations before actual welding are an important factor in the quality of a complete weld. The requirements of welding raise complexity when ensuring the quality demands set for welded products with many co-operative manufacturers in the welding network. The product requirements and quality of conformity define the demands of manufacturing which every party of the manufacturing process have to follow.

2.1. Conformable Welding Network

Companies are confronting challenges with design, manufacturing and distribution time in a highly competitive environment [7, 8, 9]. At the same time they have to improve production efficiency and ensure cost control [9]. Supply chain quality is in a significant position when expecting to achieve competitive advantage [10] and because manufacturers continuously call for improvements in supplier performance [11]. Furthermore, products are getting more complex and they have to meet customers' expectations [12]. In a welding network, the focal company of the network in the manufacture of the end product [13] is responsible for quality demands being fulfilled at every stage of the manufacturing chain.

The manufacturing failures of welded structures and products can be a result of defects in the welded joint [6] but also due to imperfections in other activities in manufacturing. It is important to define the right quality level and product specifications, and to ensure the requirements of all functions in a company that affect welding. Manufacturers rarely know the actual welding cost in their production [14]. Coordinating welding operations closely internally but also among co-operating companies in the network may decrease unnecessary defects and claims. Knowledge on requirements and possible defects has a notable effect on achieving quality. Failure to recognise weld discrepancies and nonconformity during manufacturing when fulfilling the requirements results in costly rework and lost productivity [15]. Manufacturers who understand welding economics and value added techniques are more successful in local and also global markets [14].

2.2. Impact of Design Engineering on Welding Manufacturing

Welding as a manufacturing process deeply depends on the decisions of design engineering. The design and development processes include many tools that are utilised to assess manufacturing and increase co-operation with other functions of manufacturing and have a positive impact on costs. The concurrent engineering (CC) approach shortens the time from design to delivery where many phases of the product process are running simultaneously [16]. A wider perspective on product manufacturing can be gained with product life cycle

management (PLM) which is a strategic approach to manage and support the life cycle of a product from development to withdrawal. All the information of the life cycle is determined in digital solutions. It is also an integrated approach to control and monitor the phases of product development [8.].

Usable approaches to increasing manufacturability and noticing the demands of manufacturing are the design methods from the perspective of other functions [17]. The design for X (DfX) method can be used to improve product design and the design process, for instance, manufacturability and assemblability [17]. The most commonly used DfX perspective, design for manufacturing (DFM) focuses on manufacturability in product design in the chosen manufacturing chain [17], whereas design for assembly (DFA) focuses on assembly by minimising the assembly efforts of a product [16]. Weldibng assemblies are subject to properly fitting parts and understanding the demands of welding. Design for manufacturing and assembly (DFMA) comprises both DFM and DFA [18] and enables reducing manufacturing costs while developing the product or designing a new one [19]. Fig. 1 describes the DFMA process where both aspects, manufacturing and assembly, have to be observed in a welding network where welded parts and sub-assemblies have to fit regardless of the different welding workshops where they are manufactured.

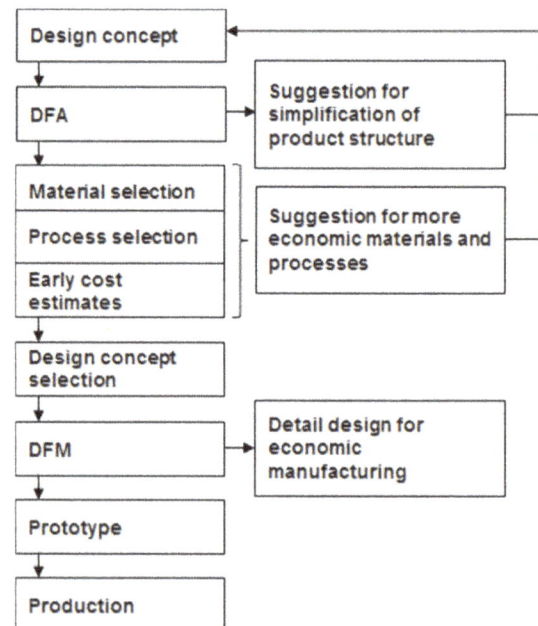

Figure 1. *Simplified DFMA process in product design engineering [17] (Adapted).*

2.3. Welding Linkages

Design engineering is not the only important function, but all departments of a company have their own specific subject field standards which define some issues of how things have to be done. Welding manufacturing includes four typical functions that have a remarkable influence on the success of producing products conforming to every demand assessed. If welding is one of the main manufacturing processes, all the other functions, like design engineering, manufacturing engineering, purchasing and quality control, also have a significant impact on welding. Therefore welding demands need to be understood in those functions of the company. Welding operations can be divided in three sections: before welding, during welding and after welding. The actual welding action can be mostly affected before welding, which is illustrated below in Fig. 2 on the important functions of welding.

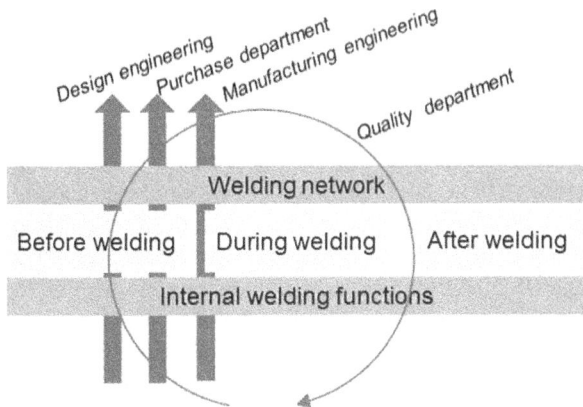

Figure 2. *Important functions of welding in welding manufacturing.*

The effectiveness of the linkage between design engineering and manufacturing mostly depends on the relations of people, employees' personal skills and capabilities, the willingness and ability to do intra-organisational co-operation, increasing knowledge and knowledge management practices and the commitment of the management to develop skills and co-operation [20]. Knowledge on welding is important in design and manufacturing [5]. Increasing knowledge and co-operation between the different functions is very important for the quality of manufacturing. Co-operative design tools, like DFM and DFMA, where manufacturing is considered at the early stage of the design process encourage co-operation with designers and manufacturing engineers and others affecting the costs of the end product at the early stages of design [17].

3. Conformity of Welding

Welding as a manufacturing process involves many different standards, guidelines and demands. Standards and technical reports are intended to help determine product specifications and quality requirements. The requirements do not, however, take into account all the demands of welded structures, the behaviour of material and the effects of the welding process. With great responsibility, design engineering and welding engineering require profound knowledge of process consequences. Product conformity assessment ensures the structure and quality requirements of the product. ISO 9000 defines the terms related to conformity: conformity, nonconformity, defect, preventive action, corrective action, correction, rework, regrade, repair,

scrap, concession, deviation permit and release [4]. These terms can be divided related to welding actions possibly affecting before welding, actions which probably follow from welding and other actions after welding as shown in Fig. 3. The manufacturer of the end product defines the demands and quality requirements of the product. However, the end product can contain other conformities by standards or other third party requirements. These usually regard safety and environmental risks. Manufacturers can use valuable tools to prove the quality of a product. A sign of the good quality of welding manufacturing, controlled welding operations decrease production costs [21].

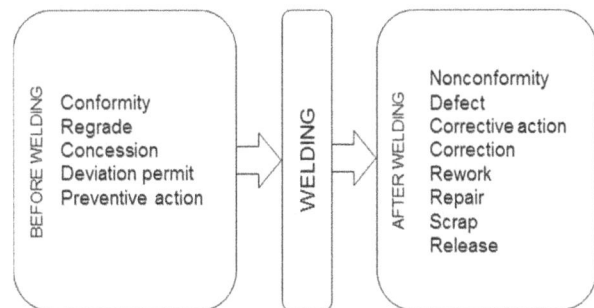

Figure 3. *The terms of conformity in welding according to ISO 9000:2005.*

3.1. Welding Quality Requirements

When a product involves complex requirements, composition or manufacturing processes, it can be defined as a complex product [22]. A welding assembly cannot be produced by choosing a fitting part for a sub-assembly or assembly, like the selective assembly technique which focuses on the fit between assembly components [22]. Therefore it is important to focus on the quality of each welded part.

Total product quality consists not only of the context of design, manufacture and post-sale service, but also of purchasing which are linked together [2]. The manufacturer, customer and third parties have many expectations with regard to the end product. Because welding processes have a significant influence on the quality of a product [23], the end result must meet all these expectations. The key to improving quality is to focus on the prevention of nonconformity [24]. Quality assurance verifies the conformity of a product and it has to reach the production process and cover the whole life cycle of a product [1]. Preventive actions can be, for example, design reviews, education, training, supplier selection, capability reviews and process improvement projects [24].

Quality can be understood in many different ways depending on the aspect. It is mostly related to product differentiation. Production quality can be understood by production efficiency [3], but it is also dependent on many functions around manufacturing. In welding production, it is important to consider the entire manufacturing process. Welding can be more effective with different tools, increased automation and fluent production. Regardless of the manufacturing technique, the product has to meet the requirements set. Control of faults and overall quality are the main things in welding design and manufacturing [25]. The quality demands of products, which are also related to the whole production efficiency, are examined in the following.

3.2. Control of Welding Operations

Table 1. ISO 3834 standard: Quality requirements for fusion welding of metallic materials.

ISO 3834 Quality requirements for fusion welding of metallic materials.	
ISO 3834-1:2005	Part 1: Criteria for the selection of the appropriate level of quality requirements
ISO 3834-2:2005	Part 2: Comprehensive quality requirements.
ISO 3834-3:2005	Part 3: Standard quality requirements.
ISO 3834-4:2005	Part 4: Elementary quality requirements.
ISO 3834-5:2005	Part 5: Documents with which it is necessary to conform to claim conformity to the quality requirements of ISO 3834-2, ISO 3834-3 or ISO 3834-4.

The ISO 3834 standard provides the basis for quality of manufacturing. It guides welding manufacturing by standards, which help organise manufacturing. It is a guideline to good welding production and continuous improvement. The standard emphasises the importance of welding coordination and control of welding operations. Adopting ISO 3834 to the course of actions can prevent critical damages because of controlled manufacturing [6]. The standard has five parts: the first one helps to choose the appropriate level of quality requirements, the subsequent

three parts define quality requirement levels and the final part is a list of documents necessary when using and conforming to the quality requirements of ISO 3834-2, ISO 3834-3 or ISO 3834-4 [23]. Table 1 presents the parts of the ISO 3834 standard.

"ISO 3834 therefore provides a method to demonstrate the capability of a manufacturer to produce products of the specified quality" [23]. ISO 3834 thus provides the basis for welding operations. It includes many standards that are important when a product is manufactured by welding. It does not take account of design engineering details but emphasises co-operation between design and manufacturing. There are also many other standards that affect actual welding. For example, design engineering has own requirements to assess the demands of a product structure, but also most of the welding decisions are made in design engineering. There are standards that define the general overview of welding and also have a direct effect on welding functions, the welding process and welding details, like ISO 5817 and ISO 13920 [26, 27], or define the details of manufacturing, for example, the welding process and welding consumables, like ISO 14341 and ISO 14175 [28, 29]. Table 2 presents examples of weld requirements according to the ISO 5817 standard. The examples illustrate the expected result of welding with limits depending on the quality grade.

Table 2. Limits for imperfections divided according to quality levels of ISO 5817 [26] (Adapted).

Imperfection designation	Remarks	t, mm	Limits for imperfections for quality levels		
			D	C	B
Continuous undercut Intermittent undercut	Smooth transition is required. This is not regarded as a systematic imperfection.	0.5 to 3	Short imperfections: $h \leq 0.2\,t$	Short imperfections: $h \leq 0.1\,t$	Not permitted
		> 3	$h \leq 0.2\,t$, but max. 1 mm	$h \leq 0,1\ t$, but max. 0.5 mm	$h \leq 0.05\ t$, but max. 0.5 mm
Spatter		≥ 0.5	Acceptance depends on application, e.g. material, corrosion protection		
Surface pore	Maximum dimension of a single pore for - butt welds - fillet welds	0.5 to 3	$d \leq 0.3\ s$ $d \leq 0.3\ a$	Not permitted	Not permitted
		> 3	$d \leq 0.3\ s$, but max. 3 mm $d \leq 0.3\ a$, but max. 3 mm	$d \leq 0.2\ s$, but max. 2 mm $d \leq 0.2\ a$, but max. 2 mm	Not permitted

The link between design engineering and manufacturing is complex because different demands affect each other. Besides ISO 5817, there can be other demands that have an effect on weld quality, e.g. finish requirements. The ISO 8501-3 standard provides requirements for painting or related

products. The requirements have to be noticed already in welding preparation and also during the welding process, such as finishing. The standard ISO 8501-3 includes preparation grades which describe the quality of product surface before painting. Preparation grades are P1 – Light preparation, P2 –

Thorough preparation and P3 – Very thorough preparation. P1 allows an unfinished surface or only minimum preparation. P2 and P3 demand more remedial efforts [30]. Table 3 shows requirements for each preparation grade of the current standard.

Table 3. Imperfections and preparation grades according to ISO 8501-3 [30] (Adapted).

Type of imperfection	P1	P2	P3
Welding spatter a) b) c)	Surface shall be free of all loose welding spatter [see a)]	Surface shall be free of all loose and lightly adhering welding spatter [see a) and b)] Welding spatter shown in c) may remain	Surface shall be free of all welding spatter
Welding slag	Surface shall be free from welding slag	Surface shall be free from welding slag	Surface shall be free from welding slag
Undercut	No preparation	Surface shall be free from sharp or deep undercuts	Surface shall be free from undercuts
Rolled edges	No preparation	No preparation	Edges shall be rounded with a radius of not less than 2 mm (see ISO 12944-3)

Companies can also have other international or national standards in use and define their own requirements for products and manufacturing. Table 4 presents one national standard on the requirements for welding. It sets extra demands for companies when usually quality grade 05 is used [31]. This particular grade assures a good base for painting, and the requirements have to be applied in every section of manufacturing, including welding. If the main supplier adopts quality level C of welding imperfections and other specific demands, like SFS 8145 offers, the same demands apply to the welding network. These supplementary demands are not necessarily known throughout the company or the whole network. The requirements of different standards can cause confusion about the total requirements of quality in products and manufacturing.

Table 4. Quality grades for mechanical preparations [31].

Object	No.	Action	Quality grade of preparation					
			01	02	03	04	05	06
Weld joints	1	Weld slag is to be removed	—	—	—	—	—	—
	2	Pieces of wire electrode are to be removed		—	—	—	—	—
	3	Welding spatters that can be loosened with a scraper are to be removed			—	—		
	4	Welding spatters are to be removed				—	—	—
	5	Open pores are to be repaired					—	—
	6	Undercuts are to be repaired					—	—
	7	Sharp peaks are to be removed					—	—

Product quality requirements have to coincide with the parts designed so that they can be manufactured without rework or extra costs. The lack of knowledge on the manufacturing challenges can cause increasing manufacturing costs due to claims and warranty costs. Standards help to determine the requirements, but designers have to understand manufacturing to satisfy the level of quality and yet achieve profitability. Steel products, commonly used in welding structures, are an example of this. The tolerance rates of raw material can be a challenge to manufacturing and have a direct influence on functionality, costs and quality of manufacturing [32]. Narrow tolerances can cause high costs [33, 34] but also problems with succeeding welds without preparation, finishing or rework. On the other hand, too wide tolerances can cause variability in the products [33, 34]. The focal company can have its own level of tolerances depending on the part, but the requirements that affect the tolerance need to be understood in manufacturing engineering or by the welding coordinator to achieve appropriate and competitive production.

The EN 10219-2 and EN 10210-2 standards define requirements for hollow section steel products [35, 36]. Corresponding international standards are ISO 10799-2 and ISO 12633-2 [37, 38]. Some common causes of unnecessary fixing or rework in welding are the concavity x1, convexity x2 (Fig. 4a), twist v (Fig. 4b) and straightness e (Fig. 4c) of these kinds of products. Standards enable imperfections in dimensions. This jeopardises compatibility when parts are meant to be fitted into each other, demonstrated in Fig. 5, or in other tightly dimensioned joining.

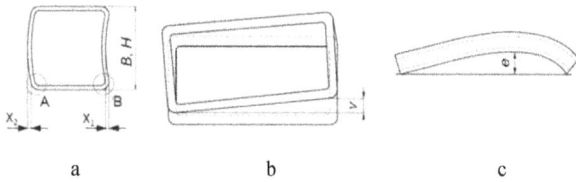

Figure 4. *Concavity (a), convexity (b), twist and straightness (c) in hollow section steel products [35, 36] (Adapted).*

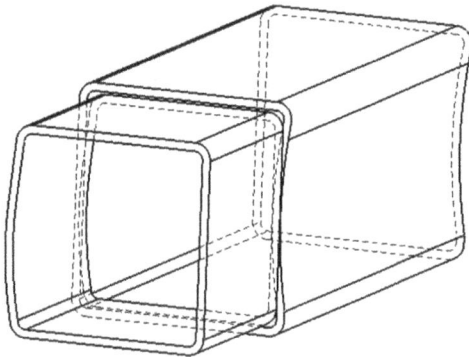

Figure 5. *An example of compatibility risked due to dimensional imperfections enabled by standards for a square tube.*

Empirical research shows that the requirements focused on products and manufacturing are insufficiently recognised at the many levels of a company. This causes deficiency of knowledge on the quality and manufacturing demands. This is one reason of nonconformity in manufacturing. Departments of design engineering and purchasing are inadequately aware of the extent of welding quality standards. The consequences of welding and preventive actions are also unknown at the management level. The management understands the importance of quality and pressures to decrease defects, but the foundation of possible welding development remains unrecognised. The control of the quality department is usually not focused on following the defect rate of welding operations in-house, but on the conformity of deliveries from collaboration partners and suppliers.

Welding is a challenging manufacturing method, and not all the challenges can be solved when applying standards and other regulations. The purpose of use of the end product can present even more requirements, for example, with regard to quality and strength, which have to be taken into account in design engineering. Also, the requirements of welding need to be understood. The welder's professional skills are primarily notable after appropriate requirements for welding. Defects can occur in actual welding which can be prevented with suitable pre-actions and the sufficient knowledge and training of welders [21, 39]. Welding coordination is in a significant role to stimulate co-operation among the departments of the company and distribute welding knowledge in every requisite stage as a response to control over quality and manufacturing demands in the welding network.

4. Costs of Conformity

Usually quality costs focus on an individual company and internal costs instead of the whole production chain [40]. In a welding network, quality costs are more closely followed by the focal company. Quite often internal quality costs are understood to include daily work rather than own countable costs. From a wider perspective, costs can be divided into different departments or other functional areas with responsibility for own departmental costs [41]. It has been known for a long time that quality costs are measurable; they can be planned, analysed and prevented and are higher when failures are detected at the end of production or by the customer [42]. Still, the focal company rarely uses this information effectively in every day work in a network. Empirical research proves that failures and defects are identifiable and known in companies, but actions to find the root of the problem are fewer, which creates continuous costs.

4.1. Quality Costs

The manufacturing process generates costs, also related to quality. Costs of quality result from not producing requisite quality or ensuring quality in accordance with requirements. Quality costs have more strategic and economic importance than earlier costs [40] by affecting profit and helping to identify the weak points in the process [43]. Many models have been developed to measure or identify quality costs. The most basic scheme is to find prevention, appraisal and failures of the process and costs.

The traditional model of developing the quality level of a company is the prevention-appraisal-failure (PAF) model [44]. It is a commonly used method for measuring quality costs [24], and it is the basic scheme in many reconstitutions of quality cost count models. Fig. 6 illustrates the PAF model. The model focuses on finding the quality level that is suitable for a company determined by specifications and the total quality costs which increase concurrently with the quality level [44]. Quality level q of a product can be defined considering a number of non-defective items, and defect rate d defective items. When increasing the quality level, it is profitable to invest in prevention and appraisal functions [44]. When total quality costs rise over the optimal quality level q, quality costs C(q) contradict with the profitability of product manufacture. Many authors divide quality costs in two parts where quality costs C(q) are a summary of prevention cost C(p) and appraisal cost C(a): $C_q = C_p + C_a$), and the total quality costs TC(q) are a summary of C(q) and failure costs N(q): $TC_q = C_q + N_q$) [24]. Another way to divide quality costs is to regroup the total quality costs into costs of conformance (prevention and appraisal costs) and costs of non-conformance (costs of internal failure and costs of external failure) [43].

The PAF model is based on the notion that higher quality causes higher costs. This view does not support the idea of continuous improvement and decreasing quality costs with higher quality. However, it has been shown that it depends on the effectiveness of the company's quality improvement program whether the quality costs are increasing or decreasing when producing higher quality and a more effective quality improvement program decreases quality costs and produces higher quality [44]. Poor design quality can also create higher production costs [3], and too narrow tolerances can generate unnecessary production costs, even though the variability of the product decreases, the quality of manufacturing improves and quality losses are reduced [34]. The balance between the requirements and manufacturing quality has to be observed.

Continuous improvement is important when the company wants to improve product quality and the flow of production. Fig. 7 describes the quality cost rates of prevention, appraisal and failure costs in continuous improvement. It has been noticed that the failure cost and total quality cost rates never reach zero because of the uneconomical aspect and because the rate turns upward at some point [42]. Figures 6 and 7 are not completely accurate for welding where qualitativeness cost more than increased quality, when quality assurance is at a sensible level with all design and manufacture demands. To maintain competitive advantage, continuous improvement of product quality is essential [1]. The main supplier has to ensure this improvement in the network.

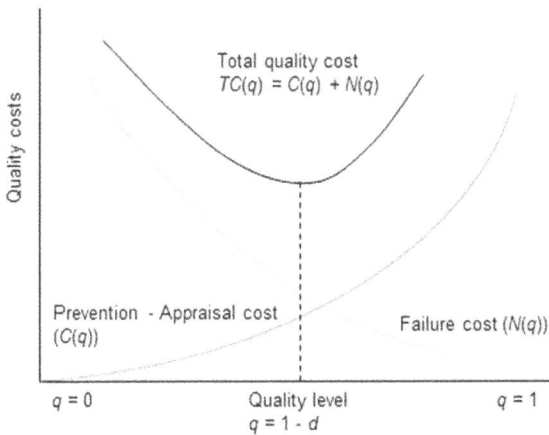

Figure 6. *PAF model for quality costs [44] (Adapted).*

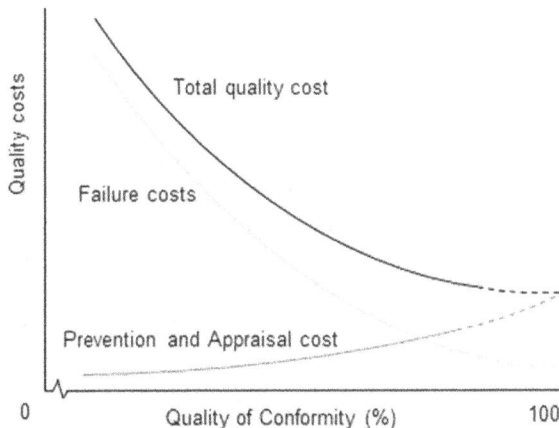

Figure 7. *Quality cost behaviour according to continuous improvement [42] (Adapted).*

Failure costs can be divided into internal and external failure costs. Internal failure costs result from a product that does not conform to the requirements before it meets the customer, whereas external failure costs occur if the product is already shipped to the customer with defects [40]. Table 5 shows examples of reasons for quality costs divided into categories prevention, appraisal, internal failure and external failure costs.

Table 5. *Example reasons for quality costs divided by different categories [44] (Adapted).*

Prevention costs	Appraisal costs	Internal failure costs	External failure costs
Process control Product and service design and redesign Process design Supplier relations, audit and screening Preventive maintenance Training and quality circles	Raw material inspection In-process inspection Final inspection Inspection material and services Quality audit	Scrap Rework Equipment repair Process downtime Re-inspection of products	Warranty charges Litigation and liability Complaint handling Returns Rework on returns Lost sales Penalties and allowances

Table 6. *Actions affecting welding quality before welding, during welding and after welding.*

ACTIONS AFFECTING QUALITY			
	Before welding	**During welding**	**After welding**
APPRAISAL	Specifications Quality requirements Manufacturing processes Training Welding knowledge Welding network control Material procurement Manufacturing details Workshop control Quality input Co-operation Design & Development	Visual inspection Welders professional skills Equipment performance Welding area control Specifications follow	Visual inspection Other quality inspections
CORRECTIVE	Quality processing	Scrap Rework	Unnecessary inspections Grinding Fine-tuning Finishing Scrap Rework

Besides considering the cost of quality, quality costs can also be assessed to manage losses. There are a lot of hidden costs that come from manufacturing loss and design loss. They are identified when quality actions are unsuccessful and generate costs. [45.] The welding manufacturing process involves unnecessary quality costs when products do not meet the requirements set on conformity. Table 6 shows actions affecting quality divided into categories before welding, during welding and after welding. The preventive actions of quality assurance create costs, but they have to be integrated into every day work and related to the level of quality and requirements. Relating quality and profitability is the most effective way to prevent failures [24]. It is also important to invest in productivity and quality knowledge to get efficient benefits to produce cost reductions and quality increase [46]. Training is one of the most important things to increase knowledge and skills in welding. Increasing welding knowledge and training is remarkably important in developing welding production and decreasing costs [21].

4.2. Influence of Nonconformity

Related to production costs, the most important decisions regarding costs and quality demands are usually made during the design stage [47]. The design phase includes the specifications of the weld structure, like the component shapes, positions of joints and also joining methods, but the whole welding network, including suppliers' own collaboration partners, has an effect on product costs by their actions in production.

Decisions made during design and manufacturing have an influence on reliability [48] and can prevent unnecessary costs caused by nonconformity. Waste can be defined in several ways. Waste losses can be related to time, motion and process flow and come from waiting, non-value added time, inappropriate layout and poor communication [49]. The waste costs of welding result from the process not working properly. Co-operation among the management, design engineering, manufacturing engineering, welding coordination, welding manufacturing and quality assurance is in an important role in profitable welding manufacturing (Fig. 8). Welding coordination links the functions together with responsibility for welding operations [13]. Each function has a specified role to achieve high quality and a profitable result.

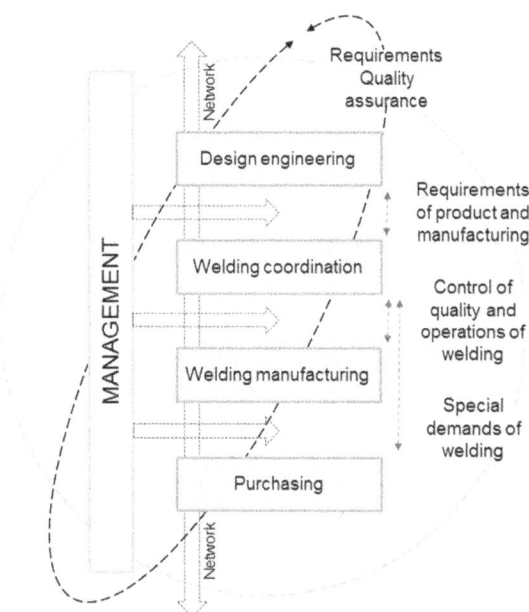

Figure 8. Links among the different functions of a company [13].

Costs arising from defects, faults, complaints and warranties are unprofitable items to a company. Cost of defects are gathered from different processes [49], and from the focal company's viewpoint, nonconformity costs from faults and defects arise not only from internal welding manufacturing, but also from the network. They are usually handled as complaints if defects are noticed by the focal company. Nonconformity is more costly than proper preventive actions in quality assurance. The work costs are only part of the total costs resulting from complaints and remanufacturing. Indirect costs come, for instance, from notice of defective processing, manufacturing engineering, welding engineering and other actions that follow from rework.

Often the closer the product is to the customer in the manufacturing chain, the greater the effect on corrective actions. Fig. 9 mirrors the cost effect from prevention to subsequent actions. The arrows present increasing quality costs during manufacturing with defects, faults, complaints and warranties. Warranties are signed between the manufacturer and the client and they oblige the manufacturer to answer for the product's operation during the warranty period [50, 51], and recovery actions create costs. It is not unambiguous how warranty costs occur, e.g. from warranty service and warranty maintenance, whereas defects, faults and complaints are connected to the manufacturing process and arise from the focal company to the network.

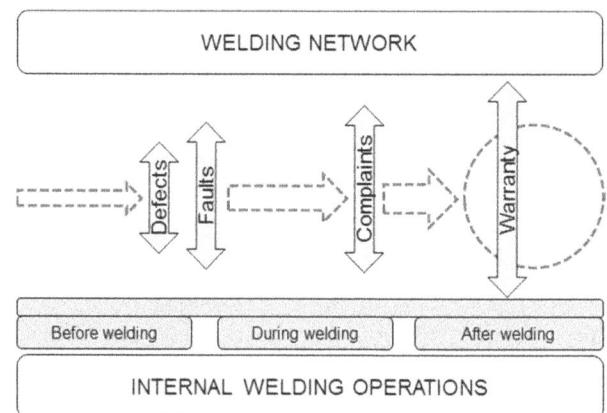

Figure 9. Cost effects from preventive to subsequent actions in welding manufacturing.

5. Conclusion

Welding is the most common joining process in the metal industry, and the customer has many demands on the end product to which the main supplier has to respond. Confronted with the challenges of a competitive environment, companies are improving their production efficiency and reducing production costs. When more than one producer is involved with a product, the whole production chain has to meet the requirements. The focal company has the responsibility to fulfil the requirements for every part of the whole final product. This also includes quality costs which come from ensuring the quality requirements. The welding process affects the costs of manufacturing and profitability; the costs have to be in control at every stage from the product's design to the manufacturing process and quality assurance. When a company manufactures products using welding in their main manufacturing process or some critical components by welding, it is important that all functions around welding are controlled. This denotes co-operation among different functions, like design engineering, purchasing, manufacturing engineering and the quality department. It is a deceptive presumption that the demands of welding need to be mastered only in the welding workshop.

Coordinating welding functions gives an advantage for profitable and quality manufacturing. For example, the ISO 3834 standard ensures welding quality requirements and gives guidelines to good welding production. It emphasises continuous improvement, controlling of welding operations and the importance of a welding coordinator. By using standards, the tool of whole production to improve welding operations, the benefits are wider than the mere focus on welding action. It is important that the company itself can answer the different quality requirements with the standards or use the standards to determine its own quality level and guidelines clarifying production to increase productivity and profitability.

Empirical research on a welding network shows that failures and defects are identifiable and known but actions to find their root cause are few. The requirements focused on product or welding manufacturing are unrecognised at the many levels of organisation, which can cause continuous nonconformity in products. One of the main obstacles to improving welding functions is the lack of co-operation and knowledge of demands on welding. This shows as deficiency in reviews on the requirements of design engineering, the purchasing department and welding coordinators and uncertainty over manufacturing demands fulfilling the quality requirements of products. Furthermore, the lack of welding knowledge among welders causes defects and also disinclination to consider the relationship of various functions affecting welding. Observing only the complaint and quality costs of network co-operation companies does not give the right idea of all the nonconformity costs of production. By increasing welding knowledge and training and clarifying the requirements of product and manufacturing among all the parties of the manufacturing chain, the demands become distinct and easier to control.

Further work will be needed to analyse the detection of defects to find the root cause of defects, faults, claims and complaints. The impact of costs of quality and complaints at the network level and the effect on the end product is an interesting area for study to further enlighten the impact of nonconformity on manufacturing.

References

[1] X. Tang and H. Yun, Data model for quality in product lifecycle, Computers in industry, 59 (2–3), 2008, pp. 167–179. doi: 10.1016/j.compind. 2007. 06. 011.

[2] D.N.P. Murthy and K.R. Kumar, Total product quality, International Journal of Production Economics, vol. 67 (3), 2000, pp. 253–267. doi: 10.1016/S0925-5273(00)00026-8.

[3] J. Freiesleben, J, Proposing a new approach to discussing economic effects of design quality, International Journal of Production Economics, vol. 124 (2), 2010, pp. 348–359. doi: 10.1016/j.ijpe.2009.11.030.

[4] ISO 9000:2005. Quality management systems. Fundamentals and vocabulary, International Organization for Standardization.

[5] P. Kah and J. Martikainen, Current trends in welding processes and materials: improve in effectiveness, Reviews on Advanced Materials Science, vol. 30, 2012, pp. 189–200.

[6] D.N. Shackleton, Reducing failure risk in welded components, Welding in the World, vol. 50 (9–10), 2006, pp. 92–97. doi: 10.1007/BF03263449.

[7] J. Váncza, L. Monostori, D. Lutters, S.R. Kumara, M. Tseng, P. Valckenaers and H. Van Brussel, Cooperative and responsive manufacturing enterprises, CIRP Annals - Manufacturing Technology, vol. 60 (2), 2011, pp. 797–820. doi: 10.1016/j.cirp.2011.05.009.

[8] V. Nosenzo, S. Tornincasa, E. Bonisoli and M. Brino, Open questions on Product lifecycle Management (PLM) with CAD/CAE integration, International Journal on Interactive Design and Manufacturing, vol. 8, 2014, pp. 91–107. doi: 10.1007/s12008-013-0184-1.

[9] F. Pan and R. Nagi, Multi-echelon supply chain network design in agile manufacturing, Omega, vol. 41 (6), 2013, pp. 969–983. doi: 10.1016/j.omega.2012.12.004.

[10] G. Xie, W. Yue, S. Wang and K.K. Lai, Quality investment and price decision in a risk-averse supply chain, European Journal of Operational Research, vol. 214 (2), 2011, pp. 403–410. doi: 10.1016/j.ejor.2011.04.036.

[11] W. Wang, R.D. Plante and J. Tang, Minimum cost allocation of quality improvement targets under supplier process disruption, European Journal of Operational Research, vol. 228 (2), 2013, pp. 388–396. doi: 10.1016/j.ejor.2013.01.048.

[12] R. Jiang and D.N.P. Murthy, Impact of quality variations on product reliability, Reliability Engineering and System Safety, vol. 94 (2), 2009, pp. 490–496. doi: 10.1016/j.ress.2008.05.009.

[13] J. Toivanen, J. Martikainen and P. Heilmann, From supply chain to welding network: A framework of the prospects of networks in welding, Mechanika, vol. 21 (2), 2015, pp. 154–160. doi: 10.5755/j01.mech.21.2.8463.

[14] D.K. Miller, Cost of Welding, Welding Design and Fabrication, vol. 3, 2004, pp. 32–37.

[15] W.C. LaPlante, How to Assure Quality in Outsourced Welded Products, Welding Journal, vol. 90 (10), 2011, pp. 42–46.

[16] F. Giudice, F. Ballisteri and G. Risitano, A Concurrent Design Method Based on DFMA-FEA Integrated Approach, Concurrent Engineering, vol. 17 (3), 2009, pp. 183–202. doi: 10.1177/1063293X09343337.

[17] T. Tomiyama, P. Gu, Y. Jin, D. Lutters, C. Kind and F. Kimura, Design methodologies: Industrial and educational applications, CIRP Annals - Manufacturing Technology, vol. 58, 2009, pp. 543–565. doi: 10.1016/j.cirp.2009.09.003.

[18] D. Dewhurst, Design First, Lean Second, Assembly, 2011, pp. 62–68.

[19] D. Hegland, DFMA Cuts Downstream Costs, Assembly, vol. 51 (6), 2008, 43–49.

[20] R. Dekkers, C.M. Chang and J. Kreutzfeldt, The interface between "product design and engineering" and manufacturing: A review of the literature and empirical evidence, International Journal of Production Economics, vol. 144 (1), 2013, pp. 316–333. doi: 10.1016/j.ijpe.2013.02.020.

[21] J.R. Barckhoff, Total Welding Management, American Welding Society, 2010, pp. 1–6, 15–36.

[22] L. Liu, F. Zhu, J. Chen, Y. Ma and Y. Tu, A quality control method for complex product selective assembly processes, International Journal of Production Research, vol. 51 (18), 2013, pp. 5437–5449. doi: 10.1080/00207543.2013.776187.

[23] ISO 3834:2005. Quality requirements for fusion welding of metallic materials - Part 1–5, International Organization for Standardization.

[24] A. Kazaz, M.T. Birgonul and S. Ulubeyli, Cost-based analysis of quality in developing countries: a case study of building projects, Building and Environment, vol. 40 (10), 2005, pp. 1356–1365. doi: 10.1016/j.buildenv.2004.11.010.

[25] G. Casalino, S.J. Hu and W. Hou, Deformation prediction and quality evaluation of the gas metal arc welding butt weld, Proceedings of the Institution of Mechanical Engineers, vol. 217 (11), 2003, pp. 1615–1622. doi: 10.1243/095440503771909999.

[26] ISO 5817:2014. Welding - Fusion-welded joints in steel, nickel, titanium and their alloys (beam welding excluded) – Quality levels for imperfections, International Organization for Standardization.

[27] ISO 13920:1996. Welding – General tolerances for welded constructions. Dimensions for lengths and angles. Shape and position, International Organization for Standardization.

[28] ISO 14341:2010. Welding consumables – Wire electrodes and weld deposits for gas shielded metal arc welding of non alloy and fine grain steels. Classification, International Organization for Standardization.

[29] ISO 14175:2008. Welding consumables – Gases and gas mixtures for fusion welding and allied processes, International Organization for Standardization.

[30] ISO 8501-3:2006. Preparation of steel substrates before application of paints and related products - Visual assessment of surface cleanliness. Part 3: Preparation grades of welds, edges and other areas with surface imperfections, International Organization for Standardization.

[31] SFS 8145:2001. Anticorrosive painting. Quality grades of mechanical surface preparations for blast cleaned or blast-cleaned and prefabrication primed steel substrates, Finnish Standards Association SFS.

[32] B. Li, X. Yang, Y. Hu and D. Zhang, Quality design of tolerance allocation for sheet metal assembly with resistance spot weld, International Journal of Production Research, vol. 47 (6), 2009, pp. 1695–1711. doi: 10.1080/00207540701644193.

[33] M-Y. Liao, Economic tolerance design for folded normal data, International Journal of Production Research, vol. 48 (14), 2010, pp. 4123–4137. doi: 10.1080/00207540902960307.

[34] N.V.R. Naidu, Mathematic model for quality cost optimization, Robotics and Computer-Integrated Manufacturing, vol. 24 (6), 2008, pp. 811–815. doi: 10.1016/j.rcim. 2008. 03. 018.

[35] EN 10210-2:2006. Hot finished structural hollow sections of non-alloy and fine grain steels - Part 2: Tolerances, dimensions and sectional properties, European Committee for Standardization.

[36] EN 10219-2:2006. Cold formed welded structural hollow sections of non-alloy and fine grain steels - Part 2: Tolerances, dimensions and sectional properties, European Committee for Standardization.

[37] ISO 10799-2:2011. Cold-formed welded structural hollow sections of non-alloy and fine grain steels – Part 2: Dimensions and sectional properties, International Organization for Standardization.

[38] ISO 12633-2:2011 Hot-finished structural hollow sections of non-alloy and fine grain steels – Part 2: Dimensions and sectional properties, International Organization for Standardization.

[39] D.C. Li, Research on Quality Management of manufacturing Equipment Welding Technology, Applied Mechanics and Materials, vol. 192, 2012, pp. 415–419. doi: 10.4028/www.scientific.net/AMM.192.415.

[40] K.K. Castillo-Villar, N.R. Smith and J.L. Simonton, A model for supply chain design considering the cost of quality, Applied Mathematical Modelling, vol. 36 (12), 2012, pp. 5920–5935. doi: 10.1016/j.apm.2012.01.046.

[41] A.I. Pettersson and A. Segerstedt, Measuring supply chain cost, International Journal of Production Economics, vol. 143 (2), 2012, pp. 357–363. doi: 10.1016/j.ijpe.2012.03.012.

[42] D.H. Besterfield, Quality control, Prentice-Hall, Inc, 1994, pp. 405–406, 420–421.

[43] O. Staiculescu, A new vision of cost: an essential optimization tool for managerial accounting, Procedia - Social and Behavioral Sciences, vol. 62, 2012, pp. 1276–1280, May 2012 [World Conference on Business, Economics and Management (BEM-2012), Antalya, Turkey]. doi: 10.1016/j.sbspro. 2012. 09. 218.

[44] S. Kim and B. Nakhai, The dynamics of quality costs in continuous improvement, International Journal of Quality & Reliability Management, vol. 25 (8), 2008, 842–859. doi: 10.1108/02656710810898649.

[45] G. Giakatis, T. Enkawa and K. Washitani, Hidden quality costs and the distinction between quality cost and quality loss, Total Quality Management, vol. 12 (2), 2001, pp. 179–190. doi: 10.1080/09544120120011406.

[46] J. Vörös, The dynamics of price, quality and productivity improvement decisions, European Journal of Operational research, vol. 170 (3), 2006, pp. 809–823. doi: 10.1016/j.ejor.2004.08.001.

[47] P.G. Maropoulos, Z. Yao, H.D. Bradley and K.Y.G. Paramor, An integrated design and planning environment for welding Part 1. Product modeling, Journal of Materials Processing Technology, vol. 107 (1–3), 2000, pp. 3–8. doi: 10.1016/S0924-0136(00)00708-1.

[48] D.N.P. Murthy and I. Djamaludin, I, New product warranty: A literature review, International Journal of Production Economics, vol. 79 (3), 2002, pp. 231–260. doi: 10.1016/S0925-5273(02)00153-6.

[49] C. Hicks, O. Heidrich, T. McGovern and T. Donnelly, A functional model of supply chains and waste, International Journal of Production Economics, vol. 89 (2), 2004, pp. 165–174. doi: 10.1016/S0925-5273(03)00045-8.

[50] M. Shafiee and S. Chukova, Maintenance models in warranty: A literature review, European Journal of Operational Research, vol. 229 (3), 2013, pp. 561–572. doi: 10.1016/j.ejor.2013.01.017.

[51] M.R. Karim and K. Suzuki. Analysis of warranty claim data: a literature review, International Journal of Quality & Reliability Management, vol. 22 (7), 2005, pp. 667–686. doi: 10.1108/02656710510610820.

14

UREAD Impact Behaviour Using Silicon Based Materials

Remi Bouttier[1], Gabriel Lopes[2], Luke Clarke[3], Rocco Lupoi[3, *]

[1]Ecole Nationale Superieure de Mechanique et D'Aerotechnique (ISAE-ENSMA), Département d'Energétique, France
[2]Federal University of Uberlandia, Engenharia Mecânica, Santa Mônica, Uberlândia - MG, Brazil
[3]Trinity College Dublin, the University of Dublin, Department of Mechanical and Manufacturing Engineering, Parsons Building, Dublin 2, Ireland

Email address:
lupoir@tcd.ie (R. Lupoi)

Abstract: Several methodologies and techniques are currently available so as to dissipate energy in engineering systems; most of them are either not re-usable, or complex in mechanism. This paper introduces an innovative re-usable energy absorption device, based upon the working principles of Equal Channel Angular Extrusion, and known as UREAD (Universal Re-usable Energy Absorption Device). This study compares the behaviour of different "low-density" deformable materials (a range of silicon rubber grades) inserted in a UREAD unit and loaded under impact condition. The energy absorbed was experimentally measured and compared against the impact energy. It was possible to dissipate levels as high as 74.91% of the impact energy when using a simple set-up, and the device re-usability was demonstrated.

Keywords: ECAE, Energy Absorption, UREAD, Non-Newtonian Materials, Impact

1. Introduction

Impacts, from road accidents to earthquakes, are threats to humans and their belongings. The scale of car, airplane and ship collisions are different than the collisions that occur naturally like in earthquakes [1]. Several technologies and methods were devised in order to prevent or minimize the damages due to these events, however the implementation of a final solution with maximum efficiency is yet far from being achieved. The development of current technologies and the engineering of new methods is currently driving the work of scientists in the field.

There are different types of energy absorbing devises based on a set number of working principles. Several methods have been used to design new and innovative systems, and a considerable research work has so far been carried out so as to find efficient ways to dissipate the energy from an impact. Materials in not traditional forms have been studied for their ability to absorb impacts, such metallic foams. Different types of foams have been explored, i.e. aluminium [2,3], and Ni/Al-hybrid [4]. It has been demonstrated possible that the engineering of objects with specific shapes can increase energy dissipation capabilities, this is the case for honeycomb and thin-wall structures. Their characteristics have been extensively explored, with interesting applications. As an example, experimental and numerical studies of honeycomb structures were made to unravel its crashworthiness parameters [5], and its absorption performance when applied to motorbike helmets [6]. However, honeycomb structures can be rather complex to manufacture and are not efficient if the impact force forms a small angle with the honeycomb cells plane. Thin-walled structures has also been explored as energy absorbers [7,8], but their typical performance under radial compression is poor. Besides that, they can only dissipate energy when they buckle, and once a first loading has occurred the unit is not re-usable. Car bumpers are also examples of energy dissipation devices based upon plastic deformation, not deemed re-usable. A design and FEA crash simulation for a composite car bumper is reported in [10], an optimized bumper systems for pedestrian lower-leg protection in [11], and an experimental study on characteristics of shock absorbers of impact energy of passenger coaches in [12]. This is also the case for a new generation of guardrails [9]; like with the others the nature of the working mechanism will not allow re-usage.

Such non-reusability characteristic is typical for "passive" shock absorbers, but a number of re-usable systems exist, and are deemed as "active". This is the case for a linear permanent-magnet motor for active vehicle suspension [13]. Examples are a permanent magnet eddy current brake for a small scaled electromagnetic launch model [14], and a system to stop roller coasters by using magnets on the train to induce eddy currents in the braking fins [15]. Other examples of active systems are advanced types of friction dampers, which can be used over a wide range of applications, i.e. for active support of piezoelectric ceramic actuators [16]. Frictional based dampers in earthquake resistant structures is another application, that has advantages over other types of energy dissipating devices. These include low cost of manufacturing and maintenance and also being less susceptible to environmental effects. Furthermore, no yielding occurs in such dampers after a severe earthquake which eliminates the need for replacement [17]. Semi-active system with friction dampers for lightweight pedestrian bridges is another example [18]. An alternative is represented by an energy-harvesting shock absorber with a mechanical motion rectifier [19]. To summarize, active or semi-active systems can consume a significant amount of energy to be efficient at impact. They are still rather complex and are currently implemented only at large o very large scale.

This paper presents experimental results obtained by using a novel passive energy absorber, known as UREAD, based upon the principles of materials plastic deformation but in such a way to allow for re-usability. It has the potential to overcome to some disadvantages of current systems. The device has been initially explored and results published in [20, 21], however primarily with Lead as deformable material. This paper will expand to the testing of alternative lighter-weight and not metallic materials.

2. Experiments set up

2.1. UREAD Working Mechanisms and Test Unit

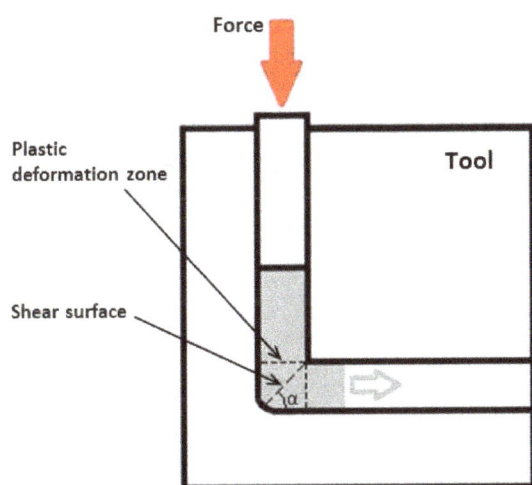

Figure 1. Shearing process of materials in 90° intersecting channels.

The Universal Reusable Energy Absorption Device (UREAD) working mechanism is based upon the principles Equal Channel Angular Extrusion (ECAE), and it is shown in Figure 1. A material is pushed through a channel of defined cross-section, with a change of direction of a determined angle. The device can be designed with different cross-

sections types, such as circular or hexagonal; while the deformation shape can be different such as with U-channels [22].

When an external force is applied to the pushing punch, the solid billet will be forced to change its moving direction at correspondence of the channel bends, in order to follow the geometry of the channel; hence energy will be dissipated by a plastic deformation shearing process [22]. While most of plastic deformations are irreversible, the fact that this one occurs inside a closed channel makes it controllable and then reversible: it only needs a push from the other side to come back to its original geometry.

This process can be explained by the diagram in Figure 1, which describes the plastic deformations mechanism occurring on a material when being moved within two intersecting channels. In order to achieve the material motion, shear has to happen at the intersection zone. As Figure 2 shows, the shearing process is modelled to develop on a shear surface, inclined at an angle α. The plastic deformation model for this process is based upon the principles of the Upper Bound Analysis technique [23]. It is clear that for this design, the material can be pushed forward and backwards through the two channels, potentially for an infinite number o times.

Figure 2(a) shows the UREAD unit which was used for the experimental work presented in this article. It is made out of tool steel (both unit and pressing punch), and has 2 channel cross-sectional geometries, however both rectangular, intersecting at 90^0 degrees, i.e. 8x8mm and 10x10mm. For the purpose of this study only the 10x10mm channel was used. The billet of deformable material is visible from the figure, positioned in this case in the 10x10mm channel. In this design the channels are covered with a flat steel plate so as to build a fully assembled UREAD, as shown in Figure 2(b).

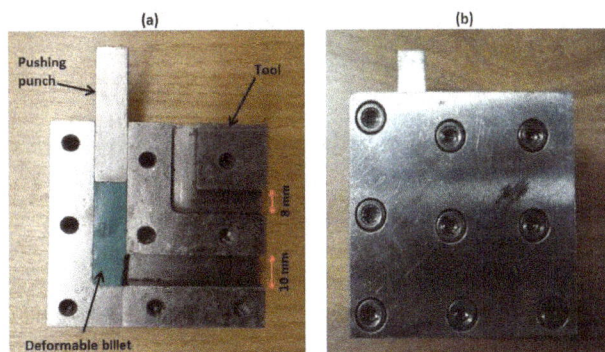

Figure 2. (a) UREAD unit with square cross-sectional channels. (b) UREAD cover plate used in this set-up.

2.1. Deformable Materials and Impact Test

In order to test the UREAD unit under impact conditions, an experimental apparatus was developed to resemble a hammer-drop test rig where the impact force is measured with a force sensor. Figure 3 shows an overview image of the set up. It consists of a hollow steel tube 965mm long; a weight is inserted in the top side of the tube, and made to impact on a UREAD unit located just above the ground. Figure 4 shows a close-up picture of the UREAD zone. It appears clear from the image the unit is bolted to a device support plate, and that the force sensor is firmly inserted between the device support plate and ground plate. The sensor used is an Integrated Circuit Piezoelectric (ICP) type purchased from PCB Piezotronics Inc. (model 208C05). A transducer is also used (signal conditioner purchased from PCB Piezotronics Inc., model 480C02). The data acquisition software is Labview.

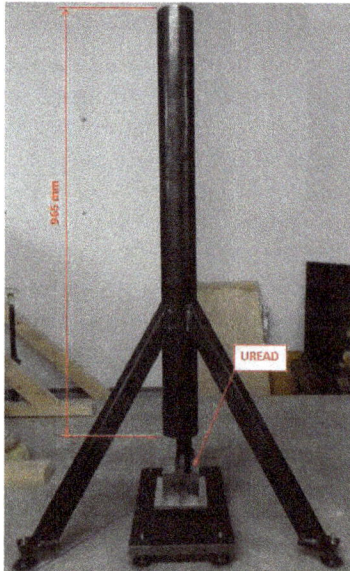

Figure 3. Impact test rig.

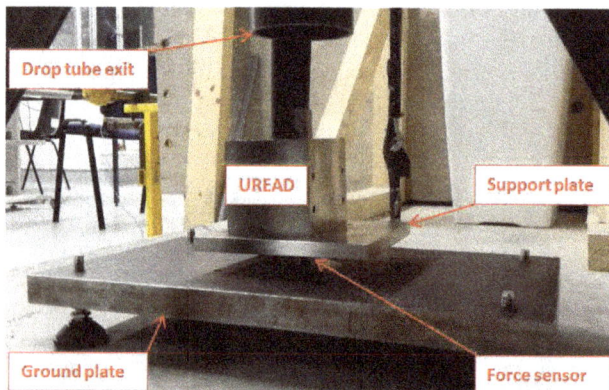

Figure 4. Close-up image of impact location.

For all impact tests a billet of deformable material measuring 30mm in height was shaped to fit the 10x10mm channel geometry, in a similar manner as it appears in Figure 1(a).

The deformable materials used are soft-dilatant polymers (silicon rubber), purchased from SportsHealth and sold under the commercial name of "Power Putty". Four commercial grades were tested known as Soft, Medium-Soft, Medium-Firm and Firm. A sample of each grade was characterized through a quasi-static compression test; a typical elasto-plastic behaviour was observed in all cases. It was possible to identify the Yield Stress and Young Modulus (E), which

have been reported in Table 1 for all the grades used in the experiments. It is shown the material mechanical characteristics enhance from Soft to Firm.

Table 1. Deformable materials characteristics.

Material	Yield Stress [Pa]	Young Modulus [MPa]
Soft	70	0,024
Medium Soft	90	0,042
Medium Firm	270	0,21
Firm	850	0,46

Following the billet preparation, the pressing punch in UREAD was placed in its location. The device is then closed and tightened with nine screws to a steel plate of 10mm thickness, as already described in Figure 2(b).

It is necessary for the drop tube to be aligned with the UREAD unit pressing punch on an axial direction, so as to make sure the weight impacts on the punch along its axis. This is to avoid the generation of bending moments, the formation of which would not be possible to take into account with the available experimental set-up. This was done, and the relative location of the UREAD unit marked so that it could be re-positioned after each test.

In order to generate a range of impact energies, 3 weights with different mass and drop heights are used. The weights are cylindrical to fit in the dropping tube and they are connected to a hook, which is used to hold each weight before being dropped. Table 2 shows the used weights masses and heights of fall, alongside with additional relevant data.

Table 2. Impact test parameters.

	Total mass [kg]	Weight length [m]	Height of fall [m]
Weight 1	1.134	0.040	0.994
Weight 2	1.689	0.060	0.984
Weight 3	2.265	0.080	0.974

2. Results and discussion

To calculate the total energy at the instant of impact for each weight the simple kinetic energy formula was used, as in Equation 1. Thus, Equation 1 and Equation 2 were used so as to calculate impact energy and velocity.

$$E_{impact} = (1/2)\, m\, v^2 \qquad \text{(Eq. 1)}$$

$$v^2 = v_o{}^2 + 2\, g\, \Delta h \qquad \text{(Eq. 2)}$$

Where m is the mass of the weight, v is the velocity of the weight just before the impact on the punch, v_o is initial velocity of the weight (which equals to zero), g is the acceleration due to gravity and Δh is the height of fall.

The values obtained are shown in Table 3:

Table 3. Generated impact energies.

	Total Impact Energy [J]
Weight 1	11.05
Weight 2	16.29
Weight 3	21.63

On the other hand, Equation 3 is used to calculate the amount of energy, E_{UREAD}, the UREAD device was able to absorb. It is therefore not assumed the full impact levels by Equation 1 are absorbed by the unit on its own.

$$E_{UREAD} = F_{ave} \cdot s \qquad \text{(Eq. 3)}$$

Where F_{ave} is the average force during the impact (by the force sensor), and s is the total punch displacement, i.e. the distance the punch has travelled (measured after the impact has occurred).

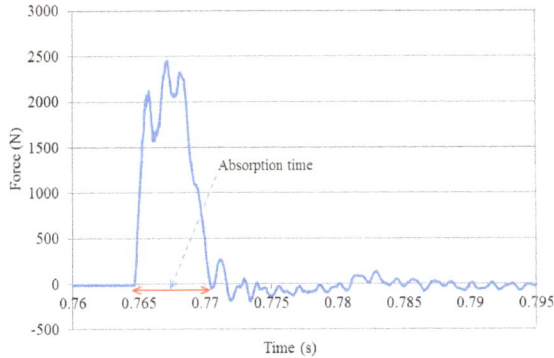

Figure 5. *UREAD measured force vs time during an impact using the Firm grade.*

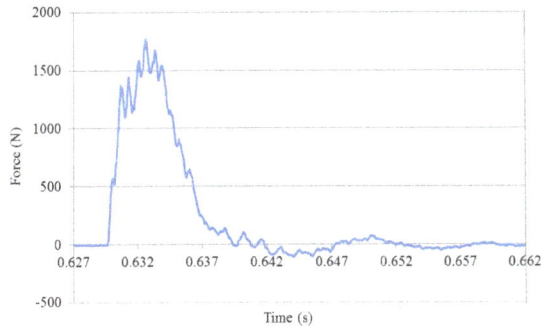

Figure 6. *UREAD measured force vs time during an impact using the Soft grade.*

Figure 5 and Figure 6 show examples of force measurements over time during an impact, where E_{impact} is the same in both cases. The data acquisition frequency was set to 30kHz. Specifically, Figure 5 is for an impact using the Firm grade in the UREAD channel, loaded with Weight 2; while Figure 6 is for an impact with the Soft grade, also using Weight 2 by following the fall parameters in Table 2. The average force, F_{ave}, has been calculated over the absorption time (highlighted in Fugure 5). Such corresponds to the period of time the channel punch is actually moving, hence energy is absorbed by the UREAD unit. In both cases the force characteristic rapidly rises as the weight lands on the channel punch, up to the reach of a peak value. This value is higher for the Firm grade, result in agreement with the material characteristics reported in Table 1. On the other hand the absorption time is longer, nearly double, for the Soft grade. This means in the latter case the punch has travelled a longer distance, despite E_{impact} has been kept constant. A residual vibrational mode was observed for periods of time beyond the actual impact, also visible in Figure 5 and Figure 6; this was attributed to a post-impact effect to the UREAD and load cell mounting/holding structures.

In the two experiments the recorded peak force was in the order of 2.5kN for the Firm grade, and 1.7kN for the Soft grade. An additional test was executed using the same impact conditions, however no deformable material was inserted in the channel, i.e. the UREAD unit was not active. In this case a peak impact force of 14.5kN was recorded, approximately 6 times higher as compared to when UREAD is functional. This clearly demonstrates it was possible using UREAD to dissipate energy and cut peak force levels considerably in the experiments. Such is a critical feature for an energy absorber.

The experimental results using the impact energies as by Table 3, and with the combination of the considered material grades, are summarized in Table 4. An average was calculated and reported in the table. The energy absorbed was estimated using Equation 3, while the frequency of data acquisition was 30kHz in all cases. Such is the maximum value as by the force sensor specifications. All experiments were repeated, giving comparable results.

Table 4. *Summary table of impact tests results.*

	Impact Average Force [N]	Absorption Time [ms]	Final Punch Displacement [mm]	Energy Absorbed [J]	Percentage of Energy Absorbed [%]
WEIGHT 1 - 1.134 KG (IMPACT ENERGY = 11.05 J)					
Very Soft	865.63	6.17	7.33	6.36	57.55
Soft Medium	1061.18	5.59	4.94	5.31	48.04
Medium Firm	1145.94	5.15	4.57	5.23	47.31
Firm	1313.68	4.74	3.37	4.42	39.96
WEIGHT 2 - 1.689 KG (IMPACT ENERGY = 16.29 J)					
Very Soft	822.32	9.37	12.05	9.96	61.09
Soft Medium	989.74	8.16	9.92	9.81	60.18
Medium Firm	1302.64	6.44	7.55	9.76	59.91
Firm	1533.89	5.62	6.43	9.87	60.54
WEIGHT 3 - 2.265 KG (IMPACT ENERGY = 21.63 J)					
Very Soft	816.25	12.40	16.60	13.59	62.80
Soft Medium	1232.66	7.75	13.23	16.21	74.91
Medium Firm	1385.62	7.18	11.38	15.78	72.92
Firm	1403.05	7.64	9.78	13.71	63.38

Figure 7 to Figure 9 show a graph of the absorbed energy, E_{UREAD}, for each material grade (from Soft to Firm) using the three levels of impact energy, E_{impact}, obtained with Weight 1, 2 and 3.

When the impact energy is the lowest (Figure 7), the Soft grade performs the best and absorbs 57.55% of the impact. The Firm grade, on the other hand, does not cross the 40% level. As the impact energy increases (Figure 8 and Figure 9), the dissipation performance trend becomes more uniform. With Weight 3 all of the grades performed very similarly by absorbing approximately 70% of the impact, with a peak of 74.91% using the Soft Medium. This behaviour emphasizes that the choice of deformable material is very critical, and careful considerations must be made in order to select the grade to suits best the application needs. Experimental results also demonstrate the versatility of UREAD is very high; it was in fact possible to dissipate a range of energy levels by interchanging the deformable material only, while the overall geometry remained unchanged. At the same time, the employment of lighter-weight materials in UREAD channels has been proved to perform efficiently, and it can be considered as a valid alternative to Lead or other metals as deformable billet so far being used in the UREAD technology.

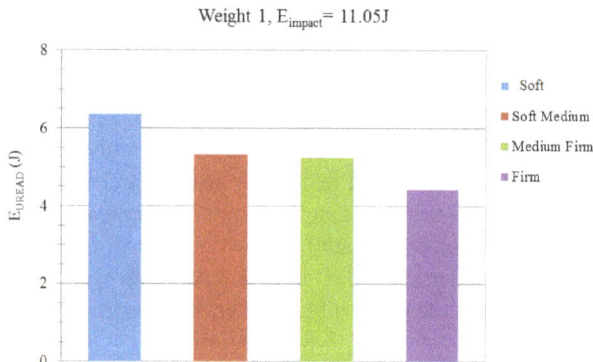

Figure 7. *Comparison of the absorbed energy for the material grades with weight 1.*

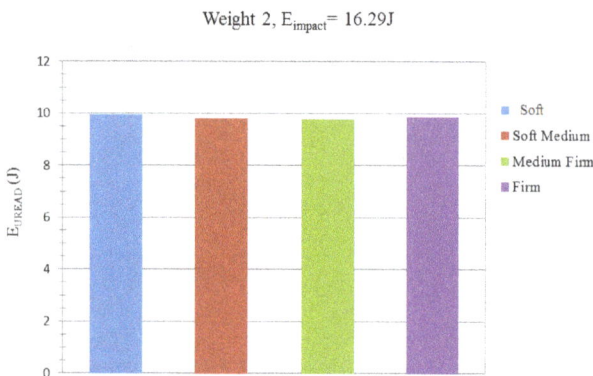

Figure 8. *Comparison of the absorbed energy for the material grades with weight 2.*

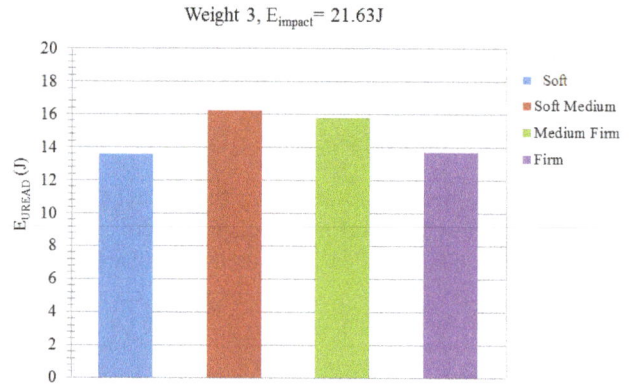

Figure 9. *Comparison of the absorbed energy for the material grades with weight 3*

Figure 10 summarises the specific behaviour for each material. The Soft grade is the most effective for low impact energies (up to just above 15J), however the effectiveness does not vary much when the actual impact energy increases (quasi-linear curve). The effectiveness of the "medium" grades (Soft Medium and Medium Firm) is rather similar, with a significant increase after approximatively 11.5J. The Firm material grade is the less effective for lower impact energies; it is however forecasted to be more suitable for higher energy tests which were not considered for the scope of this work..

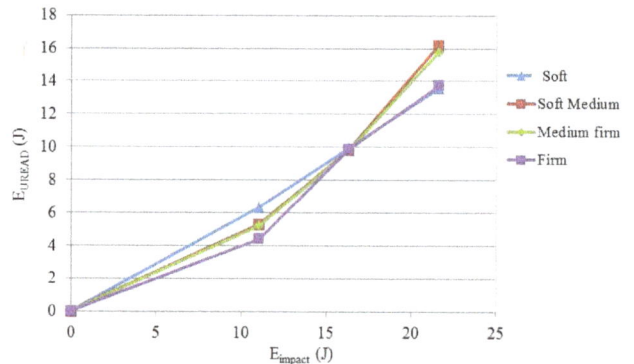

Figure 10. *Evolution of the absorbed energy vs impact energy.*

3. Conclusions

A novel technology, known as UREAD, capable of dissipating energy in engineering systems was introduced. It is based upon principles of Equal Channel Angular Extrusion (ECAE), with the ability of being re-usable after loading. This is not a typical characteristic of passive energy absorbers.

Impact tests were carried out on a UREAD device using silicon based materials as being deformed in the channel. Four different grades were tested, covering a yield stress range from 70Pa up to 850Pa measured under static compression.

A in-house drop-hammer rig was used in order to simulate different levels of impact energy. The average force during the impact period was used to calculate the energy absorbed by the device and it was compared to the impact energy. For lower impact energies (<15J), the softer Soft grade is the most efficient energy absorber as it dissipated 57.55% of it. However, for higher impact energies (between 16J and 21J), the Medium Firm and Medium Soft were shown to be able to dissipate up to 72.92% and 74.91% of the impact respectively. It was therefore possible to demonstrate the high versatility level of UREAD, as a variety of energy level was dissipated while the overall geometry remained unchanged. After each test it was possible to re-use the same UREAD unit and deformable material.

In general, all grades were able to dissipate increasing values of energy as the impact levels also increased. It is concluded such new materials represent a valid alternative and a lighter weight option to metals in UREAD applications.

Acknowledgements

The authors would like to acknowledge the support of the ERASMUS and Science Without Borders (SWB – funded by CAPES, Brazil) schemes for enabling this project. The authors are also thanking Prof. Ciaran Simms (TCD) for the support in the execution of the work presented in this article.

References

[1] Alghamdi AAA. Collapsible impact energy absorbers: An overview. Thin-Walled Structures 2001; 39:189–213.

[2] Rajendran R, Prem Sai K, Chandrasekar B, Gokhale A, Basu S. Preliminary investigation of aluminium foam as an energy absorber for nuclear transportation cask. Materials and Design 2008; 29:1732–9.

[3] Kim A, Hasan MA, Nahm SH, Cho SS. Evaluation of compressive mechanical properties of Al-foam using electrical conductivity. Composite Structures 2005; 71:191–8.

[4] Jung A, Lach E, Diebels S. New hybrid foam materials for impact protection. International Journal of Impact Engineering 2014; 64:30-8.

[5] Partovi Meran A, Toprak T, Muğan A. Numerical and experimental study of crashworthiness parameters of honeycomb structures. Thin-Walled Structures 2014; 78:87-94.

[6] Caserta GD, Iannucci L, Galvanetto U. Shock absorption performance of a motorbike helmet with honeycomb reinforced liner. Composite Structures 2011; 93:2748-59.

[7] Abramowicz W. Thin-walled structures as impact energy absorbers. Thin-Walled Structures 2003; 41:91-107.

[8] Yang Z, Yan H, Huang C, Diao X, Wu X, Wang S, Lu L, Liao L, Wei Y. Experimental and numerical study of circular, stainless thin tube energy absorber under axial impact by a control rod. Thin-Walled Structures 2014; 82:24-32.

[9] Atahan AO, Yücel AÖ, Erdem MM. Crash testing and evaluation of a new generation L1 containment level guardrail. Engineering Failure Analysis 2014; 38:25-37.

[10] Calienciug A, RADU GhN. Design and FEA crash simulation for a composite car bumper. Bulletin of the Transilvania University of Brasov, Series I: Engineering Sciences 2012; 5(1):7-12.

[11] Jiang K, Yang J. Optimization of bumper system for pedestrian lower leg protection from vehicle impact. Third International Conference on Digital Manufacturing & Automation. 2012.

[12] Simić G, Lučanin V, Tanasković J, Radović N. Experimental research of characteristics of shock absorbers of impact energy of passenger coaches. Experimental Techniques 2009; 29-35.

[13] Wang J, Wang W, Atallah K. A linear permanent-magnet motor for active vehicle suspension. IEEE Transactions On Vehicular Technology 2011; 60(1):55-63.

[14] Zhou S, Yu H, Hu M, Huang L. Design of permanent magnet eddy current brake for a small scaled electromagnetic launch model. Journal of Applied Physics III 2012; 07A738:1-3.

[15] Pendrill AM, Karlsteen M, Rödjegård H. Stopping a roller coaster train. Physics Education 2012; 47(6):728-35.

[16] Mingfu L, Mingbo S, Siji W. Active Elastic Support/Dry Friction Damper with Piezoelectric Ceramic Actuator. Shock and Vibration 2014; ID 712426:1-10.

[17] Samani HR, Mirtaheri M, Zandi AP, Bahai H. The Effects of Dynamic Loading on Hysteretic Behavior of Frictional Dampers. Shock and Vibration 2014; ID 181534:1-9.

[18] Wieczorek N, Gerasch WJ, Rolfes R, Kammerer H. Semiactive Friction Damper for Lightweight Pedestrian Bridges. Journal of Structural Engineering 2014; 140:1-13.

[19] Li Z, Zuo L, Kuang J, Luhrs G. Energy-harvesting shock absorber with a mechanical motion rectifier. Smart Materials and Structures 2013; 22:1-10.

[20] Lupoi R, Osman FH. Loading behaviour of 900 UREAD energy channels. International Journal of Crashworthiness 2008, 13, 2: 195-203.

[21] Osman F.H., Lupoi R. Application of "UREAD" for the energy dissipation in engineering structures. Key-Engineering Materials, 486, 2011, 1-4.

[22] Lupoi R. Investigation into Energy Dissipation in Equal Channel Angular Extrusion. PhD Thesis. University of Bath (UK), Department of Mechanical Engineering, 2008.

[23] Hosford W.F., Caddell R.M. Metal Forming: Mechanics and Metallurgy. Cambridge University Press, 4th edition, 2014.

15

Development Research of a New Welding Manufacturing Quality Control and Management System

Xiaochen Yang, Paul Kah, Jukka Martikainen

Laboratory of Welding Technology, Lappeenranta University of Technology, Lappeenranta, Finland

Email address:

xiaochen.yang@lut.fi (Xiaochen Yang), paul.kah@lut.fi (P. Kah), jukka.martikainen@lut.fi (J. Martikainen)

Abstract: The Chinese welding industry is growing every year. Companies around the world are looking to use Chinese enterprises as their cooperation partners. However, the Chinese welding industry also has relatively low quality and weak management system. A modern, advanced welding management system appropriate for local socio-economic conditions is required to enable Chinese enterprises to enhance further their business development. This article designed and implemented a new welding quality management system in China. This new system is called 'welding production quality control management model system in China' (WQMC). Constructed on the basis of surveys and in-company interviews, the welding management system comprises the following different elements and perspectives: a 'Localized congenital existing problem resolution strategies ' (LCEPRS) database, a 'human factor designed training system' (HFDT) training strategy, the theory of modular design, ISO 3834 requirements, total welding management (TWM), and lean manufacturing (LEAN) theory. The paper also describes the design and implementation of a HFDT strategy in Chinese welding companies. Such training is an effective way to increase employees' awareness of quality and issues associated with quality assurance. The study identified widely existing problems in the Chinese welding industry and constructed a LCEPRS database that can be used to mitigate and avoid common problems. The work uses the theory of modular design, and TWM as tools for the implementation of the WQMC system. Analysis of the WQMC system effects indicates that its adoption has resulted in improved quality and reduced costs.

Keywords: Welding Quality Management, Training, WPS, ISO 3834, Total Welding Management

1. Introduction

The current situation in China as in many other developing and newly industrialized countries demands the development and implementation of a locally-appropriate, modern welding quality management and welding personnel training system that enables the attainment of an ideal balance of welding product quality, welding process costs and sustainable development. Such a training and management system will help staff to acquire appropriate and efficient working attitudes and procedures needed to increase production quality.

The aim of this paper is to construct a localized, sustainable, reliable and broadly applicable 'welding production quality control management system model for China' (WQMC) that can be used in small, medium, and large enterprises and is suitable for most industries in which welding is an inherent part. The system is based on the ISO 3834 standard and combines personnel training strategies, processes adaptations, and the implementation of existing problem-solving methods. The items listed below constitute the basic concepts and elements of the model, and each individual company can use the model to replace or enhance their daily work or management strategies:

- ISO 3834 and the documents with which it is necessary to conform to be able to claim conformity to the quality requirements. ISO 3834 forms the foundation of the model (WQMC) and defines its programmatic approach.
- A localized, human factor and psychology theory based welding personnel (welders, welding specialists, welding technicians, and welding engineers) training system for China. The training system is the origin of change and the guarantor of sustainable development of welding enterprises, factories and workshops.
- A long-term guidance and supervision system for welding processes. This system demands not only

administrators but also staff to continue improving their tasks. Fast error correction guarantees the dynamism of enterprise development.

- A series of recommendations for improvements based on surveys, case studies, interviews and careful observation in over 30 Chinese welding factories or workshops. This series of recommendations are practical in nature and applicable also to welding enterprises in other developing and newly industrialized countries.[6, 7, 8, 9, 10, 11]

Together, the four items above constitute a comprehensive improvement model for welding enterprises in China and newly industrialized countries. This research is related to the research fields of welding technology, quality and knowledge management, psychology, and pedagogy and training.

Every enterprise requires its own management system to monitor internal affairs. An enterprise management system can be divided based on business functions into eleven parts: 1) plan management, 2) production management, 3) procurement management, 4) market management, 5) quality management, 6) logistic management, 7) financial management, 8) project management, 9) human resources management, 10) statistic management, and 11) information management [1]. Welding production enterprises often have the same above-mentioned eleven parts. Within any one enterprise, the listed management aspects are integrated into a whole management system in order to guarantee normal daily operations.

One management system for one company can include a number of different management sub systems. For example, a company may utilize the ISO 9001 quality management system, ISO 14001 environment management system, OHSAS 18001 occupational safety & health management system, and ISO/IEC 27001 information security management system at the same time [2]. For a welding production or manufacturing company, the ISO 3834 welding quality management system should be integrated into the whole enterprise management system in order to ensure relatively high productivity, high quality, and low production costs. Within the context of this work, the theory of total welding management (TWM), the LEAN theory, the human factors designed training system, and the localization factors for China can be integrated into the management system in the same manner and thus the 'welding production quality control management system model for China (WQMC)' can be generated.

Fig. 1 illustrates the overall structure of the WQMC system, which comprises the following key elements:

- Normal enterprise management systems (NEMS): ISO 9001, ISO 14001, OHSAS 18001, ISO/IEC 27001 and other related management systems. The function of the NEMS is to provide the most basic foundation of the WQMC system;
- ISO 3834 welding quality management system;
- Human factor designed training system (HFDT);
- Total welding management (TWM) system, LEAN theory, and modular design model;
- Strategies for resolution of localized congenital existing problems (LCEPRS).

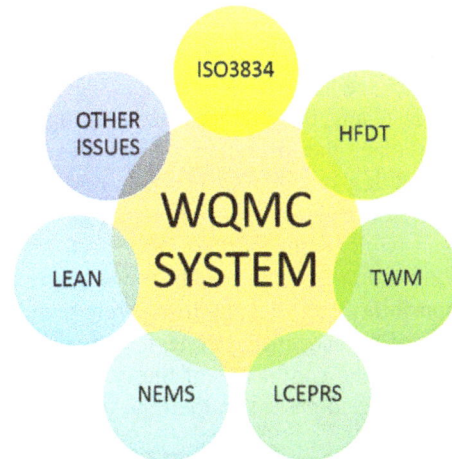

Fig. 1. *Structure of WQMC and relationship to other management systems.*

Before a company can implement the WQMC system, the NEMS should already be built inside company operations. While the nature of the NEMS and their implementation is clearly of importance, research of NEMS is beyond the scope of this study. Normally, if a manufacturing company is qualified by a system form NEMS, the 3rd party certification body must issue a certification document to the manufacturing company. However, ISO 3834 is of immediate relevance to the subject of this work and implementation steps for integration of ISO 3834 in the WQMC system are discussed in chapter 2.

2. Welding Quality System Implementation Based on the Tools Using in WQMC System

The implementation steps of the ISO 3834 quality management system (the core of the WQMC system) are:

- Setting goals. The whole personnel participate in this process. During this period, the welding personnel must set out the difficulties and focus points for changes to the work processes;
- Placing the welding engineer or welding specialist in the right place in the personnel organization structure in order to give real power to the welding coordination teams to carry out necessary rectification work.
- Building an organizational structure between different departments inside the enterprise to ensure effective co-operation patterns and guarantee smooth communication channels.
- Communication of the goals. The whole personnel participates in a meeting to announce the goals of the activities in order to demonstrate the determination of the company;
- Setting of the quality policy. The quality policy should be set during the meeting, and the quality policy should be announced to customers;
- Documentation analysis and preparation. The welding coordination team do the documentation analysis work.

This work involves checking which documents are missing and adding the necessary documents;

- Initiation of the inspection work. The welding coordination team, the welders and the inspection department launch the inspection work related to every aspect of welding quality. Surface treatment, welding consumable storage, the welding station, and maintenance work are the most crucial areas to check;

- Orientation of top management. The top managers have the responsibility to be familiar with and have knowledge of management strategies regarding the welding work. The confidence of the team is built by the top manager and passed on to all participants in the team;

- Continuous improvement process. A continuous improvement system is a requirement of the ISO 3834 standard. A PDCA cycle plan needs to be made and followed in order to ensure increasing productivity, increasing quality, and decreasing cost.

Welding factory in China which wants to implement the WQMC system needs to build the ISO 3834 welding quality management system as the core of WQMC. This is the first step to building the whole WQMC system. The LCEPRS strategies at the beginning steps of WQMC before ISO 3834 implementation and the HFDT strategy during the WQMC system building are useful tools to guarantee the effectiveness of the WQMC system implementation.

The modular design methodology is a useful tool in the WQMC system. In welding production processes, advanced methodology for design and manufacture is needed in order to increase quality and productivity, and decrease cost. The idea of modular design is to try to use similar design approaches or methods to meet different customer requirements for almost identical products. The products may have differences in dimensional aspects, but the main function should be the same. If modular design theory is used, a number of changes in the design and manufacture departments are usually required:

a) Revised design of jigs and fixtures for welding;
b) New arrangement of the layout of the welding workshop and its upstream and downstream processes;
c) Improved modular design awareness among designers;
d) Changes to the manufacturing habits of welders.

During the case project research, some commonly existing problems were found in the Chinese welding industry that clearly influenced quality, productivity and cost. These problems to some extent do not occur because of the welding industry itself but mainly because of the local Chinese culture, human factors and other related reasons. In the WQMC system, such problems are called 'localized congenital existing problems (LCEP)', and the solving method is called 'Localized congenital existing problem resolution strategies' (LCEPRS).

LCEP should be solved before the ISO 3834 implementation, or even before the NEMS implementation. On the other hand, some LCEPs may be recognized only during the implementation of ISO 3834 and later. In the WQMC system, the identification and implementation of mitigation actions for LCEP, LCEPRS, will be largely complete after the survey and interview work, and before the implementation of ISO 3834.

More than 20 LCEPs were identified during the case project and on the basis of previous research. The LCEPs were noted through the use of surveys, interviews, discussions, observations, and the author's experiences. Some LCEPs will be similar to the problems discussed in previous chapters, but in the WQMC system, iterative improvement is a core concept. Knowledge of repeated problems will help enterprises prioritize areas requiring improvement.

The LCEPRS database is ready to update. Along with the development of the Chinese industry, and the deep research of this topic, more and new LCEPs is also can be found. The database will update according to the changing of the real situation. Fig. 2 illustrates the problem solving approach.

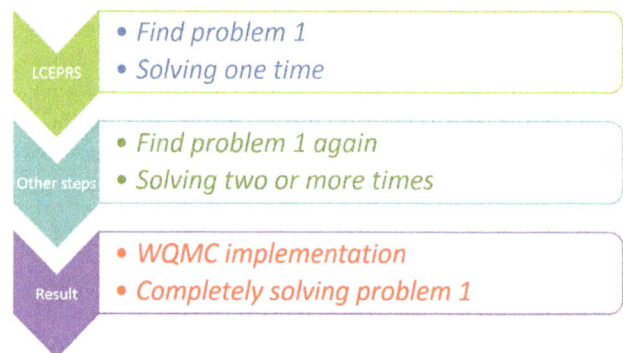

Fig. 2. Solving processes for repeated problems.

This process is also a concept from PDCA cycle. Problem 1 can be found in later steps, so the whole system ensure the problem will not be skipped and will be solved.

3. Training System - HFDT

Training of company staff is required in the WQMC system. TWM theory and ISO 3834 also require that business entities develop training activities for their personnel (SFS-EN ISO 3834-2). In the WQMC system, the novelty is to build an up-to-date and localized welding personnel training strategy for China. In addition to its foundations on welding technology and production theory, this strategy employs the knowledge of psychology, pedagogy and behavioral science in order to establish training modules appropriate for contemporary circumstances in China, for welders as well as welding specialists, welding technicians, welding engineers, design engineers and welding inspectors. To some extent, managers and administrators in other roles are also considered for participation in the training in order to ensure that the same concept of welding and the WQMC system exists throughout the enterprise [12]. In this study, the training strategy is abbreviated to HFDT (human factor designed training system).

Training content and lecture formats are discussed in this thesis. In the training method, the training modes are teaching and lectures, discussions, and audiovisual and case study methods. This study recommends adoption of 'small class' teaching methods in order to achieve better training quality. The small group size help to avoid distractions and improve focus. The passivity of large group lecture-based learning can induce fatigue and boredom in welding personnel. Furthermore, having a class with a large number of trainees means that it becomes difficult for the trainer to monitor and assist each trainee properly and respond to individual needs. Consequently, the objectives of the lectures are seldom totally met. Furthermore, small class teaching methods promote discussion and case study. Normally, the number of trainees in a small-group class ranges from 5 to 6.

The content of the training builds and expands on welding theory; the following topics, among others, are addressed:

- Welding symbols and engineering drawing;
- Welding terms;
- Welding standards: ISO 3834 series, ISO 5817, ISO 9606, ISO 13920, ISO 15607, ISO 15609, ISO 15610-ISO 15614, ISO 4136, ISO 5173, ISO 9015, ISO 9016, ISO 4063, ISO 6947, ISO 6520-1, GBT/324, GBT/3375, etc.;
- Welding defects and imperfections;
- Welding stress and deformations;
- Welding cracks and their causes;
- WPS concept and application;
- Welding inspections and experiments: destruction testing (DT) and non-destruction testing (NDT);
- Welding management strategies;
- Welding joints and welding structures production;
- Welding production knowledge for designers.

Table 1. *The HFDT training lecture design content [13, 14, 15, 16, 17, 18, 19, 20, 21, 22, 23, 24, 25].*

Topics	Content
Welding symbols and drawings	Welding drawing and mechanical drawing reading; Welding symbols explanation; GB/T 324 and ISO 2553, welding symbol representation on drawings; Link to drawings used in daily work to give further explanation.
Welding technology terms	Explanation of different terms used in daily work; Explanation of GB/T 3375, welding terms and Chinese national standards; Explanation of ISO 4063 and ISO 6947.
ISO 3834 and related knowledge	The main structure of ISO 3834; Welding quality management; Different parts of ISO 3834 (part2, part5 are the most significant parts); The balance between cost and quality; implementation of ISO 3834; ISO 3834 and WQMC, the core of WQMC system.
ISO 9606	What is ISO 9606 and EN 287; The test selection principle and criteria; Test requirements and steps; Reading the welder designation after the welder test.
Welding deformation	Reasons for welding deformation; Relationship between residual stress and welding deformation; Judging welding deformation in current practices; Avoiding welding deformation during the processes.
Welding imperfections and defects	Concepts of welding imperfection and welding defects; The different kinds of welding defects; Explanation of each welding defect: the reason and origin of the defect; How to avoid welding defect during the production steps; The importance of ISO 5817; VT knowledge for welders and inspectors.
WPS knowledge and related ISO standards	The idea of WPS, WPQR, pWPS, and welding instructions; Explanation of ISO 15607 and ISO 15609; Explanation of ISO 15610 to ISO 15613; Explanation of ISO 15614.
ISO 5817	The concept and explanation of ISO 5817; The concept of ISO 6520-1; The different requirements for level B, C, and D welding seams; Use ISO 5817 as the judgment document between welders and inspectors.
Welding cracks and its origins	The importance of avoiding welding cracks; The origins of cold cracking, hot cracking, reheat cracking, and lamellar tearing; Methods to avoid different cracks; Importance of paying attention to the hydrogen and cleaning before welding.
Welding seam DT and NDT to inspectors	Explanation of DT and NDT tests; Show the test specimen photos for DT tests; Short explanation of ISO 4136, ISO 5173 ISO 9015 (part 1 and 2), and ISO 9016; Short explanation of PT, RT, UT, and MT.
ISO 13920	The requirement of welding seam tolerance in ISO 13920; The different welding seam tolerance levels in ISO 13920.
Welding strength knowledge for designers	Short explanation to the designers about welding seam mechanics; The strength of the welding joints; How to check the welding handbook to ensure the strength.
Welding production knowledge for designers	The main steps of welding processes; Explanation of different welding processes (MIG, MAG, SAW, and TIG); Importance of paying attention to the welding accessibility during the drawing design.
Welding management strategy	Layout of welding workshop; Welding environment issues; Total welding management concepts; LEAN theory; WQMC system concepts; PDCA, 5S, 6sigma, and related concepts; Welding consumable and parent material storage.
New welding technologies and new concepts introduction	Laser welding and hybrid welding; Welding robots and semi-automatic welding processes; New technologies of welding machines; Ergonomics during welding works; Advanced shielding gas information.

Clearly, the training content is not permanent and needs to change with developments in welding science and technology. The skills demanded of the trainer is an important research topic in this study. A knowledgeable, dutiful and well prepared trainer is required to ensure the success of the WQMC system and HFDT strategies. This research topic is expanded with reference to knowledge of training activities and training psychology.

The HFDT strategy is a tool which is built inside the WQMC system. Training, in this case, is defined as an implementation activity by the enterprise in order to promote the knowledge and working abilities of employees according to a well prepared, detailed plan [3]. The aim of the training is to update employees' knowledge, skills and activities as regards the topics emphasized in the training project, and encourage the application of this knowledge, skills and activities in their daily work. Furthermore, continuous learning requires that the employees understand the whole operation, and the steps and structures of the working system, which includes the connections between the different tasks, different departments, and even outsourced or subcontractor activities [4]. Welding training should follow the lead of modern manufacturing and incorporate information about modern innovations in both welding technology and training theory into courses at all levels [5]. The HFDT strategy exhibits the same characteristics as given in modern training theory. For example, in welding production, the HFDT strategy requires that the welders, designers, coordination team, managers, inspectors, and the welding upstream processes workers become familiar with each other and each other's work in order to fulfill the requirements of the continuous learning.

The welding knowledge training lectures are the most significant part in the HFDT strategy. In the case project, the training lectures are divided into 15 main topics. Table 1 lists the topics and key related content.

4. Conclusion

The complete implementation of the WQMC system includes several steps. Before the system implementation, the enterprise should make the decision to adopt the WQMC system, hold mobilization meetings for the employees, prepare the NEMS systems, and make a financial plan for the activities.

During the implementation steps of the WQMC system, the personnel structure analysis, the survey and interview work, the documentation work, the rectification of the manufacturing processes, the LCEPRS stage, the HFDT stage, the TWM and LEAN theory implementation, the modular design methodology implementation and the ISO 3834 requirements implementation should be all well planned and conducted.

After implementation of the WQMC system, the results discussion forums, the PDCA cycle and the future improvement plan should be established. Awareness among the personnel of the importance of welding quality and how welding quality is achieved is the most significant aim of the implementation of the WQMC system.

The flow chart below depicts the main steps in implementation of the WQMC system (Fig. 3): The WQMC system implemented in three different companies had positive results according to company statistics and feedback from welders, inspectors, and welding coordinators. The results indicate that the WQMC system is suitable for Chinese welding enterprises to utilize in their company management. The WQMC system will help Chinese welding enterprises to improve product quality and productivity, and decrease welding manufacturing cost.

The HFDT strategy helped the Chinese welding enterprises to improve the level of knowledge of their staff. Welding manufacturing and design personnel can get improved training from the implementation of this strategy.

The design of the training program, the qualifications of the trainer and the arrangement of the classroom were seen as effective drivers for the training developed in the three companies. Based on the results of this research, the HFDT strategy is suitable for use in Chinese welding enterprises.

The LCEPRS database is a concept that can help Chinese welding enterprises to improve their daily work not only from the production aspects, but also from the human factor, culture, and management aspects. The LCEPRS database can help Chinese welding enterprises to find and solve their inherent problems efficiently and easily based on the case project results and data collection.

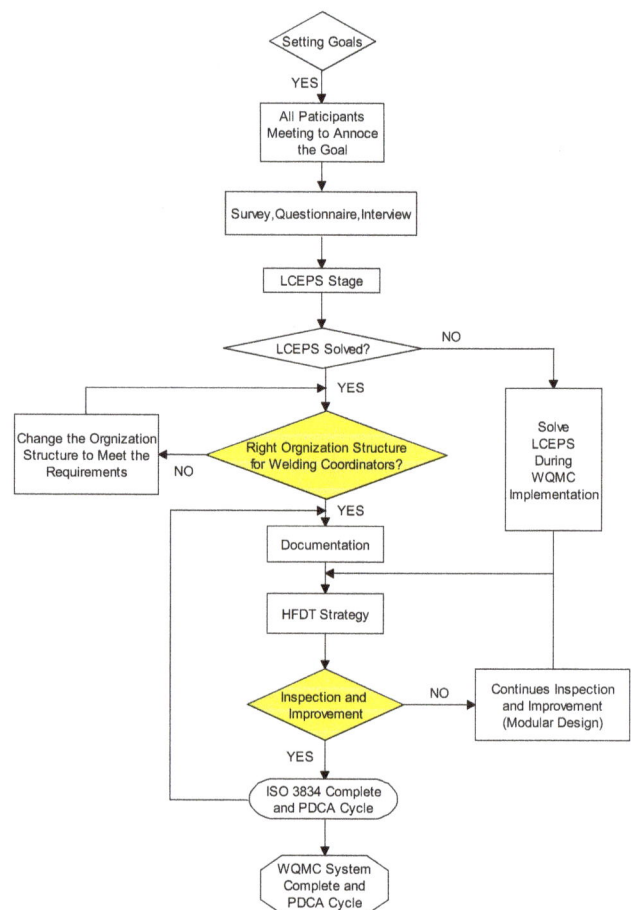

Fig. 3. Flow chart of WQMC system implementation.

In this research, only one company had launched the 'modular design' methodology. More examples of this approach in the future will help to verify the effects of the 'modular design' methodology.

Data collection and statistics is not very sufficient because some of the data is the confidential information in the company. More statistical data would definitely help to confirm the results of the WQMC system implementation. The data collected from company A and B is sufficient to illustrate the positive effect of the WQMC system.

Based on statistical data from the quality department in the enterprise, Figure 62 (a, b) presents results of WQMC system implementation in Company B.

From the data of Figure 4, it can be seen that the total

welding defect rate decreased from 9.3% to 5% during one year. And the wrong dimension, porosity, and other defects decreased from 5.4% to 3.1%, 2.8% to 1.4%, and 1.1% to

0.5% respectively.

Figure 5 presents the welding repair rate and the welding imperfection rate.

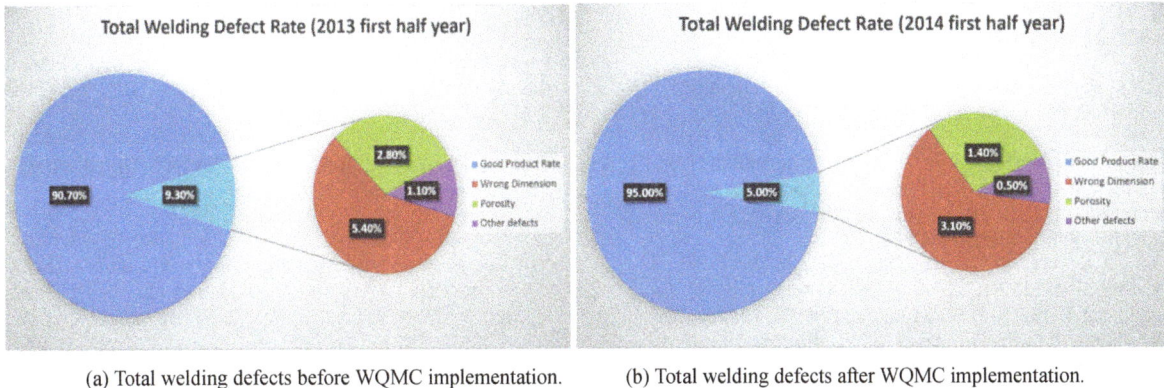

(a) Total welding defects before WQMC implementation.   (b) Total welding defects after WQMC implementation.

Fig. 4. *Total welding defect rate of Company B.*

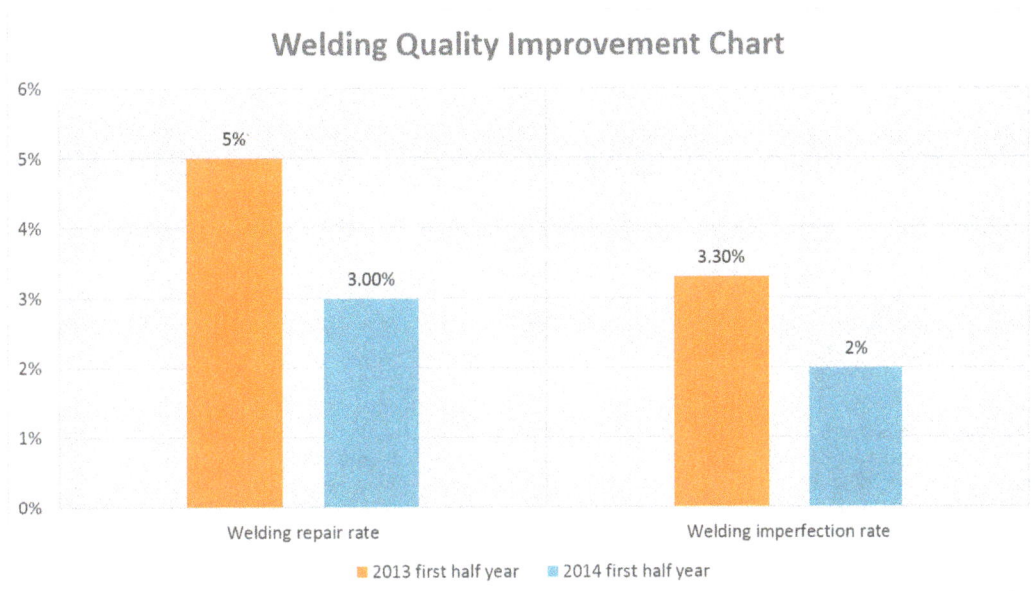

Fig. 5. *Welding quality improvement of Company B.*

From Figure 5, it can be seen that the welding repair rate decreased from 5% to 3%, and the welding imperfection rate from 3.3% to 2% one year after implementation of the WQMC system.

According to the assessments, the basic research results are reached. The implementation of WQMC system gives the positive result for the companies in the case project.

The contribution of this paper is the finding, designing and building of the new welding manufacturing quality control management system implementing in China (name: welding production quality control management system in China: for short, WQMC system). This research is related to the management science, welding technology, and training technology and methodology. The new system will help the Chinese welding manufacturing enterprises (in different geographical areas, in different industrial areas, and in different company scales) to control their welding production quality in a certain high level, and management their welding manufacturing related personnel cooperate

and work in an effective way by training. This research is important and have contribution because that the Chinese welding industry and foreign companies now all need the increasing quality and high productivity in the welding productions in China. Only when the Chinese welding factories have and implement a useful quality management control system now, that the two sides, China and foreign countries, can gain a win-win result in the welding production industries.

References

[1] Miao Chengdong, Yao Weimin, Zhang Jianzhong, The Theory of Modern Enterprise Management. – Peking University Press, ISBN 978-7-301-21032-1/F. 3269, 2012, 265 p.

[2] Chai Bangheng, ISO 9000 Quality Management System (Second Edition). – China Machine Press, ISBN 978-7-111-31292-5, 2010, 401 p.

[3]   Noe A. Raymond, X. Fang, Employee Training and Development. – China Renmin University Press, ISBN 978-7-300-08186-1, 2007, 428 p.

[4]   Rosow J. M. and Zager R. Training the Competitiveness Edge. – San Francisco, Jossey-Bass, 1988 p.

[5]   Polanin W. Richard The future of welding education. – Welding journal, 2014, vol. 93, no. 4, p. 38-41.

[6]   SFS-EN ISO 3834-6: en. 2007. Quality requirements for welding. Fusion welding for metallic materials. Part 6: implementation of ISO 3834-1 to - 5. Brussels: European Committee for Standardization (CEN). 21 p.

[7]   SFS-EN ISO 3834-1: en. 2006. Quality requirements for fusion welding of metallic materials. Part 1: criteria for the selection of the appropriate level of quality requirements. Brussels: European Committee for Standardization (CEN). 7 p.

[8]   SFS-EN ISO 3834-2: en. 2006. Quality requirements for fusion welding of metallic materials. Part 2: comprehensive quality requirements. Brussels: European Committee for Standardization (CEN). 10 p.

[9]   SFS-EN ISO 3834-3: en. 2006. Quality requirements for fusion welding of metallic materials. Part 3: standard quality requirements. Brussels: European Committee for Standardization (CEN). 9 p.

[10] SFS-EN ISO 3834-4: en. 2006. Quality requirements for fusion welding of metallic materials. Part 4: elementary quality requirements. Brussels: European Committee for Standardization (CEN). 3 p.

[11] SFS-EN ISO 3834-5: en. 2006. Quality requirements for fusion welding of metallic materials. Part 5: documents with which it is necessary to conform to claim conformity to the quality requirements of ISO 3834-2, ISO 3834-3, or ISO 3834-4. Brussels: European Committee for Standardization (CEN). 19 p.

[12] Barckhoff R. J., 2007, Total Welding Management, China Machine Press, ISBN 978-7-111-20887-7, 167 p.

[13] Chen Qiang, 2007, Welding Manual volume 1 (Third Edition), China Machine Press, ISBN 978-7-111-22263-7, 1154 p.

[14] Lei Shiming, 2010, Welding Processes and Equipment, China Machine Press, ISBN 978-7-111-13617-0, 254 p.

[15] Liu Huijie, 2007, Welding Metallurgy and Weldability, China Machine Press, ISBN 978-7-111-20921-8, 246 p.

[16] Li Yajiang, 2005, Welding crystal structure performance and quality control, Chemical Industry Press, ISBN 7-5025-6468-3/TH.281, 356 p.

[17] Pan Jimin, 2010, Guidebook of Welder Operation Quality Assurance, ISBN 978-7-111-28350-8, 223 p.

[18] SFS-EN ISO 4063: en. 2010. Welding and allied processes. Nomenclature of processes and reference numbers (ISO 4063: 2009, corrected version 2010-03-01). Brussels: European Committee for Standardization (CEN). 21 p.

[19] SFS-EN ISO 5817: en. 2006. Welding. Fusion- welded joints in steel, nickel, titanium and their alloys (beam welding excluded). Quality levels for imperfection (ISO 5817: 2003, corrected version: 2005, including technical corrigendum 1: 2006). Brussels: European Committee for Standardization (CEN). 25 p.

[20] SFS-EN ISO 6520-1: en. 2008. Welding and allied processes. Classification of geometric imperfections in metallic materials. Part 1: fusion welding. Brussels: European Committee for Standardization (CEN). 49 p.

[21] SFS-EN ISO 6947: en. 2011. Welding and allied processes. Welding positions. (ISO 6947: 2001). Brussels: European Committee for Standardization (CEN). 17 p.

[22] SFS-EN ISO 9606-1: en. 2013. Qualification testing for welders. Fusion welding. Part 1: steels (ISO 9606-1: 2012 including COR 1: 2012). Brussels: European Committee for Standardization (CEN). 33 p.

[23] Wang Fengqing, 2009, The Development Research of the Internationalization of the Chinese Certification and Accreditation, China Standard Press, ISBN 978-7-5066-5545-3, 115 p.

[24] Yin Shike, 2014, Welding Features and Welding Consumables for Low Carbon Steel, Chemical Industry Press, ISBN 978-7-122-18703-1, 230 p.

[25] Zhao Xihua, 2009, Welding Inspection, China Machine Press, ISBN 978-7-111-03695-1, 201 p.

16

Heat release rate in a reduced-scale model of a subway car on fire

Won-Hee Park[1, 2]

[1]Transportation Environmental Research Team, Korea Railroad Research Institute, #176, Choeldo bangmulgwan-ro, Uiwang-si, Gyeonggi-do, 437-757 Korea
[2]Railway System Engineering, Korea University of Science and Technology (UST), 217 Gajeong-ro Yuseong-gu, Daejeon, 305-350 Korea

Email address:
whpark@krri.re.kr

Abstract: The heat release rate of a subway car on fire is measured for various positions of an ignition source and simulated materials differing in combustibility by using a reduced-scale model. Although the maximum HRR value is nearly independent of the position of the ignition source, the time required to reach this value varies greatly for different ignition positions. The open area of a subway car is a key factor that determines the maximum HRR value, although material combustibility also has an effect. Finally, the HRR curve is compared with that of a fire test in a real-scale subway car.

Keywords: Heat Release Rate (HRR), Subway Car, Reduced-Scale Model

1. Introduction

Various fire accidents including the Daegu subway fire in 2003 and the Austria mountain tunnel fire in 2001 have shown that fire accidents in underground tunnels are more dangerous than those above ground. Many studies have been conducted to anticipate and analyze tunnel fire phenomena. However, research on railcar fires as fire sources that determine the type and characteristics of a train tunnel fire remains insufficient. In particular, the characteristics of a passenger railcar fire vary depending on the structures and interior materials of the railcar. Previous studies on railcar fires include Amtrak's full-scale passenger railcar fire test [1] by the NBS (National Bureau of Standards) in 1984, the EUREKA 499 test [2] program in 1991, and the reduced-scale and full-scale fire research by Peacock et al. [3] in 1994. Ingason [4] has studied the change in the heat release rate (HRR) according to the conditions of ventilation by using a reduced-scale model railcar. Recently, full-scale fire tests for subway and inter-city railcars have been conducted [5] and analyzed [6]. Although full-scale experiments are ideal for collecting and analyzing accurate data, they incur considerable amounts of expense and effort. Therefore, reduced-scale tests have been actively conducted. Although these tests have limitations such as turbulence intensity, thermal inertia of materials, radiation heat effects,

and fire propagation, they allow researchers to forecast quantitatively characteristics including the HRR and temperature more easily.

In this study, fire propagation phenomena were analyzed by measuring the HRR, mass loss rate, and temperature according to the location of ignition source and combustibility of interior materials by using a 1/10-scale model of an actual subway car. In addition, the HRR was compared with that measured in an actual fire test conducted with a subway car.

2. Tests Using Reduced-Scale Model

2.1. Froude Scaling

In this research, Froude scaling [7] was used for the relation between a reduced-scale and an actual-scale subway car. The HRR measured in the reduced-scale model is converted using the following equation:

$$Q_R = Q_M \left(L_R / L_M \right)^{5/2}, \qquad (1)$$

where Q is the HRR, L is the character length, and the subscripts R and M denote real and model, respectively. Time as measured in the reduced-scale model is revised according to the following equation:

$$t_R = t_M \left(L_R / L_M \right)^{1/2} , \qquad (2)$$

where t is time. The actual total energy emitted is calculated with the following equation by using the energy emitted from the reduced-scale model:

$$E_R = E_M \left(L_R / L_M \right)^3 \left(\Delta H_{c,M} / \Delta H_{c,R} \right) , \qquad (3)$$

where E is the total heat content and ΔHc is the heat of combustion.

2.2. Reduced-Scale Subway Car

(a) Inside of the reduced-scale model.

(b) Outside of the reduced-scale model.

Figure 1. Reduced-scale model.

The model in this study was a 1/10-scale model of an actual subway car operated in Seoul (Figure 1). The dimensions of the reduced-scale model are 1.9 m in length, 0.29 m in width, and 0.235 m in height. The length of the actual train is 19.7 m. LR/LM, the reciprocal of the scale ratio in eqs. (1)–(3), is 10.37. Generally, when a fire breaks out in a subway car and the train is stopped at the platform, all four doors at the sidewall near the platform will open for passenger evacuation. Therefore, in the reduced-scale model, all four doors at the sidewall were open. Each door was 0.13 m in width, 0.185 m in height, and 0.02405 m2 in area. The total ventilation area when the four doors were open was 0.0962 m2. The plaster board, which was used as the structural material for the reduced-scale car, weighed 28 kg and had a density of 860 kg/m3, a thickness of 0.016 m, and a thermal conductivity of 0.069 W/mK. The interior materials were manufactured based on the size of the ceiling, floor, and

ends of an actual subway car, and then attached to the plaster boards. As shown in Figure 1(a), interior materials with dimensions of 0.045 m × 1.9 m and 0.09 m × 1.9 m were attached to the sidewalls with and without open doors, respectively. The ceiling and floor were completely covered by interior materials. At both ends, small seats were installed as in an actual subway car. Long seats were installed at the sidewall sections between doors. For the long seat, the size of the seating area was 0.055 m × 0.312 m; the back of the seat and support measured 0.04 m × 0.312 m. For the small seat, the size of the seating area was 0.055 m × 0.134 m, and that of the back of the seat and support measured 0.04 m × 0.134 m. All interior materials were attached using bolts without adhesive bond. The total surface area of the interior materials used for the ceiling and floor was 0.551 m2, that for both ends was 0.06815 m2, and that for the sidewalls with or without open doors was 0.0855 or 0.171 m2, respectively. The total cross-sectional areas of the four small seats and six large seats were 0.07236 m2 and 0.25434 m2, respectively.

2.3. Materials of Model Car

Corrugated cardboard 7 mm thick and plywood 3 or 5 mm thick were used as the interior materials. To determine the combustibility of each interior material, the combustion behaviors of the 7-mm corrugated cardboard and the 3- or 5-mm thick plywood under incident heat fluxes of 10, 20, and 30 kW/m2 were measured using a ISO5660-compliant cone calorimeter [8]. The oxygen consumption calorimeters employed in this study, such as the cone calorimeter and the large-scale calorimeter of ISO9705 [9], were used to measure the HRR from the reduced-scale model. The HRR is calculated by measuring the amount of emitted flow and concentration of oxygen in the combustion gas to quantify the consumption of oxygen by combustion. The HRR, or Q, from the fire is calculated as [10],

$$Q = E \cdot \dot{m} \cdot \frac{M_{O_2}}{M_{air}} (1 - X_{H_2O}^0) \left/ \frac{\alpha - 1}{X_{O_2}^0} + \frac{1 - X_{O_2}/1 - X_{CO_2}}{X_{O_2}^0 - X_{O_2}(1 - X_{CO_2}^0)/X_{CO_2}} \right. \qquad (4)$$

Here, E [kJ/kg] is the amount of energy per consumed unit of oxygen, $\dot{m}$ [kg/s] is the mass flow in the exhaust duct, M is the molecular weight, X is the mole fraction, and α is the ratio between the number of moles of combustion products including nitrogen and the number of moles of reactants including nitrogen. The superscript 0 denotes the value in ambient air.

Test results for the cardboard and plywood specimens are shown in Table 1. The combustion behavior under a certain incident heat flux was measured three times under each heat flux; the averages are summarized in Table 1. Only the corrugated cardboard was ignited under a heat flux of 10 kW/m2; in contrast, neither plywood was ignited. A higher incident heat flux caused specimens to ignite faster and show higher maximum HRR values. The effective heat of combustion for the cardboard and the plywood of two thicknesses were 8.01, 7.90, and 6.56 kJ/g, respectively. The cardboard gave the largest heat of combustion among the tested specimens in a fire.

Table 1. *Cone calorimeter results for materials of the model car*

Thickness (mm)	Cardboard			Plywood					
	7 (case A)			3 (case B)			5 (case C)		
Heat flux (kW/m2)	10	20	30	10	20	30	10	20	30
Max. HRR (kW/m2)	106	110	144	-	242	274	-	175	190
Time to ignition (s)	253	36	13	-	112	39	-	188	61
Effective heat of combustion (kJ/g)	8.01			7.90			6.56		

2.4. Ignition Source

The fire source was placed at 1 of the 3 positions shown in Figure 2. The first location (case 1) was between the left end of the subway car and the seat, and the second (case 2) was between the second seat on the left and the second door on the left. The third location (case 3) was at the center of the

subway car. The test conditions are summarized in Table 2. The fire source in this test was a square pool pan (0.04 m × 0.04 m) 0.02 m in height with n-heptane as the fuel. The HRR of the fire source used for the reduced-scale model was measured 6 times by using a cone calorimeter. The average HRR during 300 s was 1 kW, and the HRR of the full-scale model as calculated by eq. (1) was 350 kW.

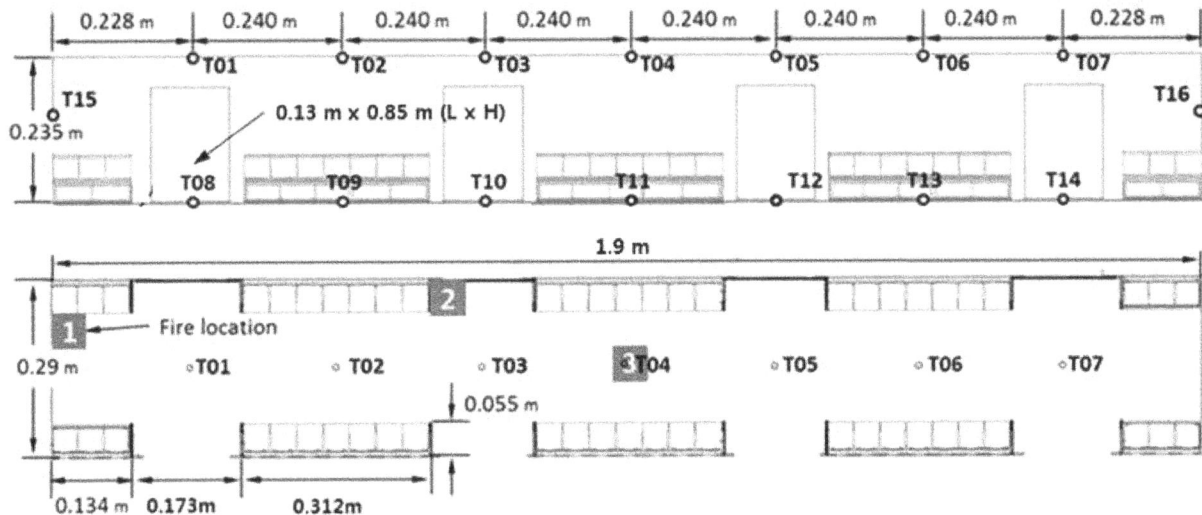

Figure 2. *Positions of ignition sources and thermocouples.*

Table 2. *Test conditions for various materials and fire source locations*

	Corrugated cardboard	Plywood (3 mm)	Plywood (5 mm)
at 1 in Fig. 2	case A1	case B1	case C1
at 2 in Fig. 2	case A2	case B2	case C2
at 3 in Fig. 2	case A3	case B3	case C3

Table 3. *Maximum HRR and its time of occurrence for all cases in the reduced-scale model*

Case	A			B			C		
	1	2	3	1	2	3*	1	2	3*
Peak HRR	71.43	83.78	82.37	49.82	51.32	9.06	70.81	64.30	12.43
Time to reach peak HRR	123	204	258	426	564	351	501	1023	351
Ratio of max. peak HRR for each case to peak HRR	0.85	1	0.98	0.97	1	0.19*	1	0.91	0.18

*Fire spread out only near the source

2.5. Measurements in Model

Figure 3. Hood of the fire test room and the model.

To measure the HRR of the reduced-scale model subway car on fire, the model was placed under the exhaust hood in the fire test room, in compliance with ISO 9705 [9]. Figure 3 shows the exhaust hood (with a 3-m square area) and the model placed under it. The combustion gas from the ignited model and the entrainment air, which are collected by the exhaust hood, are mixed along the duct. Therefore, the mixture takes time to reach the sensors in the duct. This time lag in measurement is made up for by comparing with results of the standard propane burner. With the heat of combustion for the plywoods and cardboard from the cone calorimeter tests, the HRR can be obtained by measuring the mass loss rate, which in turn is measured by an electronic scale (QA60FEG-S of Sartorius) with a resolution of 1 g/s. To investigate flame propagation, 16 k-type thermocouples were installed in the model (Figure 2) to measure the temperature in the fire tests.

3. Results of Model Test

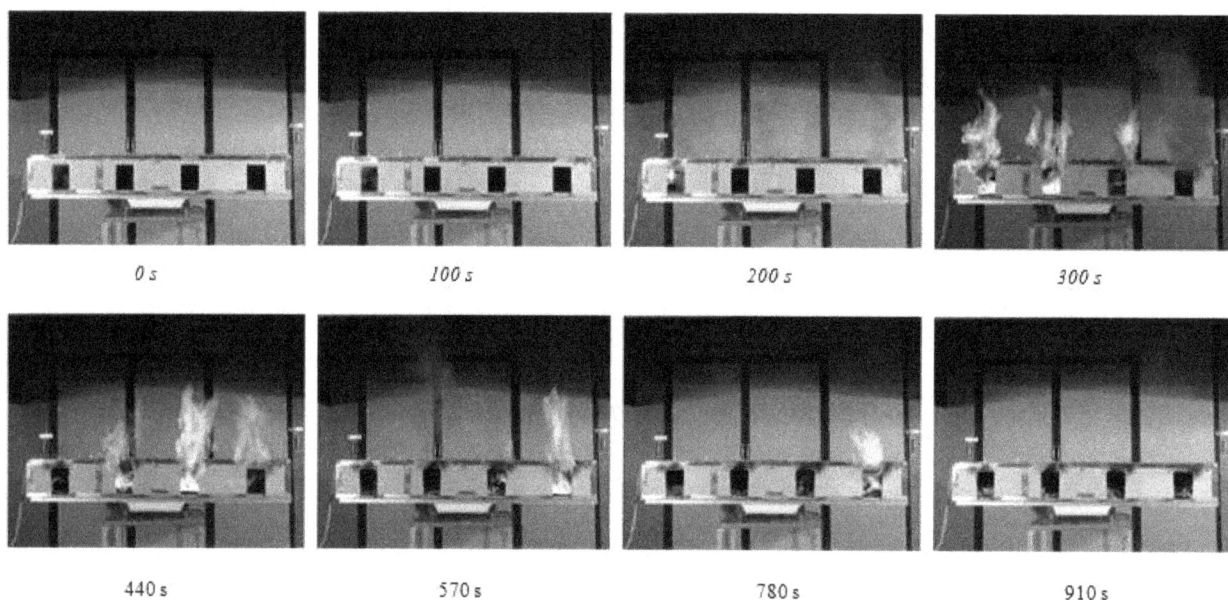

Figure 4. Fire development on the subway car for case B1.

Figure 4 presents photographs of the time course of fire propagation for case B1, in which 3-mm plywood was used as the interior material. Unless otherwise mentioned, time is represented in the reduced scale. The small seat near the fire source caught fire owing to radiative heat flux after about 100 s. The fire started growing after 120 s and grew rapidly after 280 s. The flames propagated beneath the ceiling of the model subway car, after which the floor was ignited because of the incident radiative heat flux. The difference in propagation time between the ceiling and floor at the same vertical position was about 70 to 90 s, where the times were determined from both video clips and measured temperatures. Flames and smoke were subsequently emitted through the first three doors and the last door, respectively, after 300 s. Table 3 shows the maximum HRR and the time to reach it for each case. For cases in which corrugated cardboard was used, the fire reached its peak after 123, 204, and 258 s for source locations 1 (case A1), 2 (case A2), and 3 (case A3), respectively. Interior materials burned more quickly near the fire source (case A1); hence, less time was required to reach the maximum HRR than in other cases (cases A2 and A3). Plywood materials showed the same location-dependent characteristics for the peak HRR. The fire spread fastest when the fire source was at the end wall. This demonstrates that the most vulnerable location during a subway car fire is at the end, because it is easily exposed to the fire source. For cases B3 and C3, in which a fire source of the same size was used, the fire burned only a portion of the interior materials around the fire source and died naturally. The fire did not grow; it was confined to the vicinity of the fire source because only the interior material of the ceiling was exposed directly to the source. Therefore, the nozzles for the water mist and sprinkler, which are installed to suppress a fire in a subway car, should ideally be installed toward the seats, sidewall, and end wall, rather than the center of the floor, to prevent flashover or fire propagation. Except for cases B3 and C3, wherein the fire did not spread to the entire area, the peak HRR did not vary greatly with the location of fire source (Table 3). For case A, the ratio of the lowest peak HRR (case A1) to the highest peak HRR (case A2) was relatively small at 0.85; the ratio was 0.97 for case B and 0.91 for case C. Except for cases wherein the fire did not spread to the entire area, the difference in peak HRR among all interior materials was only 15%. In short, the location of the fire in a subway car did not affect its size (HRR) greatly but considerably influenced the time to reach the peak HRR. Figures 5–7 show the HRRs for all cases. The difference in the time to reach the peak HRR increased according to the combustibility of interior materials (Table 3). Evaluation of combustibility by using the cone calorimeter revealed that among the interior materials, corrugated cardboard (case A) had the lowest combustibility, followed successively by 3-mm plywood (case B) and 5-mm Plywood (case C). Similar to the result of the specimen test, in the reduced-scale model, the time to reach the peak HRR was shortest in case A1, followed successively by cases B1 and C1. In the reduced-scale model test, the peak HRR was highest in case A, followed successively by cases C and B. This ordering was the reverse of that for the maximum HRR measured in the specimen test; presumably, case A had the highest HRR in the reduced-scale model test because of rapid spreading of the flame. Figure 8 shows the average temperature in the reduced-scale model. The filled gray plots represent the average temperature of the thermocouples on the ceiling, and the open black plots show the average temperature of all thermocouples installed in the model. Case C (5-mm plywood) had a higher peak HRR than case B (3-mm plywood) because many interior materials burned simultaneously owing to the greater increase in temperature (Figure 8).

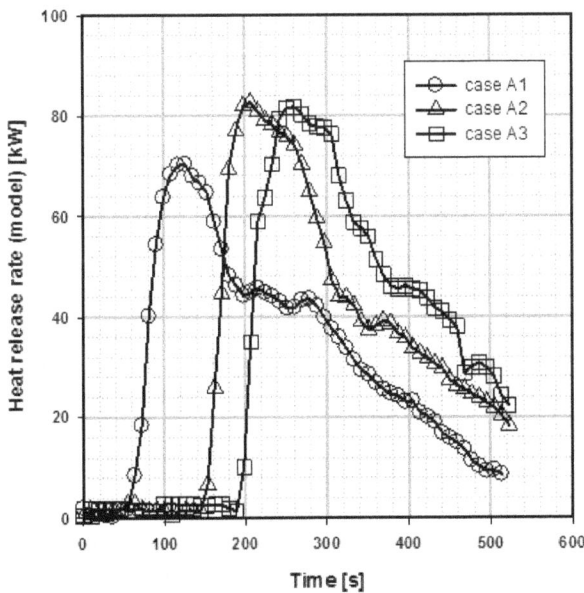

Figure 6. Heat release rates for case B.

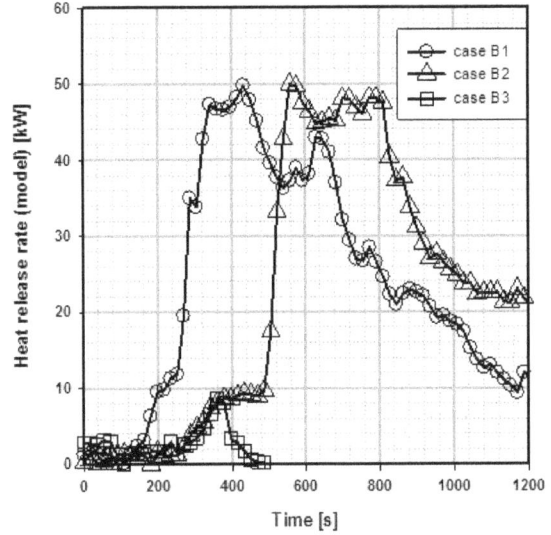

Figure 7. Heat release rates for case C.

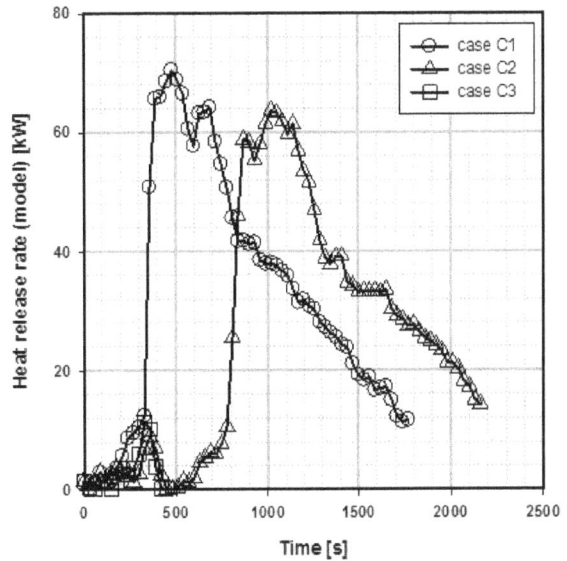

Figure 5. Heat release rates for case A.

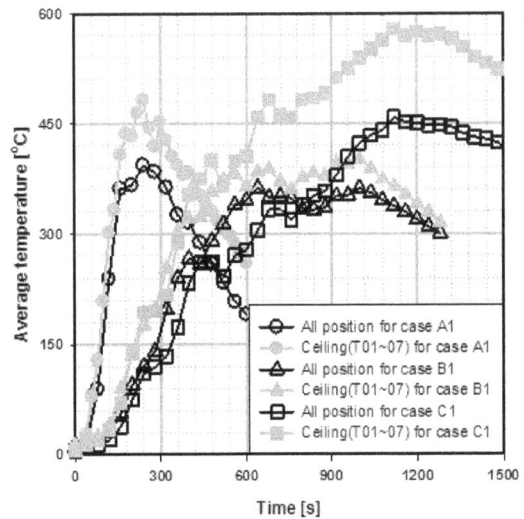

Figure 8. Average temperatures of all positions and ceiling.

4. Comparison with Full-Scale Test

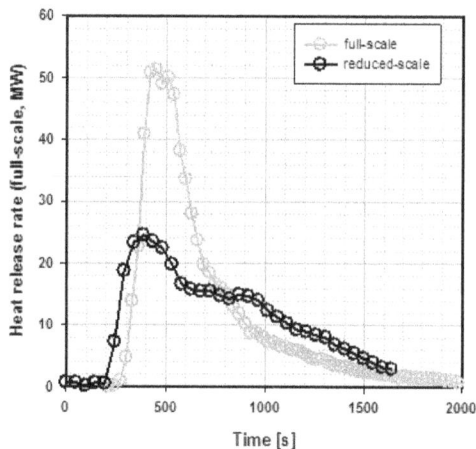

Figure 9. Heat release rate in the full- and reduced-scale tests.

In the full-scale test [5], the fire source was installed in the corner, in the same place as in cases A1, B1, and C1 of this study. In each reduced-scale test, the time to reach the peak HRR converted using eq. (2) was 396, 1372, and 1614 s, respectively. The time to reach the peak HRR in the full-scale test was 430 s, similar to the result of case A1. Hence, we selected case A1 as the reference. The HRRs of the reduced-scale and full-scale subway cars are compared in Figure 9. The peak HRR of the reduced-scale model was 24.7 MW, or 50% that of the full-scale model (52.5 MW). Although there are many reasons for the discrepancy between the reduced- and real-scale tests, the amount of interior materials used in the reduced-scale model is the likely reason for this large difference. The total fire load of the trailer car used in the full-scale test was 21.4 GJ [5]. The effective heat of combustion of a subway car with similar materials was 19.1 MJ/kg [11], and that of the reduced-scale model measured in the specimen test was 8.01 MJ/kg. The total fire load of the real-scale subway car, converted to the reduced-scale by using eq. (3), was 45.5 MJ. This value was much larger than that of the actually used reduced-scale model (17.0 MJ). Therefore, the peak HRR of the reduced-scale model was measured to be smaller than that of the real-scale subway car. Measuring the peak HRR more accurately will require consideration of the similarity of the fire load when installing combustibles in the reduced-scale model fire test.

5. Conclusions

In sum, we used a 1/10-scale model of an actual subway car in a fire test and obtained the following results.

For a subway car, the peak HRR did not significantly differ according to the location of the fire, but the time of fire growth varied greatly.

The HRR changed rapidly in regions where the mass loss rate of the interior materials in the reduced-scale model trailer car changed rapidly. Measuring a more accurate peak HRR will require installing interior materials of the reduced-scale

model by considering the similarity of the fire load.

The most vulnerable location during the fire was the end of the subway car, followed successively by the sidewall close to the seats and the center of the car.

When a subway car with many doors stops in an emergency at a nearby platform or tunnel, the open area becomes much wider than that for normal railcars. This is because the doors on one or both sides are open for passenger evacuation, and hence air (oxygen) is sufficient (the railcar is well ventilated). Therefore, the fire size of this type of railcar (having many open doors or windows) can be determined by the amount of interior materials (fire load) and their combustibility.

Acknowledgement

The author gratefully acknowledge the financial support from Disaster Safety Technology Development & Infrastructure Construction Research Group and National Emergency Management Agency. (NEMA-Infra -2013– 103).

References

[1]   Peacock R, Braun E, Fire tests of amtrak passenger rail vehicle interiors, National Bureau of Standards Technical Note 1193, 1984

[2]   Fires in Transport Tunnels: Report on Full-Scale Tests. Edited by Studiensgesellschaft Stahlanwendung e. V., EUREKA-ProjectEU499: FIRETUN, Du¨ sseldorf, Germany, 1995

[3]   Peacock R, Bukowski R, Jones W, Reneke P, Babrauskas V, Brown J, Fire safety of passenger Trains, National Bureau of Standards Technical Note 1406, 1994

[4]   Ingason H, Model scale railcar fire tests, Fire Safety J, 2007, 42(4): 271-282

[5]   George H, Lee D -H and Park W -H, Full-scale experiments for heat release rate measurements of railcar fires, The 5th International Symposium on Tunnel Safety and Security, New York, USA, 14-16th March, 2012

[6]   Lee D., Park W., Jung W, Yang S., Kim H., Hadjishophocleous G, Hwang J., Estimations of heat release rate curve in case of railcar fire, J Mech Sci Technol, 2013, 27(6):1665-1670

[7]   Quintiere J G, Scaling Application in Fire Research, Fire Safety J 15(1): 3-29, 1989

[8]   ISO 5660 - 1, Reaction to fire tests - Heat release, smoke production and mass loss rate - Part 1 : Heat release rate(cone calorimeter method, 2002

[9]   ISO 9705, Fire tests-Full-scale room test for surface products, 1996.

[10] Huggett C, Estimation of rate of heat release by means of oxygen consumption measurements, Fire Mater, 1980, 4(2): 61-65

[11] Korea Land, Transport and Maritime Affairs, Technology development of safety evolution and accident prevention for railway fire, 2010.

17

The elastic scattering of an electron from the target BY absorbing a photon via free- free scattering theory

Kishori Yadav, Jeevan Jyoti Nakarmi[*]

Central Department of Physics, T. U., Kirtipur, Nepal

Email address:

nakarmijj@gmail.com (J. J. Nakarmi)

Abstract: This paper intended to the elastic scattering of an electron from the target by absorbing a photon from the laser field has been studied for the polarized potential. Since the solution of the Schrödinger equation of whole three-body system has not been found, we consider such intensities of electromagnetic field (Laser field) that the electron field coupling is the dominant process and the target is transparent to the field such that photon- target coupling can be ignored. Therefore the internal structure of target can be ignored and represented just as a scattering potential. For number of photon, l=-1 i.e, absorption of a photon (inverse Bremsstrahlung), we have concluded that the differential scattering cross section of an electron depends upon the fourth power of the wavelength (λ^4) and the intensity of the Laser field. From this work we see that at certain values of laser parameters the differential scattering cross section of scattered electron decreases with increase in scattering angle and attains a minimum value of 0.1 *barn* and further increase in scattering angle also increases in differential scattering cross section and attains a maximum value of 0.3 *barn*.

Keywords: Scattering, Bremsstrahlung, Volkov Wave Function, Polarized Potential

1. Introduction

Electron atom interaction in the presence of a laser field attracted considerable theoretical attention in the recent year not only because of the importance in applied areas (such as plasma heating or laser driven fusion), but also in view of their interest in fundamental atomic theory. The problem of this process, is in general, very complex, since in addition to the difficulties associated with the treatment of electron atom collision, the presence of the laser introduces new parameters (for example, the laser photon energy $\hbar\omega$ and intensity I) which may influence the collision. Moreover, the laser photon can play the role of a "third body" during the collision, and "dressed" the atomic states. It is therefore of interest to begin the theoretical analysis by considering the simpler problem of the scattering of an electron by a potential in the presence of a laser field. A fully realistic description of the target atom is quite difficult. We shall represent it here by a potential model.

Mason and Newell (1982) reported experimental evidence of simultaneous electron-photon excitation of atoms. However most experimental studies have been performed with noble gases (Wallbank et al. 2009), a recent on being with a Nd:YAG laser (Luan et al 2011). On the other hand, theoretical studies are not easy to perform with these atoms and the hydrogen atom has been studied extensively (Rahman ans Faisal 1978, Jetzek et al 1988, Bhattacharya et al 1993). Hydrogen as a one electron atom is a simple to deal with and it is often interesting information concerning the main features of the problem.

The free process can theoretically be studied at various levels. As the target does not change states in this process, its own energy spectrum can be ignored and a simple potential can mimic the electron atom interaction. The collision process can then be treated either classically or quantum mechanically by means of the simple scattering theories . Furthermore, the collision can be treated as occurred at such intensities of electromagnetic field that the electron-field coupling is the dominant process and the target is transparent to the field such that photon-target coupling can be ignored. If, however, the frequency of the photon is such as to couple two stationary states of the target, then the target-field interaction becomes extremely important. Here we discussed such intensity of the electromagnetic field where the photon-field interaction can be neglected.

We want to show the effect of various collision and laser parameters on the collision process. This is the motive of the paper. Furthermore, most theoretical studies for scattering of electron by atoms in an intense radiation field are based on perturbation theory (Gersten and Mittleman 1976; Byron and Joachain 1984; Garvila et al 1990) starting with the well known Kroll and Waston (1976) work on the soft photon approximation, there exist only a few non-perturbative approaches for this problem. Shakeshaft (1983) formulated a non- perturbative method of coupled integral equations for calculating the scattering cross section by assuming the potential to be separable. Rosenberg (2000) applied the variation method for coulomb scattering in a laser field using a low frequency approximation.

The reaction studied in the present work is,

$$e_{k_i}^- + H(i) + N(\omega, \hat{e}) \rightarrow e_{k_f}^- + H(j) + (N \pm l)\gamma(\omega, \hat{e})$$

Representing the collision of an incoming electron with momentum k_i with hydrogen atom initially in the state i in the presence of a single mode laser beam moving to the excited state j with exchange of l photons between the electron and the laser field. We have used the Born approximation, to treat the electron- atom interaction as it is simple enough to allow calculation of a larger number of transitions, and it becomes exact at high energies.

2. Materials and Methods

We consider a collision between an electron and hydrogen atom in the presence of a laser field. We begin this section by considering the simple case in which the target atom is modeled by a center of force and hence does not interact with the laser field. The field is assumed to be purely monochromatic with angular frequencyω, linearly polarized with linear polarization vector$\hat{e}$. We also assume that the dipole approximation is valid. The Hamiltonian of the electron-atom system in the presence of a laser beam can be written as,

$$H = H_f + H_t + V(r_o, r_a) \qquad (1)$$

Where

$H_f \rightarrow$Hamiltonian of the free electron

$H_t \rightarrow$Hamiltonian of the target atom

And $V(r_o, r_a)$ is the interaction between the incident electron and atom, defined as,

$$V(r_o, r_a) = \frac{-Z}{|r_o|} + \sum_{j=1}^{Z} \frac{1}{|r_{oj}|} \qquad (2)$$

Where,

$r_o \rightarrow$ Position coordinate of the projectile electron

$Z \rightarrow$ Atomic number of target or charge of the target atom

$r_a \rightarrow$ Position of an atomic electron

For H-atom, Z=1 (i.e, one electron) equation (2) becomes

$$V(r_o, r_a) = \frac{-Z}{|r_o|} + \frac{1}{|r_o - r_1|}$$

In equation (2) Z is the charge of the target atom. Working in the coulomb gauge we have the electric field

$$\vec{E}(t) = \vec{E}_o \sin \omega t$$

And the corresponding vector potential is

$$\vec{A}(t) = \vec{A}_o \cos \omega t \quad \text{with} \quad \vec{A}_o = \frac{c\vec{E}_o}{\omega}$$

In the presence of a laser field the incidence electron of momentum k is represented as

$$\chi(\vec{r_o}, t) = (2\pi)^{\frac{-3}{2}} exp \left[(i\vec{k}.\vec{r} - i\vec{k}.\vec{\alpha}_o \sin \omega t) - \frac{iE_k t}{\hbar} \right] \qquad (3)$$

Where,

$$E_k = \frac{\hbar^2 k^2}{2m}$$

$$\vec{\alpha}_o = \frac{e\vec{A}_o}{mc\omega}$$

Also, $\vec{\alpha}_o = \frac{e\vec{E}_o}{m\omega^2}$ is the measure of the coupling between the field and the projectile.

Equation (3) is also known as volkov wave function. This equation gives the states of a free electron in a laser field. Since there is no conservation of energy in free state of an electron. So the states repersented by the volkov wave functiion is also called as virtual state, to make the energy conservation there must be another parameter. Here in our case we take the static potential as the third parameter for energy conservation of electron.

The S-matrix (scattering matrix) element for the transition from initial state (i) to final state (f) $(i.e\ i \rightarrow f)$ is given by, [ref]

$$S = \frac{-i}{\hbar} < \chi_f V \psi_i^+ >$$

The angle bracket denotes both space and time integration. Where,

χ_f is the volkov wave function for the final state of an electron

ψ_i^+ is the exact formal solution of the following schrödinger wave equation,

$$\frac{1}{2m} \left(\vec{P} + \frac{e\vec{A}(t)}{c} \right)^2 \psi(\vec{r}, t) + V(\vec{r})\psi(\vec{r}, t) = i\hbar \frac{\partial \psi(\vec{r}, t)}{\partial t}$$

where $\psi(\vec{r}, t)$ is the time dependent wave function in presence of the scattering potential $V(\vec{r})$ in the presence of laser field.

the solution this equation (4.65) can be written as [ref],

$$\psi^+(\vec{r}, t) = \psi_i(\vec{r}, t) + \int_{-\infty}^{\infty} G_+(t, t')V\psi^+(\vec{r}, t')dt'$$

Now taking electron atom interaction upto the first Born approximation, we get scattering matrix (S-matrix) as,

$$S = \frac{-i}{\hbar} < \chi_f |V| \chi_i > \qquad (4)$$

χ_i is the Volkov wave function for the initial state of an electron

And V is the interaction potential defined by equation (2).

Initially the electron is free from potential and take part in interaction for certain time and after that it scattered from the potential to the Free State. So this interaction is also known as free- free scattering process.

Now we can write equation (4) as,

$$S_{k_f k_i} = \frac{-i}{\hbar} \iint_{-\infty}^{t} (2\pi)^{\frac{-3}{2}} exp\left[(i\vec{k_f}.\vec{r} - i\vec{k_f}.\vec{\alpha_o} \sin \omega t) - \frac{iE_{k_f} t}{\hbar} \right] V(\vec{r}) (2\pi)^{\frac{-3}{2}} exp\left[(i\vec{k_i}.\vec{r} - i\vec{k_i}.\vec{\alpha_o} \sin \omega t) - \frac{iE_i t}{\hbar} \right] d^3 r dt$$

Where,

$k_i \rightarrow$ Initial wave vector of projectile particle

$k_f \rightarrow$ Final wave vector of the scattered particle

$$S_{k_f k_i} = \frac{-i}{\hbar} (2\pi)^{-3} \iint_{-\infty}^{t} e^{-i\Delta.\vec{r}} e^{i\Delta.\vec{\alpha_o} \sin \omega t} e^{i(E_{k_f} - E_{k_i})\frac{t}{\hbar}} V(\vec{r}) d^3 r dt$$

Here, $\Delta \rightarrow (\vec{k_f} - \vec{k_i})$

(momentum transfer during scattering process)

$$S_{k_f k_i} = \frac{-i}{\hbar} \hat{V}(\Delta) \int_{-\infty}^{t} e^{i\Delta.\vec{\alpha_o} \sin \omega t} e^{i(E_{k_f} - E_{k_i})\frac{t}{\hbar}} dt$$

where,

$$\hat{V}(\Delta) = (2\pi)^{-3} \int e^{-i\Delta.\vec{r}} V(\vec{r}) d^3 r$$

This is the Fourier transformation of the potential in the momentum space.

Since $V(\vec{r})$ is independent of time (t), we can seperate time and space integration so that $\hat{V}(\Delta)$ can be taken outside of time integration.

Using the generating function of the Bessel Polynomial We get,

$$e^{i\Delta.\alpha_o \sin \omega t} = \sum_{l=-\infty}^{\infty} J_l(\Delta.\alpha_o) e^{il\omega t}$$

So the S-matrix element becomes

$$S_{k_f k_i} = \frac{-i}{\hbar} \hat{V}(\Delta) \int_{-\infty}^{t} \sum_{l=-\infty}^{\infty} J_l(\Delta.\alpha_o) e^{il\omega t} e^{i(E_{k_f} - E_{k_i})\frac{t}{\hbar}} dt$$

where l is the no of photons exchange during the scattering process. It may take the values,

$$l = 0, \pm 1, \pm 2, \pm 3, \dots \dots \dots \dots \dots$$

Where, positive l describes the photon emission stimulated

Bremsstrahlung and negative l describes the photon absorption or inverse Bremsstrahlung and $l = 0$ corresponds to the pure elastic scattering in the presence of the laser field.

$$S_{k_f k_i} = \frac{-i}{\hbar} T^l_{k_f k_i} \int_{-\infty}^{t} e^{i(E_{k_f} - E_{k_i} + l\hbar\omega)\frac{t}{\hbar}} dt \qquad (5)$$

where,

$$T^l_{k_f k_i} = \hat{V}(\Delta) \sum_{l} J_l(\Delta.\alpha_o)$$

Using the idea of integral form of Dirac delta function

$$(\delta(x - x') = \frac{1}{2\pi} \int_{-\infty}^{\infty} e^{ik(x-x')} dk)$$

We reduce equation (5) in the following form,

$$S_{k_f k_i} = -2\pi i T^l_{k_f k_i} \delta(E_{k_f} - E_{k_i} + l\hbar\omega) \qquad (6)$$

The delta function in above equation ensures energy conservation,

$$i.e, \qquad E_{k_f} = E_{k_i} - l\hbar\omega$$

$$\frac{k_f}{k_i} = \left(1 - \frac{l\hbar\omega}{E_{k_i}} \right)^{\frac{1}{2}}$$

Here $\hbar\omega$ is the photon energy and E_{k_i} is the kinetic energy of a projectile electron.

Now the differential scattering cross section in terms of transiton matrix is given by,

$$\frac{d\sigma^{B1}}{d\Omega} = \frac{m^2}{(2\pi)^2 \hbar^4} \frac{k_f}{k_i} \left| T^l_{k_f k_i} \right|^2$$

This is the required relation of differential cross section of electron with transfer of l photon.

where ,

$$\left| T^l_{k_f k_i} \right|^2 = \sum_{l} J_l^2(\Delta.\alpha_o) |\hat{V}(\Delta)|^2$$

$$\hat{V}(\Delta) = (2\pi)^{-3} \int e^{-i\Delta.\vec{r}} V(\vec{r}) d^3 r$$

For spherically symmetric potential,

$$V(\vec{r}) = V(r)$$

So that,

$$\hat{V}(\Delta) = (2\pi)^{-3} \int_{0}^{\infty} V(r) r^2 dr \int_{0}^{\pi} e^{-i\Delta r \cos \theta} \sin \theta d\theta \int_{0}^{2\pi} d\phi$$

This gives,

$$|\hat{V}(\Delta)|^2 = \frac{e^4}{4\pi^4 \Delta^4}$$

To calculate $\sum_l J_l(\Delta.\alpha_o)$ we have the expression for the Bessel function which is, [ref]

$$J_n(x) = \frac{x^n}{2^n \Gamma(n+1)} \left[1 - \frac{x^2}{2.2(n+1)} + \frac{x^4}{2.4.2(n+1)(n+2)} - \cdots \right]$$

Here we take

$$n = l \text{ and } x = \Delta.\alpha_o \text{ then}$$

$$J_l(x) = \frac{(\Delta.\alpha_o)^l}{2^l \Gamma(l+1)} \left[1 - \frac{(\Delta.\alpha_o)^2}{2.2(l+1)} + \frac{(\Delta.\alpha_o)^4}{2.4.2(l+1)(l+2)} - \cdots \right]$$

For low intensity and low frequency we get,

$$[J_l(\Delta.\alpha_o)]^2 = \frac{\Delta^2 {\alpha_o}^2}{4}$$

Thus we get square modulus of the transition matrix element as,

$$\left| T_{k_f k_i}^l \right|^2 = \frac{\Delta^2 {\alpha_o}^2}{4} \frac{e^4}{4\pi^4 \Delta^4}$$

Substituting these values we get the value of the differential cross section as,

$$\frac{d\sigma^{B1}}{d\Omega} = \frac{e^6}{(2\pi)^6 \hbar^4} \frac{k_f}{k_i} \frac{{E_o}^2}{\omega^4 (k_f - k_i)^2}$$

$$\frac{d\sigma^{B1}}{d\Omega} = \frac{e^6}{(2\pi)^6 \hbar^4} \frac{k_f}{k_i} \frac{{E_o}^2}{\omega^4 (\Delta)^2} \tag{7}$$

We have,

$$\Delta = \overrightarrow{k_f} - \overrightarrow{k_i}$$

$$\Delta^2 = k_f^2 + k_i^2 - 2\overrightarrow{k_f}\overrightarrow{k_i} \cos\theta$$

Where, θ is the angle between k_f and k_i, that means θ is the scattering angle between initial and final momentum of the projectile particle.

Thus we have,

$$\Delta^2 = k_i^2 \left[\left(1 - \frac{l\hbar\omega}{E_{k_i}}\right) - 2 \left(1 - \frac{l\hbar\omega}{E_{k_i}}\right)^{\frac{1}{2}} \cos\theta + 1 \right]$$

Substituting this value in equation (7) we get,

$$\frac{d\sigma^{B1}}{d\Omega} = \frac{e^6}{(2\pi)^6 \hbar^4} \frac{k_f}{k_i} \frac{{E_o}^2}{\omega^4 k_i^2 \left[\left(1 - \frac{\hbar\omega}{E_{k_i}}\right) - 2\left(1 - \frac{\hbar\omega}{E_{k_i}}\right)^{\frac{1}{2}} \cos\theta + 1 \right]}$$

$$\frac{d\sigma^{B1}}{d\Omega}$$
$$= \frac{e^6}{\hbar^2 (2\pi)^6} \left(1 - \frac{\hbar\omega}{E_{k_i}}\right)^{\frac{1}{2}} \frac{{E_o}^2}{\omega^4 2m E_{k_i} \left[\left(1 - \frac{\hbar\omega}{E_{k_i}}\right) - 2\left(1 - \frac{\hbar\omega}{E_{k_i}}\right)^{\frac{1}{2}} \cos\theta + 1 \right]}$$

$$\frac{d\sigma^{B1}}{d\Omega}$$
$$= \frac{e^6}{(2)^8 m\hbar^2 (\pi)^{10} c^5} \left(1 - \frac{l\hbar\omega}{E_{k_i}}\right)^{\frac{1}{2}} \frac{\lambda^4 I_o}{E_{k_i} \left[\left(1 - \frac{l\hbar\omega}{E_{k_i}}\right) - 2\left(1 - \frac{l\hbar\omega}{E_{k_i}}\right)^{\frac{1}{2}} \cos\theta + 1 \right]}$$

We have the relation

$$\omega = \frac{2\pi c}{\lambda} \text{ and } E_o^2 = \frac{8\pi I_o}{c}$$

where,

I_o is the intensity of the laser field.
λ is the wave length of the radiation.
c is the velocity of light.

$$\frac{d\sigma^{B1}}{d\Omega} = C \left(1 - \frac{l\hbar\omega}{E_{k_i}}\right)^{\frac{1}{2}} \frac{\lambda^4 I_o}{E_{k_i} \left[\left(1 - \frac{l\hbar\omega}{E_{k_i}}\right) - 2\left(1 - \frac{l\hbar\omega}{E_{k_i}}\right)^{\frac{1}{2}} \cos\theta + 1 \right]}$$

Where, $C = \frac{e^6}{(2)^8 m\hbar^2 (\pi)^{10} c^5}$ is the constant quantity.

Now, for $l = 1$ i.e, one photon emission (stimulated emissioin)

$$\frac{d\sigma^{B1}}{d\Omega} = C \left(1 - \frac{\hbar\omega}{E_{k_i}}\right)^{\frac{1}{2}} \frac{\lambda^4 I_o}{E_{k_i} \left[\left(1 - \frac{\hbar\omega}{E_{k_i}}\right) - 2\left(1 - \frac{\hbar\omega}{E_{k_i}}\right)^{\frac{1}{2}} \cos\theta + 1 \right]} \tag{8}$$

This is the differential cross section for the scattering of an electron with one photon emission (stimulated Bremsstrahlung) in the presence of potential .

Also for $l = -1$ i.e, one photon absorption (inverse Bremsstrahlung),

$$\frac{d\sigma^{B1}}{d\Omega} = C \left(1 + \frac{\hbar\omega}{E_{k_i}}\right)^{\frac{1}{2}} \frac{\lambda^4 I_o}{E_{k_i} \left[\left(1 + \frac{\hbar\omega}{E_{k_i}}\right) - 2\left(1 + \frac{\hbar\omega}{E_{k_i}}\right)^{\frac{1}{2}} \cos\theta + 1 \right]} \tag{9}$$

This is the differential cross section for the scattering of an electron with one photon absorption (inverse Bremsstrahlung) in the presence of potential.

3. Result and Discussion

In the present work, we have studied the elastic scattering of an electron-atom interaction by absorbing a photon from the laser field. For simplicity, we have neglected the dressing effect of an atom i.e., we have neglected the field- atom interaction so that we can neglect the internal structure of an atom because we choose the intensity of the field in such a way that the target atom is transparent for the field.

Here, the differential cross-section equations (8) and (9) is found to be proportional to λ^4 (where λ is the wavelength of the radiation) which explains why it has been easier to observe laser –assisted cross section using infrared laser (high wave length and low frequency) than those operating in the visible or ultraviolet spectral regions.

Equation (9) is plotted as the functions of scattering angle, kinetic energy of an electron, wavelength and intensity of the laser field as shown below:

1: Variation of differential scattering cross section with kinetic energy of the incident electron

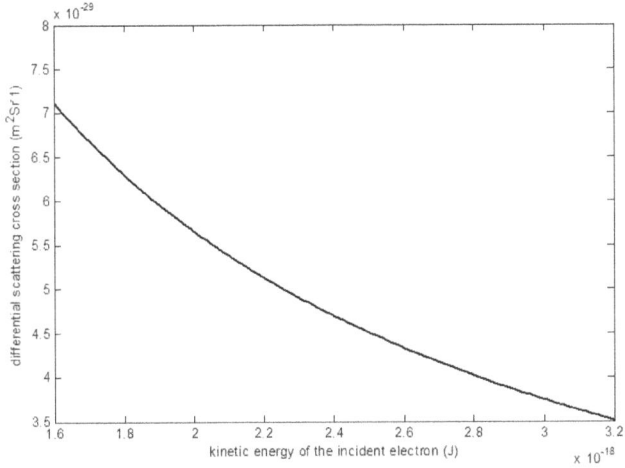

Fig. 1. *Variation of differential scattering cross section with kinetic energy of the incident electron*

From the plot, we see that the differential scattering cross section decrease as the kinetic energy of the incident electron increases. Going on increasing the kinetic energy of an incident electron we will get zero value of scattering cross section.

2: Variation of differential cross section with the wave length of the laser field

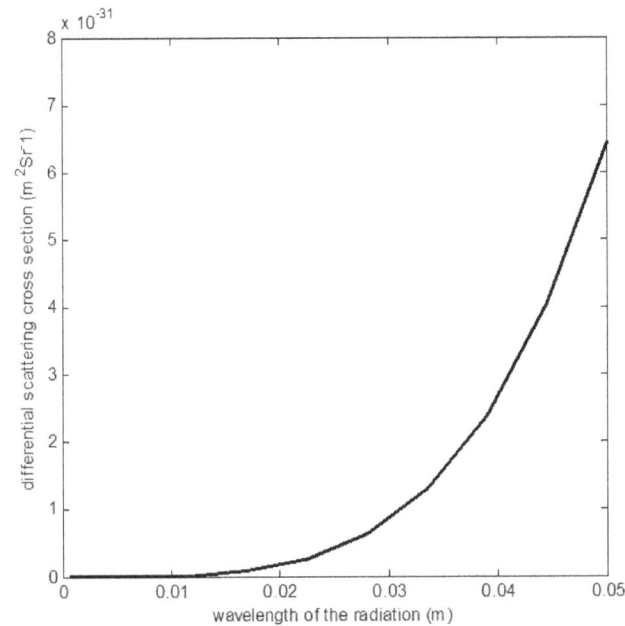

Fig. 2. *Variation of differential cross section with wavelength of the field*

From the plot, it is found that the differential scattering cross section increases with increase in wavelength of the laser field. It shows that the differential scattering cross section is zero below the wavelength 0.01m.

3: Variation of differential cross section with intensity of laser field

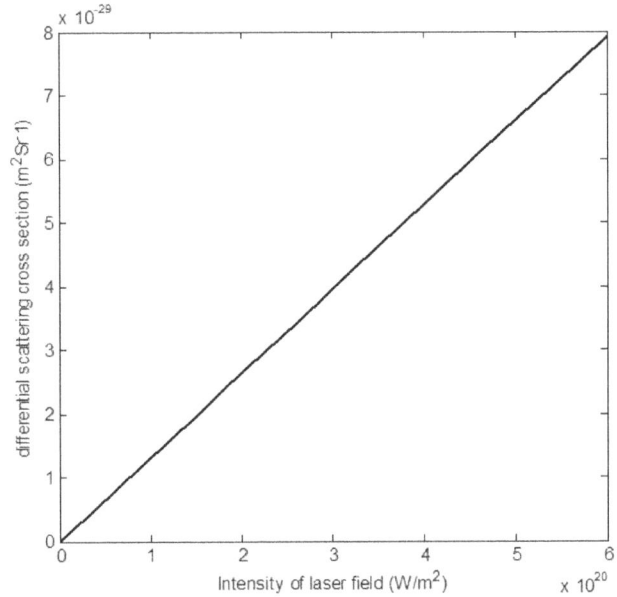

Fig. 3. *Variation of differential cross section with intensity of the laser field*

From the figure, it is clear that differential scattering cross section increase with increase in the intensity of the laser field. It shows that the differential scattering cross section varies linearly with intensity.

4: Variation of differential cross section with scattering angle

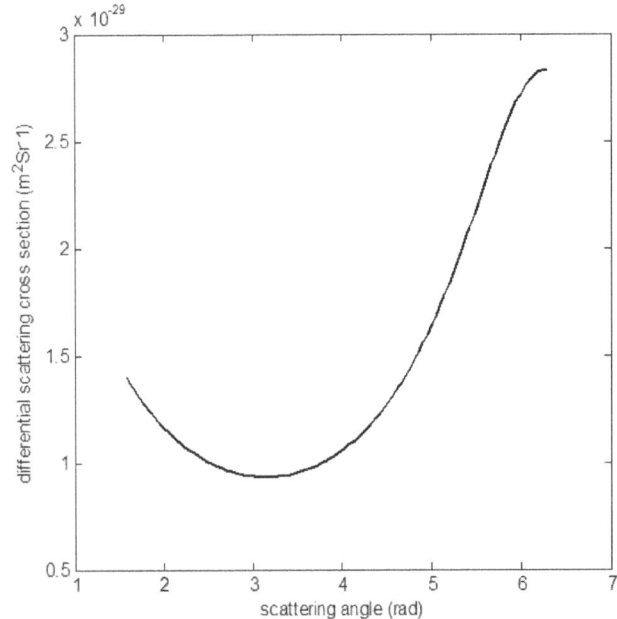

Fig. 4. *Variation of differential cross section with scattering angle*

From the plot, it is clear that differential cross section decreases as the scattering angle increase and attains the minimum value $1 \times 10^{-29} m^2$ and further increase in scattering angle also increase in differential scattering cross section which is clearly shown in polar plot.

5: Polar plot of differential cross section with scattering angle

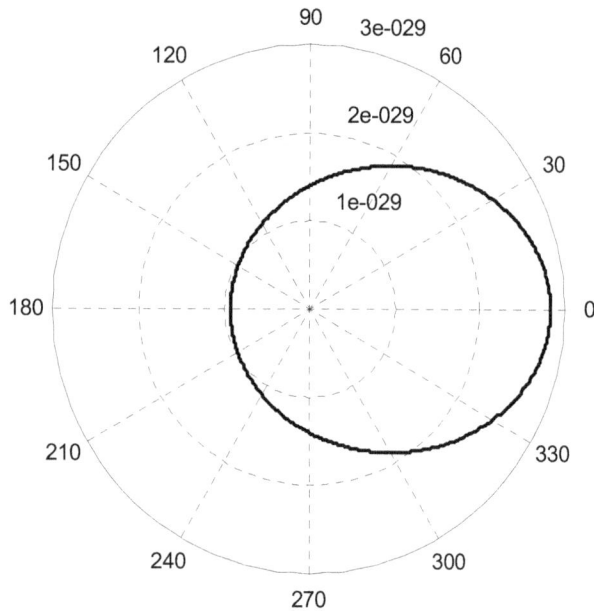

Fig. 5. *Polar plot of differential cross section with scattering angle*

From the polar plot, it is clear that the differential scattering cross section of an electron decreases with increase in scattering angle and attains minimum value and further increase in scattering angle also increase in cross section and attains maximum value $3 \times 10^{-29} m^2$.

It is generally observed that when electrons are scattered from the atom in the presence of a laser field, a new effect is observed which are not accessible in ordinary electron –atom scattering. This collision have the basic peculiarity of being processes in which three subsystem are present (i) the electron (ii) the target atom (iii) the radiation field. The last one provided energy and momentum and is characterized by the polarization of its electric field, which introduces in this collision process a new physical axis.

In this work, we have studied the scattering of an electron from the target atom by absorption of a photon from the laser field. From this study we concluded that the differential scattering cross section of an electron depends upon the intensity and wavelength of the laser field where as in ordinary electron -atom scattering, cross section only depends upon the scattering angle. The differential scattering cross section increases with increase in wavelength and intensity of the laser field. For a fixed value of a laser parameters and kinetic energy of an incident electron, the differential scattering cross section of an scattered electron decreases with increase in scattering angle and attains a minimum value of $1 \times 10^{-29} m^2 = 0.1 barn$ and further increase in scattering angle also increase in differential scattering cross section and attains a maximum value of $3 \times 10^{-29} m^2 = 0.3 barn$. Also from this study we see that, the differential scattering cross section for the electric field perpendicular to the direction of momentum transfer is zero. So from this study we concluded that the differential scattering cross section is also greatly depends upon the polarization of the laser field.

4. For Polarized Potential

If we choose $V(r)$ as polarized potential
i.e, $V(r) = -\frac{\alpha_p}{2(r^2+d^2)^2}$
Where,

$$d^4 = \frac{\alpha_p}{2z^{\frac{1}{3}}}$$

$\alpha_p \rightarrow$ dipole polarizability
$Z \rightarrow$ atomic number
Now,
Fourier Transformation of polarized potential,

$$\hat{V}(\Delta) = -\frac{1}{2\pi^2} \int_0^\infty \frac{\sin(\Delta r)}{\Delta} V(r) dr$$

$$\hat{V}(\Delta) = \frac{1}{2\pi^2} \int_0^\infty \frac{\sin(\Delta r)}{\Delta} \frac{\alpha_p}{2(r^2+d^2)^2} dr$$

$$= \frac{\alpha_p}{8\pi^2 i\Delta} \frac{\pi i\Delta}{2d} e^{(-\Delta d)} \quad [20]$$

$$\hat{V}(\Delta) = \frac{\alpha_p}{16\pi d} e^{(-\Delta d)}$$

So that,

$$|\hat{V}(\Delta)|^2 = |\frac{\alpha_p}{16\pi d} e^{(-\Delta d)}|^2$$

$$= \frac{\alpha_p^2}{256\pi^2 d^2} e^{(-2\Delta d)}$$

$$= \frac{\alpha_p^2}{256\pi^2 d^2}\left[1 - \frac{2\Delta d}{1!} + \frac{(2\Delta d)^2}{2!} - \cdots\right]$$

$$|\hat{V}(\Delta)|^2 = \frac{\alpha_p^2}{256\pi^2 d^2}\left[1 - 2\Delta d + \frac{(2\Delta d)^2}{2!}\right]$$

Higher order terms can be neglected for small momentum transfer

$$|\hat{V}(\Delta)|^2 = \frac{\alpha_p^2}{256\pi^2 d^2} - \frac{2\alpha_p^2\Delta}{256\pi^2 d} + \frac{\alpha_p^2}{256\pi^2 d^2}\frac{(2\Delta d)^2}{2}$$

To calculate $\sum_l J_l(\Delta. \alpha_o)$ we have the expression for the Bessel function which is, [15]

$$J_n(x) = \frac{x^n}{2^n\Gamma(n+1)}\left[1 - \frac{x^2}{2.2(n+1)} + \frac{x^4}{2.4.2(n+1)(n+2)} - \cdots\right]$$

Here we take

$$n = l \text{ and } x = \Delta. \alpha_o \text{ then}$$

$$J_l(x) = \frac{(\Delta.\alpha_o)^l}{2^l\Gamma(l+1)}\left[1 - \frac{(\Delta.\alpha_o)^2}{2.2(l+1)} + \frac{(\Delta.\alpha_o)^4}{2.4.2(l+1)(l+2)} - \cdots\right] \quad (**)$$

for $l = 1$ i.e, for one photon emission (stimulated Bremsstrahlung) then we get

$$J_l(\Delta.\alpha_o) = \frac{(\Delta.\alpha_o)}{2\Gamma(2)}\left[1 - \frac{(\Delta.\alpha_o)^2}{2.2.2} + \frac{(\Delta.\alpha_o)^4}{2.4.2.2.3} - \cdots\right]$$

$$J_l(\Delta.\alpha_o) = \frac{(\Delta.\alpha_o)}{2} - \frac{(\Delta.\alpha_o)^2}{2.2.2.2} + \frac{(\Delta.\alpha_o)^4}{2.4.2.2.3.2} - \dots \dots \quad (10)$$

If we take ξ as the angle between electric field and momentum transfer i.e,

$$\Delta.\alpha_o = \Delta\alpha_o \cos\xi$$

and for high frequency and low intensity the higher order terms of equation (4.97) can be neglected because $\alpha_o = \frac{eE_o}{m\omega^2}$.

Then we get,

$$J_l(\Delta.\alpha_o) = \frac{\Delta\alpha_o \cos\xi}{2}$$

If the direction between electric field and momentum transfer is parallel (i.e, $\xi = 0$) then $\cos\xi = \cos 0 = 1$

Thus , $J_l(\Delta.\alpha_o) = \frac{\Delta\alpha_o}{2}$ and also,

$$[J_l(\Delta.\alpha_o)]^2 = \frac{\Delta^2\alpha_o^2}{4} \quad (11)$$

If the direction between electric field and momentum transfer is perpendicular (i.e, $\xi = 90°$) then $\cos\xi = \cos 90° = 0$

Thus, $J_l(\Delta.\alpha_o) = 0$.

for $l = -1$ one photon absorption (inverse Bremsstrahlung) the Bessel function is $J_{-1}(\Delta.\alpha_o)$.

We have the relation,

$$J_{-l}(x) = (-1)^l J_l(x)$$

Using this relation, for $l = 1$ we get,

$$J_{-1}(\Delta.\alpha_o) = -J(x)$$

$$J_{-1}(\Delta.\alpha_o) = -\frac{\Delta\alpha_o}{2}$$

$$[J_{-1}(\Delta.\alpha_o)]^2 = \frac{\Delta^2\alpha_o^2}{4} \quad (12)$$

We have,

$$|T^l_{k_f k_i}|^2 = \sum_l J_l^2(\Delta.\alpha_o) |\hat{V}(\Delta)|^2$$

$$= \frac{\Delta^2\alpha_o^2}{4}\left(\frac{\alpha_p^2}{256\pi^2 d^2} - \frac{2\alpha_p^2\Delta}{256\pi^2 d} + \frac{\alpha_p^2}{256\pi^2 d^2}\frac{(2\Delta d)^2}{2}\right)$$

$$= \left(\frac{\Delta^2\alpha_o^2\alpha_p^2}{1024\pi^2 d^2} - \frac{2\alpha_o^2\alpha_p^2\Delta^3}{1024\pi^2 d} + \frac{\alpha_o^2}{4}\frac{\alpha_p^2}{256\pi^2 d^2}\frac{(2d)^2\Delta^4}{2}\right)$$

Here the higher order term of momentum transfer is also neglected.

So that we get,

$$|T^l_{k_f k_i}|^2 = \frac{\alpha_o^2\alpha_p^2}{1024\pi^2 d^2}\Delta^2$$

Thus we get,

$$T^l_{k_f k_i} = \frac{\alpha_o\alpha_p}{32\pi d}\Delta$$

$$T^l_{k_f k_i} = \frac{\alpha_{pe E_o}^2}{32\pi m\omega^2 d}\Delta$$

$$\alpha_o = \frac{eE_o}{m\omega^2}$$

Now,

$$\frac{d\sigma}{d\Omega} = \frac{m^2}{(2\pi)^2\hbar^4}\frac{k_f}{k_i}|T^l_{k_f k_i}|^2$$

$$\frac{d\sigma}{d\Omega} = \frac{m^2}{(2\pi)^2\hbar^4}\left(1 - \frac{l\hbar\omega}{E_{k_i}}\right)^{\frac{1}{2}}\frac{\alpha_o^2\alpha_p^2}{1024\pi^2 d^2}\Delta^2 \quad (13)$$

$$\frac{d\sigma}{d\Omega} = \frac{m^2\alpha_p^2}{4096\pi^4\hbar^4 d^2}\left(1 - \frac{l\hbar\omega}{E_{k_i}}\right)^{\frac{1}{2}}\alpha_o^2\Delta^2 \quad (14)$$

Here,

$$\Delta = k_f - k_i \text{ thus}$$

$$\Delta^2 = (k_f - k_i)^2$$

$$= k_f^2 + k_i^2 - 2k_f k_i \cos\theta$$

Where, θ is the angle between k_f and k_i, that means θ is the scattering angle between initial and final momentum of the projectile particle (electron).

$$\left(\frac{\Delta}{k_i}\right)^2 = \left(\frac{k_f}{k_i}\right)^2 + 1 - 2\left(\frac{k_f}{k_i}\right)\cos\theta$$

$$\left(\frac{\Delta}{k_i}\right)^2 = \left(1 - \frac{l\hbar\omega}{E_{k_i}}\right) - 2\left(1 - \frac{l\hbar\omega}{E_{k_i}}\right)^{\frac{1}{2}}\cos\theta + 1$$

$$\Delta^2 = k_i^2\left[\left(1 - \frac{l\hbar\omega}{E_{k_i}}\right) - 2\left(1 - \frac{l\hbar\omega}{E_{k_i}}\right)^{\frac{1}{2}}\cos\theta + 1\right] \quad (15)$$

Substitute this value in equation (4.100) we get,

$$\frac{d\sigma}{d\Omega} = \frac{m^2\alpha_p^2}{4096\pi^4\hbar^4 d^2}\left(1 - \frac{l\hbar\omega}{E_{k_i}}\right)^{\frac{1}{2}}\left(\frac{eE_o}{m\omega^2}\right)^2 k_i^2\left[\left(1 - \frac{l\hbar\omega}{E_{k_i}}\right) - 2 1 - l\hbar\omega E_{ki} 1 2\cos\theta + 1\right] \quad (16)$$

we have from equation (4.83),

$$\frac{k_f}{k_i} = \left(1 - \frac{l\hbar\omega}{E_{k_i}}\right)^{\frac{1}{2}}$$

for $l = 1$ it is

$$\frac{k_f}{k_i} = \left(1 - \frac{\hbar\omega}{E_{k_i}}\right)^{\frac{1}{2}}$$

for $l = 1$ i.e, for one photon emission (stimulated Bremsstrahlung)

$$\Delta^2 = k_i^2 \left[\left(1 - \frac{\hbar\omega}{E_{k_i}}\right) - 2\left(1 - \frac{\hbar\omega}{E_{k_i}}\right)^{\frac{1}{2}} \cos\theta + 1 \right]$$

Substitute this value in equation (4.102) we get,

$$\frac{d\sigma}{d\Omega} = \frac{m^2 \alpha_p{}^2}{4096\pi^4\hbar^4 d^2} \left(1 - \frac{\hbar\omega}{E_{k_i}}\right)^{\frac{1}{2}} \left(\frac{eE_o}{m\omega^2}\right)^2 k_i^2 \left[\left(1 - \frac{\hbar\omega}{E_{k_i}}\right) \right.$$
$$\left. - 2\left(1 - \frac{\hbar\omega}{E_{k_i}}\right)^{\frac{1}{2}} \cos\theta + 1 \right]$$

$$\frac{d\sigma}{d\Omega} = \frac{m^2 \alpha_p^2}{4096\pi^4\hbar^6 d^2} \left(1 - \frac{\hbar\omega}{E_{k_i}}\right)^{\frac{1}{2}} \frac{e^2 E_o^2}{m^2\omega^4} \hbar^2 k_i^2 \left[\left(1 - \frac{\hbar\omega}{E_{k_i}}\right) - \right.$$
$$2 1 - \hbar\omega E k i 1 2 \cos\theta + 1 \qquad (17)$$

$$\frac{d\sigma}{d\Omega} = \frac{me^2 \alpha_p{}^2}{32768\pi^8\hbar^6 d^2 c^4} \left(1 - \frac{\hbar\omega}{E_{k_i}}\right)^{\frac{1}{2}} E_o{}^2 E_{k_i} \lambda^4 \left[\left(1 - \frac{\hbar\omega}{E_{k_i}}\right) \right.$$
$$\left. - 2\left(1 - \frac{\hbar\omega}{E_{k_i}}\right)^{\frac{1}{2}} \cos\theta + 1 \right]$$

$$\frac{d\sigma}{d\Omega} = C \left(1 - \frac{\hbar\omega}{E_{k_i}}\right)^{\frac{1}{2}} E_o{}^2 E_{k_i} \lambda^4 \left[\left(1 - \frac{\hbar\omega}{E_{k_i}}\right) - 2\left(1 - \frac{\hbar\omega}{E_{k_i}}\right)^{\frac{1}{2}} \cos\theta + 1 \right]$$

Where, $C = \frac{me^2 \alpha_p{}^2}{32768\pi^8\hbar^6 d^2 c^4}$ (constant)

This is the differential cross section for the scattering of an electron with one photon emission (stimulated Bremsstrahlung) in the presence of potential V(r).

Again from equation (14) we have,

$$\frac{k_f}{k_i} = \left(1 - \frac{l\hbar\omega}{E_{k_i}}\right)^{\frac{1}{2}}$$

for $l = -1$,one photon absorption (inverse Bremsstrahlung) it is

$$\frac{k_f}{k_i} = \left(1 + \frac{\hbar\omega}{E_{k_i}}\right)^{\frac{1}{2}}$$

Also, $\Delta^2 = k_i^2 \left[\left(1 + \frac{\hbar\omega}{E_{k_i}}\right) - 2\left(1 + \frac{\hbar\omega}{E_{k_i}}\right)^{\frac{1}{2}} \cos\theta + 1 \right]$

Then equation (4.102) becomes

$$\frac{d\sigma}{d\Omega} = \frac{m^2 \alpha_p^2}{4096\pi^4\hbar^6 d^2} \left(1 + \frac{\hbar\omega}{E_{k_i}}\right)^{\frac{1}{2}} \frac{e^2 E_o^2}{m^2\omega^4} \hbar^2 k_i^2 \left[\left(1 + \frac{\hbar\omega}{E_{k_i}}\right) - \right.$$
$$2 1 + \hbar\omega E k i 1 2 \cos\theta + 1 \qquad (18)$$

This is the differential cross section for the scattering of an

electron with one photon absorption (inverse Bremsstrahlung) in the presence of potential V(r).

Where,

m = mass of an electron

k_i = initial momentum vector of an electron

E_{k_i} = initial kinetic energy of the incident an electron

$\hbar\omega$ = photon energy of the laser

l = no. of the photon transfer during interaction

θ = scattering angle

Δ = momentum transfer

E_o = amplitude of the electric field of the laser

5. Result and Discussion

5.1. Result and Discussion

In the present thesis work, we have studied the elastic scattering of an electron-atom interaction by absorbing a photon from the laser field. For simplicity, we have neglected the dressing effect of an atom [21] i.e., we have neglected the field- atom interaction so that we can neglect the internal structure of an atom because we choose the intensity of the field in such a way that the target atom is transparent for the field.

When an electron scatters elastically from the target by absorbing a photon (inverse Bremsstrahlung, i.e., $l = -1$) from the fields, then the differential scattering cross-section is given by equation (17) as,

$$\frac{d\sigma}{d\Omega} = \frac{m^2 \alpha_p{}^2}{4096\pi^4\hbar^6 d^2} \left(1 + \frac{\hbar\omega}{E_{k_i}}\right)^{\frac{1}{2}} \frac{e^2 E_o{}^2}{m^2\omega^4} \hbar^2 k_i^2 \left[\left(1 + \frac{\hbar\omega}{E_{k_i}}\right) \right.$$
$$\left. - 2\left(1 + \frac{\hbar\omega}{E_{k_i}}\right)^{\frac{1}{2}} \cos\theta + 1 \right]$$

From this relation, it is clear that the frequency ω of the laser field is not only important but also the field amplitude E_o is important for the differential cross section.

Now we changed this equation in terms of wavelength and the intensity of the laser field.

We have the relations,

$$E_o^2 = \frac{8\pi I_o}{c}$$

where, I_o is the intensity of the laser field.

and also, $\omega = \frac{2\pi c}{\lambda}$

where

λ is the wave length of the radiation.

c is the velocity of light.

Substituting the values of E_o^2 and ω in equation (17), then we get the following form for the differential cross- section,

$$\frac{d\sigma}{d\Omega} = \frac{2me^2 \alpha_p{}^2}{4096\pi^4\hbar^6 d^2} \left(1 + \frac{\hbar\omega}{E_{k_i}}\right)^{\frac{1}{2}} \frac{\left(\frac{8\pi I_o}{c}\right) E_{k_i} \lambda^4}{(2\pi c)^4} \left[\begin{array}{c} \left(1 + \frac{\hbar\omega}{E_{k_i}}\right) - \\ 2\left(1 + \frac{\hbar\omega}{E_{k_i}}\right)^{\frac{1}{2}} \cos\theta + 1 \end{array} \right]$$

$$\frac{d\sigma}{d\Omega} = \frac{me^2\alpha_p^2}{32768\pi^8\hbar^6 d^2 c^4}\left(1+\frac{\hbar\omega}{E_{k_i}}\right)^{\frac{1}{2}}\left(\frac{8\pi I_o}{c}\right)E_{k_i}\lambda^4\left[\left(1+\frac{\hbar\omega}{E_{k_i}}\right)-\right.$$
$$21+\hbar\omega E k i 12\cos\theta+1 \qquad (19)$$

Similarly, when an electron scatters elastically from the target by emitting a photon (stimulated Bremsstrahlung, i.e., $l = 1$), then the differential scattering cross-section is given by equation (18) as,

$$\frac{d\sigma}{d\Omega} = \frac{m^2\alpha_p^2}{4096\pi^4\hbar^6 d^2}\left(1-\frac{\hbar\omega}{E_{k_i}}\right)^{\frac{1}{2}}\frac{e^2 E_o^2}{m^2\omega^4}\hbar^2 k_i^2\left[\left(1-\frac{\hbar\omega}{E_{k_i}}\right)-\right.$$
$$21-\hbar\omega E k i 12\cos\theta+1 \qquad (20)$$

Substituting the values of E_o^2 and ω in equation (5.3), then we get the following form for the differential cross- section,

$$\frac{d\sigma}{d\Omega} = \frac{me^2\alpha_p^2}{32768\pi^8\hbar^6 d^2 c^4}\left(1-\frac{\hbar\omega}{E_{k_i}}\right)^{\frac{1}{2}}\left(\frac{8\pi I_o}{c}\right)E_{k_i}\lambda^4\left[\left(1-\frac{\hbar\omega}{E_{k_i}}\right)-\right.$$
$$21-\hbar\omega E k i 12\cos\theta+1 \qquad (21)$$

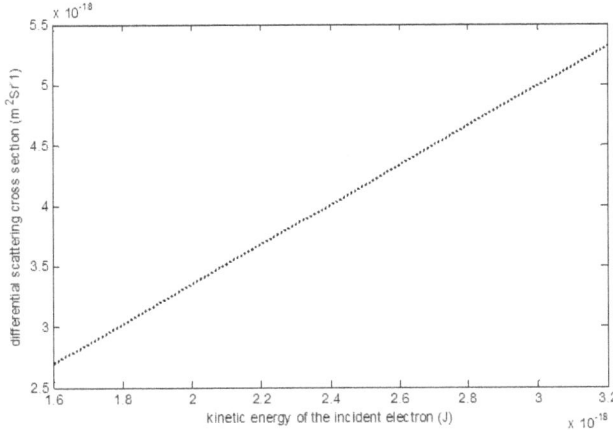

Fig 6. Variation of differential scattering cross section with kinetic energy of the incident electron

Here, the differential cross-section equations (18) and (19) is found to be proportional to λ^4 (where λ is the wavelength of the radiation) which explains why it has been easier to observe laser –assisted cross section using infrared laser (high wave length and low frequency) than those operating in the visible or ultraviolet spectral regions[22].

Equation (18) and (19) are plotted as the functions of scattering angle, kinetic energy of an electron, wavelength and intensity of the laser field as shown below:

From the plot, we see that the differential scattering cross section increases as the kinetic energy of the incident electron increases. Going on increasing the kinetic energy of an incident electron we will get infinite value of scattering cross section.

5.2. Variation of Differential Cross Section with the Wave Length of the Laser Field

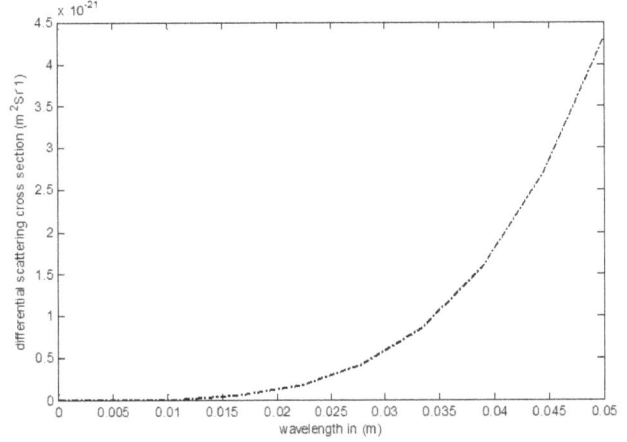

Fig 7. Variation of Differential Cross Section with the Wave Length of the Laser Field

From the plot, it is found that the differential scattering cross section increases with increase in wavelength of the laser field. It shows that the differential scattering cross section is zero below the wavelength 0.015m.

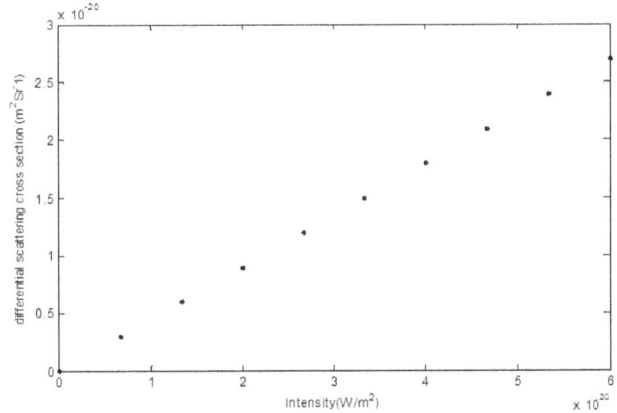

Fig 8. Variation of differential cross section with intensity of the laser field

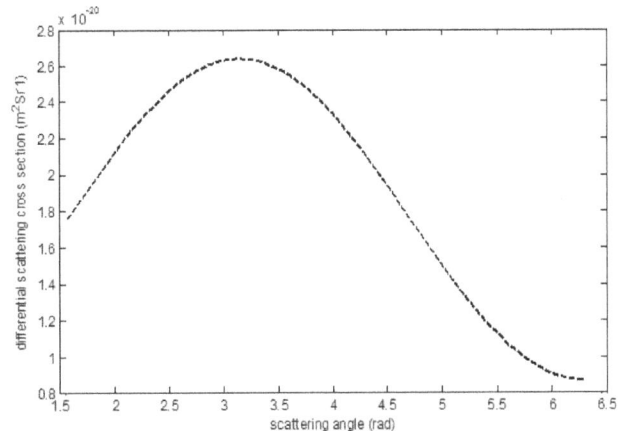

Fig 9. Variation of differential cross section with scattering angle

From the figure, it is clear that differential scattering cross section increase with increase in the intensity of the laser field. It shows that the differential scattering cross section varies linearly with intensity.

From the plot, it is clear that differential cross section increases as the scattering angle increase and attains the maximum value $2.6 \times 10^{-20} m^2$ and further increase in scattering angle decreases in differential scattering cross section which is clearly shown in polar plot[23].

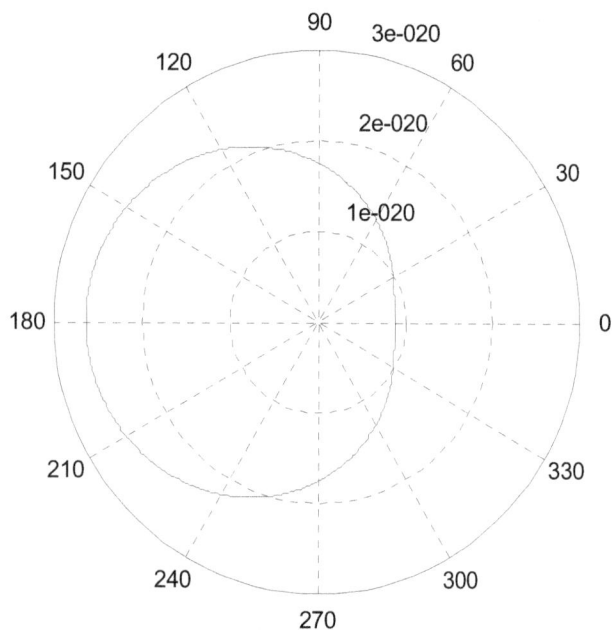

Fig 10. Polar plot of differential cross section with scattering angle

From the polar plot, it is clear that the differential scattering cross section of an electron increase with increase in scattering angle and attains maximun value of $2.6 \times 10^{-20} m^2$ and further increase in scattering angle, decreases in cross section and attains minimum value of $0.98 \times 10^{-20} m^2$.

5.3. Conclusion

It is concluded that from the above discussion that when electrons are scattered from the atom in the presence of a laser field, a new effect is observed which are not accessible in ordinary electron atom scattering. This collision have the basic peculiarity of being processes in which three subsystem are present (i) the electron (ii) the target atom (iii) the radiation field. The last one provided energy and momentum and is characterized by the polarization of its electric field, which introduces in this collision process a new physical axis.

This work reflected that the scattering of an electron from the target atom by absorption of a photon from the laser field. From this study we concluded that the differential scattering cross section of an electron depends upon the intensity and wavelength of the laser field where as in ordinary electron - atom scattering, cross section only depends upon the scattering angle. The differential scattering cross section increases with increase in wavelength and intensity of the laser field. For a fixed value of a laser parameters and kinetic energy of an incident electron, the differential scattering cross section of an scattered electron increases with increase

in scattering angle and attains a maximum value of $2.6 \times 10^{-20} m^2$ and further increase in scattering angle , decreases in differential scattering cross section and attains a minimum value of $0.98 \times 10^{-20} m^2$.

References

[1]   J. Mathews and R.L. Walker, Mathematical Methods of Physics, Second Edition, California Institute of Technology, copyright © 1970 by Pearson Education (2005).

[2]   C. Harper, Introduction to Mathematical physics, Department of Physics California State University, Hayward, Prentice Hall of India Pvt. Ltd, New Delhi-110 001 (2006)

[3]   J. D. Jackson, Classical Electrodynamics, Third Edition, John Wiley and Sons, 709 (1971) .

[4]   A.A Balakin and G.M.Frainman, Bremsstrahlung in a strong Laser field JEPT, 93, 695, (2001).

[5]   G.M Frainman, V.A.Mironov and A.A Balakin , Representative Electrons and Energy Exchange in Strong Laser field, Phys.Rev.Lett, 82, 319 (1999).

[6]   F.W. Byron, P.Francken and C.J. Joachain, Laser assisted elastic electron atom collision, J. Phys: B: At. Mol. Phys. 20 5487-5503 (1987).

[7]   N.M Kroll and KM Waston, Phys. Rev. A 8, 804(1973).

[8]   J.R. Reitz, F.J. Milford, R. W. Christy, Foundation of electromagnetic theory, Third Edition, Narosa Publishing House, New Delhi (1998).

[9]   H. Goldstein, C. Poole, J. Safko, Classical Mechanics, third edition, Pearson Education, India (2007).

[10] F. Ehlotzky – Fundamentals of Laser Interaction –Springer (1985).

[11] Marvin H. Mittleman, Introduction to the Theory of Laser – Atom Interaction, second edition, Printed in the United State of America (1993).

[12] A.G.Sitenko and P.J. Shepherd, Lectures in Scattering Theory.

[13] J.J. Sakurai- Modern Quantum Mechanics Revised Edition, Addition-Wesley Publishing Company (1994).

[14] B.H Bransden, W.A. Benjamin, New York - Atomic Collision Theory, standard book number 8053-1180-7 (C) (1970).

[15] G.N. Watson- Theory Of Bessel Function, Cambridge at the University Press 1922.

[16] M. Hannachi, Z. Rouabah, C. Champion, N. Bouarissam *Journal of Electron Spectroscopy and Related Phenomena, Volume 195, August 2014, Pages 155-159*

[17] A. Bekzhanov, S. Bondarenko, V.urov*Nuclear Physics B - Proceedings Supplements, Volume 245, December 2013, Pages 65-68*

[18] H. Aouchiche, F. Medegga, C. Champion   *Nuclear Instruments and Methods in Physics Research Section B: Beam Interactions with Materials and Atoms, Volume 333, 15 August 2014, Pages 113-119*

[19] E. Merzbacher-Quantum Mechanics, Third edition, John Wiley and Sons, Inc (1998).

[20] N. Zettili - Quantum Mechanics Concepts and Application, second edition, John Wiley and Sons, Ltd (2009).

[21] M. L. Goldberger- Collision Theory, John Wiley and Sons, Inc, Third Printing (1967).

[22] Laser-Assisted Elastic Electron scattering From Argon, Theor. Phys.(Beijing, China) 51 (2009) pp (131-134).

[23] S. Balasubramian, Oxford and IBH Publishing Co. New Delhi, Bombay, Calcutta, 1985.

18

On some problems of synthesis of spatial five-bar hinged mechanisms with two degrees of freedom

Nodar Davitashvili[*], Otar Gelashvili

Department of Transport and Mechanical Engineering of Georgian Technical University, Tbilisi, Georgia

Email address:

nodav@pam.edu.ge (N. Davitashvili), gelashviliotari@mail.ru (O. Gelashvili)

Abstract: Solution of the problems of synthesis of spatial five-bar hinged mechanisms with two degrees of freedom task considering the angle of transmission and definition of conditions of existence of cranks is given in the paper. A possibility of movement of the designed mechanism without seizure that is depended on shape and sizes of the mechanism links is envisaged. At solution of the problem of synthesis first are determined sizes of the mechanism couples considering the angle of transmission and then are determined the limits of variation of the angle of transmission depending on sizes of the mechanism links. The conditions are ascertained which must be met by sizes of the spatial five-bar mechanism links so that the two links adjoining to the frame are crank (theorem on existence of cranks in the spatial five-bar mechanism).

Keywords: Synthesis, Angle of Transmission, Spatial Five-Bar Mechanism, Coupler, Conditions of Existence of Crank

1. Introduction

Spatial mechanism with two degrees of freedom rank high among numerous hinged mechanisms used in practice, which in contrast to four-bar mechanisms, can realize more complex laws of motion favouring at the same time rise of the machines efficiency and optimization of the technological processes.

A distinctive feature of these mechanisms is their capacity for variation of the coupler point locus from the simple to the more complex form in the course of operation at variation of angular velocities of the input links considering directions of motions and their mutual initial inter-location at constant lengths of the links.

These circumstances are directly related with selection of length of the output links of mechanisms and predetermine necessary conditions for creation of compact, lightweight mechanisms, especially in the fields of space and robotics industry, decreasing at this consumption of materials and energy.

The spatial five-bar hinged mechanisms with two degrees of freedom are widely used in various devices, robotics systems as well as in industrial practice as operating members [1, 2] and so on.

The researches into hinged mechanisms with two degrees of freedom started at the end of XIX and beginning of XX

centuries by first studying planar five-bar mechanisms with two input links.

Solution of the problem on locations for the spatial mechanisms with several degrees of freedom and spherical five-bar mechanisms with two degrees of freedom was first considered by A.G. Ovakimov [3] only in the 70-ies of the XX century.

The problems of kinematical analysis of the spatial five-bar mechanism of type RCRCR with one degree of freedom were considered in the work of J. Duffi and G.I. Habib-Olahi. [4]. The analytical formulae were derived with the use of spherical trigonometry. The positions of output links of the mechanism are determined by the forth order equation and in the work of J. Duffi [5] for determination of displacements of the spatial five-bar mechanism were used the dual equations.

Much attention was given to the problems of analysis, synthesis, kinetostatics and dynamics of mechanisms with two degrees of freedom by Georgian scientists having carried out deep analysis of the mentioned mechanisms.

In should be noted in this aspect the works [6, 7] of N.S. Davitashvili. In the work [6] is given analysis of influence of parameters of the five-bar hinged mechanism on trajectories of the couplers points and in the work [7] – are studied problems of kinematics and precision of these mechanisms.

Besides the mentioned works numerous researches into analysis, synthesis, dynamics and precision of hinged mechanisms with two degrees of freedom are carried out by first author of these articles [8, 9, 10]. In the work [11] of N.S. Davitashvili and O. Gelashvili is given synthesis of the five-bar spatial mechanism with two degrees of freedom according to the trajectory.

In the presented work is carried out synthesis of the spatial five-bar hinged mechanism with two degrees of freedom taking into account the angle of transmission and conditions of existence of two cranks.

2. Synthesis of the Spatial Five-Bar Mechanism Taking into Account the Angle of Transmission

At solution of the problem of synthesis of spatial five-bar mechanisms with two degrees of freedom it is necessary to consider the following requirements: correct structure of the mechanism to be designed, kinematical precision of the movement to be performed, possibility of realization of the given movement by mechanism to be designed from the standpoint of dynamics, possibility of movement of the mechanism under action of applied forces with maximum possible efficiency (taking into account the angle of transmission to avoid their seizure) and finally, ensuring existence of cranks and reproduction of the given movement by the sizes of links of the synthesized mechanism.

2.1. Dimensioning of Couplers of the Five-Bar Spatial Mechanism Taking into Account the Given Value of the Angle of Transmission

Let's consider spatial five-link hinged leverage RSSRR type mechanism ABCDE (Fig. 1) with two degrees of freedom and determine the dimensions $BC=\ell_3$ and $CD=\ell_4$ of couplers with taking into account the given values of angle of transmission γ ($\gamma_{min}, \gamma_{max}$).

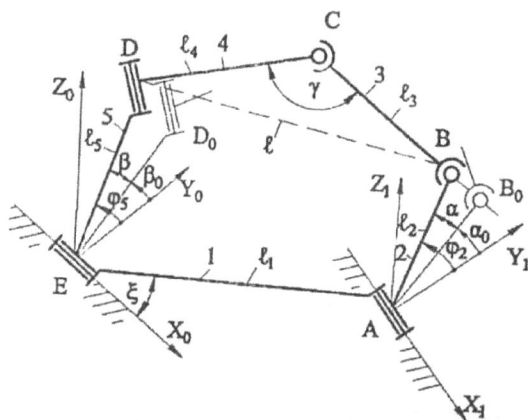

Figure 1. Spatial five-bar mechanism with two degrees of freedom.

For solution of this problem are given: dimensions of links 1, 2, 5, designated as ℓ_1, ℓ_2, and ℓ_5; angles α_0 and β_0, specifying the initial position of the input AB_0 and ED_0 links 2 and 5, with respect to the frame 1.

The laws of motion of the input links 2 and 5 are given as:

$$\varphi_2 = \varphi_2(t) \text{ and } \varphi_5 = \varphi_5(t), \qquad (1)$$

where

$$\varphi_2 = \alpha_0 + \alpha(t) \text{ and } \varphi_5 = \beta_0 + \beta(t). \qquad (2)$$

Here α and β – are the variable angles characterizing the motion of input links.

The gear ratios from link 5 to link 2 is defined as

$$u_{52} = \frac{\varphi_5}{\varphi_2} = \frac{\omega_5}{\omega_2} = \frac{\beta}{\alpha} = \pm \frac{p}{q}, \qquad (3)$$

where the sign (+) is taken for the case when input links 2 and 5 are rotating in the same direction, and the sign (–) when they are rotating in the opposite directions..

It should be noted that by giving functions of displacements of the input links, gear ratio is also

$$u_{52} = u_{52}(t). \qquad (4)$$

We select two fixed coordinate systems for the input links 2 and 5 so that point A is located in the plane X_0EY_0. The system $EX_0Y_0Z_0$ is selected for link 5, and system $AX_1Y_1Z_1$ - for link 2.

Accordingly are given the coordinates of point $A(x_0 y_0 z_0)$ in the system $EX_0Y_0Z_0$ considering angle ξ between axis EX_0 and frame 1:

$$x_0 = l_1 \cos \xi; \quad y_0 = l_1 \sin \xi; \quad z_0 = 0. \qquad (5)$$

The transformation of system $AX_1Y_1Z_1$ in the system $EX_0Y_0Z_0$ is carried out due the matrix:

$$\begin{vmatrix} b_{11} & b_{21} & b_{31} \\ b_{12} & b_{22} & b_{32} = 0 \\ b_{31} & b_{32} & b_{33} \end{vmatrix}, \qquad (6)$$

where $b_{11}, b_{12},...., b_{33}$ – are directing cosines of axes of the system $AX_1Y_1Z_1$ with respect to the $EX_0Y_0Z_0$ system.

It is possible at first to set the four values of directing cosines, for example, b_{11}, b_{12}, b_{31} and $b_{32} = 0$, then for $b_{33}, b_{12}, b_{23}, b_{12}$ we will have:

$$b_{33} = \sqrt{1 - b_{31}^2}; \qquad b_{13} = -\frac{b_{11}b_{31}}{b_{33}};$$
$$b_{23} = -\frac{b_{21}b_{31}}{b_{33}}; \qquad b_{12} = \sqrt{1 - b_{11}^2 - b_{31}^2}. \qquad (7)$$

The variable distance ℓ between movable joints B and D, the value of that is depended on dimensions of frame ℓ_1 and input links 2 (ℓ_2) and 5 (ℓ_5), initial position of input links (α_0, β_0), gear ratio (u_{52}) between input links with taking into account directions of their movement can be determined in the system $EX_0Y_0Z_0$ with the help of formula for the distance between two points in space:

$$\ell^2 = (x_B - x_D)^2 + (y_B - y_D)^2 + (z_B - z_D)^2, \qquad (8)$$

where

$$x_B = 0; \quad y_B = \ell_2 \cos\varphi_2; \quad z_B = \ell_2 \sin\varphi_2. \qquad (9)$$

$$x_D = 0; \quad y_D = \ell_5 \cos\varphi_5; \quad z_D = \ell_5 \sin\varphi_5. \quad (10)$$

Considering the location of point A at the origin of the system, , the coordinates of point B CAN be written in the system $EX_0Y_0Z_0$ using matrix (6);

$$x_B = b_{12}\ell_2 \cos\varphi_2 + b_{13}\ell_2 \sin\varphi_2 + \ell_1 \cos\zeta;$$
$$y_B = b_{22}\ell_2 \cos\varphi_2 + b_{23}\ell_2 \sin\varphi_2 + \ell_1 \sin\zeta; \qquad (11)$$
$$z_B = b_{33}\ell_2 \sin\varphi_2.$$

Considering values (2), (10) and (11) expression (8) will take a form:

$$\ell = \{\ell_1^2 + \ell_2^2 + \ell_5^2 + 2\ell_1\ell_2[\cos(\alpha_0 + \alpha) \cdot$$
$$\cdot (b_{12}\cos\xi + b_{12}\sin\xi) + \sin(\alpha_0 + \alpha) \cdot$$
$$\cdot (b_{31}\cos\xi + b_{23}\sin\xi)] -$$
$$-2\ell_1\ell_5 \sin\xi \cos(\beta_0 + u_{52}\alpha) -$$
$$-2\ell_2\ell_5[b_{22}\cos(\alpha_0 + \alpha)\cos(\beta_0 + u_{52}\alpha) +$$
$$+b_{23}\sin(\alpha_0 + \alpha)\cos(\beta_0 + u_{52}\alpha) +$$
$$+b_{33}\sin(\alpha_0 + \alpha)b_{23}\sin(\beta_0 + u_{52}\alpha)]\}^{1/2}, \qquad (12)$$

where

$$\beta = u_{52}\alpha.$$

Here α is a variable value.

Analysis of expression (12) shows that distance ℓ depends on $\ell_1, \ell_2, \ell_5, \alpha_0, \beta_0, u_{52}, \alpha$ and directing cosines.

FROM triangle BCD taking into account the given maximal (γ_{max}) and minimal (γ_{min}) values of the angle of transmission can written:

$$\ell_3^2 + \ell_4^2 - 2\ell_3\ell_4 \cos\gamma_{max} = \ell_{max}^2;$$
$$\ell_3^2 + \ell_4^2 - 2\ell_3\ell_4 \cos\gamma_{min} = \ell_{min}^2, \qquad (13)$$

where ℓ_{max} and ℓ_{min} – are the extreme values of variable distance ℓ.

It is known that if derivative $d\ell/d\alpha$ exists, then function ℓ will have maximum or minimum only if

$$\frac{d\ell}{d\alpha} = 0. \qquad (14)$$

If second derivative $d^2\ell/d\alpha^2$ exists, then function ℓ will have maximum when

$$\frac{d\ell}{d\alpha} = 0, \quad \frac{d^2\ell}{d\alpha^2} < 0, \qquad (15)$$

and minimum at

$$\frac{d\ell}{d\alpha} = 0, \quad \frac{d^2\ell}{d\alpha^2} > 0. \qquad (16)$$

Accordingly we can write

$$\frac{d\ell}{d\alpha} = \frac{k\prime}{2\sqrt{k}}; \qquad (17)$$

$$\frac{d^2\ell}{d\alpha^2} = \frac{2kk'' - (k\prime)^2}{4\sqrt{k}}, \qquad (18)$$

where

$$k = \ell_1^2 + \ell_2^2 + \ell_5^2 + 2\ell_1\ell_2[\cos(\alpha_0 + \alpha) \cdot$$
$$\cdot (b_{12}\cos\xi + b_{12}\sin\xi) + \sin(\alpha_0 + \alpha) \cdot$$
$$\cdot (b_{13}\cos\xi + b_{23}\sin\xi)] -$$
$$-2\ell_1\ell_5 \sin\xi \cos(\beta_0 + u_{52}\alpha) -$$
$$-2\ell_2\ell_5[b_{22}\cos(\alpha_0 + \alpha)\cos(\beta_0 + u_{52}\alpha) +$$
$$+b_{23}\sin(\alpha_0 + \alpha)\cos(\beta_0 + u_{52}\alpha) +$$
$$+b_{33}\sin(\alpha_0 + \alpha)\sin(\beta_0 + u_{52}\alpha)];$$

$$k' = 2\ell_1\ell_2[\cos(\alpha_0 + \alpha)(b_{13}\cos\xi + b_{23}\sin\xi) -$$
$$- \sin(\alpha_0 + \alpha)(b_{12}\cos\xi + b_{22}\sin\xi)] +$$
$$+2\ell_1\ell_5(u_{52}'\alpha + u_{52})\sin\xi \sin(\beta_0 + u_{52}\alpha) -$$
$$-2\ell_2\ell_5[b_{23}(\cos(\alpha_0 + \alpha)\cos(\beta_0 + u_{52}\alpha) -$$
$$-(u_{52}'\alpha + u_{52})\sin(\beta_0 + u_{52}\alpha)\sin(\alpha_0 + \alpha)) -$$
$$- b_{22}(\sin(\alpha_0 + \alpha)\cos(\beta_0 + u_{52}\alpha) +$$
$$+(u_{52}'\alpha + u_{52})\sin(\beta_0 + u_{52}\alpha)\cos(\alpha_0 + \alpha)) +$$
$$+b_{33}(\cos(\alpha_0 + \alpha)\sin(\beta_0 + u_{52}\alpha) +$$
$$+(u_{52}'\alpha + u_{52})\cos(\beta_0 + u_{52}\alpha)\sin(\alpha_0 + \alpha))].$$

The values k" will be found by differentiation of k'. In turn

$$u_{52}' = \frac{\beta'\alpha - \beta}{\alpha^2}. \qquad (19)$$

For determining of such value of α that gives the extreme values of distance ℓ it is necessary that condition (14) be met. From the relations (17) we will have

$$2\ell_5 \cos u_{52}\alpha \{\ell_1(u_{52}'\alpha + u_{52})\sin\xi \sin\beta_0 +$$
$$+\ell_2 \sin(\alpha_0 + \alpha)[b_{22}\cos\beta_0 +$$
$$+b_{23}(u_{52}'\alpha + u_{52})\sin\beta_0 -$$
$$-b_{33}(u_{52}'\alpha + u_{52})\cos\beta_0] +$$
$$+\ell_2 \cos(\alpha_0 + \alpha)[b_{22}(u_{52}'\alpha + u_{52})\sin\beta_0 -$$
$$-b_{23}\cos\beta_0 - b_{33}\sin\beta_0]\} +$$
$$+2\ell_5 \sin u_{52}\alpha \{\ell_1(u_{52}'\alpha + u_{52})\sin\xi \sin\beta_0 +$$
$$+\ell_2 \sin(\alpha_0 + \alpha)[b_{23}\cos\beta_0 (u_{52}'\alpha + u_{52}) -$$
$$-b_{22}\sin\beta_0 + b_{33}(u_{52}'\alpha + u_{52})\sin\beta_0] +$$
$$+\ell_2(\alpha_0 + \alpha)[b_{22}(u_{52}'\alpha + u_{52})\cos\beta_0 -$$

$$-b_{23}\sin\beta_0 - b_{33}\cos\beta_0]\} -$$

$$-2\ell_1\ell_2[\sin(\alpha_0 + \alpha)(b_{12}\cos\xi + b_{22}\sin\xi) +$$

$$+ \cos(\alpha_0 + \alpha)(b_{13}\cos\xi + b_{23}\sin\xi)] = 0. \quad (20)$$

On the base of known formulae of the Moivre $\cos u_{52}\alpha$ and $\sin u_{52}\alpha$ included in the equation (20) can be presented as

$$\cos u_{52}\alpha = \cos^{u_{52}}\alpha - C_{u_{52}}^2\cos^{u_{52}-2}\alpha\sin^2\alpha +$$

$$+ C_{u_{52}}^4\cos^{u_{52}-4}\alpha\sin^4\alpha - \cdots \quad (21)$$

$$\sin u_{52}\alpha = u_{52}\cos^{u_{52}-1}\alpha\sin\alpha -$$

$$- C_{u_{52}}^2\cos^{u_{52}-3}\alpha\sin^3\alpha +$$

$$+ C_{u_{52}}^5\cos^{u_{52}-5}\alpha\sin^5\alpha - \cdots \quad (22)$$

Taking into account values (21) and (22) equation (20) will take a form:

$$a_1 \pm a_2 - a_3 = 0, \quad (23)$$

where

$$a_1 = 2\ell_5\Big(\cos^{u_{52}}\alpha - C_{u_{52}}^2\cos^{u_{52}-2}\alpha\sin^2\alpha +$$

$$+ C_{u_{52}}^4\cos^{u_{52}-4}\alpha\sin^4\alpha - \cdots\Big)\cdot$$

$$\cdot \{\ell_1(u_{52}'\alpha + u_{52})\sin\xi\sin\beta_0 +$$

$$+ \ell_2\sin(\alpha_0 + \alpha)[b_{22}\cos\beta_0 +$$

$$+ b_{23}(u_{52}'\alpha + u_{52})\sin\beta_0 - b_{33}(u_{52}'\alpha + u_{52})\cos\beta_0] +$$

$$+ \ell_2\cos(\alpha_0 + \alpha)[b_{22}(u_{52}'\alpha + u_{52})\sin\beta_0 -$$

$$- b_{23}\cos\beta_0 - b_{33}\sin\beta_0]\};$$

$$a_2 = 2\ell_5(u_{52}\cos^{u_{52}-1}\alpha\sin\alpha -$$

$$- C_{u_{52}}^2\cos^{u_{52}-3}\alpha\sin^3\alpha +$$

$$+ C_{u_{52}}^2\cos^{u_{52}-5}\alpha\sin^5\alpha - \cdots)$$

$$\cdot \{\ell_1(u_{52}'\alpha + u_{52})\sin\xi\sin\beta_0 + \ell_2\sin(\alpha_0 + \alpha)\cdot$$

$$\cdot [b_{23}\cos\beta_0(u_{52}'\alpha + u_{52}) - b_{22}\sin\beta_0 +$$

$$+ b_{33}(u_{52}'\alpha + u_{52})\cos\beta_0] + \ell_2\cos(\alpha_0 + \alpha)\cdot$$

$$\cdot [b_{22}(u_{52}'\alpha + u_{52})\cos\beta_0 -$$

$$- b_{23}\sin\beta_0 - b_{33}\cos\beta_0]\};$$

$$a_3 = 2\ell_1\ell_2[\sin(\alpha_0 + \alpha)(b_{12}\cos\xi + b_{22}\sin\xi) +$$

$$+ \cos(\alpha_0 + \alpha)(b_{13}\cos\xi + b_{23}\sin\xi)].$$

Binomial coefficients in turn will have forms

$$C_{u_{52}}^2 = \frac{u_{52}(u_{52} - 1)}{2!};$$

$$C_{u_{52}}^3 = \frac{u_{52}(u_{52} - 1)(u_{52} - 2)}{3!};$$

$$C_{u_{52}}^4 = \frac{u_{52}(u_{52} - 1)(u_{52} - 2)(u_{52} - 3)}{4!};$$

$$C_{u_{52}}^5 = \frac{u_{52}(u_{52} - 1)(u_{52} - 2)(u_{52} - 3)(u_{52} - 4)}{5!}$$

and so on

$$C_{u_{52}}^n = \frac{u_{52}(u_{52} - 1)(u_{52} - 2)(u_{52} - 3)\cdots\big(u_{52} - (n-1)\big)}{n!}.$$

In equations (23) a sign (+) is taken when gear ratios u_{52} is a positive number, and sign (−) − when u_{52} is a negative number.

The series of binomial coefficients is finite when u_{52} is a positive or negative integer, and infinite when u_{52} is a positive or negative fractional number.

By solution of equations (23) are determined the values $\alpha_i(\alpha_1, \alpha_2, ..., \alpha_n)$, with the help of which we determine extreme values ℓ_{max} and ℓ_{min} of distance ℓ accordingly to formula (12).

In practice are mainly encountered cases when the input links are moving with constant angular velocity (u_{52}=const). In such case expressions (17), (18) and equation (23) will be significantly simplified.

Thus knowing values ℓ_{max} and ℓ_{min}, and transforming the system (13) we obtain a value of coupler 4 (DC=ℓ_4) length:

$$\ell_4 = \pm\sqrt{\frac{A_2 \pm \sqrt{A_2^2 - 4A_1A_3}}{2A_1}}, \quad (24)$$

where

$$A_1 = 4(\cos\gamma_{min} - \cos\gamma_{max})^2;$$

$$A_2 = 4[\ell_{max}^2(\cos\gamma_{min} - \cos\gamma_{max})^2 +$$

$$+ \cos\gamma_{max}(\ell_{max}^2 - \ell_{min}^2) - (\cos\gamma_{min} - \cos\gamma_{max})];$$

$$A_3 = (\ell_{max}^2 - \ell_{min}^2)^2.$$

Accordingly a length of coupler 3 (BC=ℓ_3) will be determined from system (13):

$$\ell_3 = \frac{\ell_{max}^2 - \ell_{min}^2}{2\ell_4(\cos\gamma_{min} - \cos\gamma_{max})}. \quad (25)$$

2.2. Determination of Limits of Angles of Transmission Depending on Dimensions of Links of the Spatial Five-Bar Mechanism

At solution of this problem a condition is envisaged according to which an angle of transmission at any position of the mechanism must not be less that certain given minimum value ensuring normal functioning of the mechanism.

Selection of allowable angles of transmission is determined by dynamical conditions of the mechanism operation and in the first place by absence of seizure of output links.

Considering a polygon of links ABCDE and triangle BCD (Fig. 1) for distance ℓ we can write:

$$\ell^2 = \ell_1^2 + \ell_2^2 + \ell_5^2 + 2\ell_1\ell_2[\cos(\alpha_0 + \alpha) \cdot$$
$$\cdot (b_{12}\cos\xi + b_{22}\sin\xi) +$$
$$+ \sin(\alpha_0 + \alpha)(b_{31}\cos\xi + b_{23}\sin\xi)] -$$
$$-2\ell_1\ell_5\sin\xi\cos(\beta_0 + u_{52}\alpha) -$$
$$-2\ell_2\ell_5[b_{22}\cos(\alpha_0 + \alpha)\cos(\beta_0 + u_{52}\alpha) +$$
$$+ b_{23}\sin(\alpha_0 + \alpha)\cos(\beta_0 + u_{52}\alpha) +$$
$$+ b_{33}\sin(\alpha_0 + \alpha)\sin(\beta_0 + u_{52}\alpha)] =$$
$$= \ell_3^2 + \ell_4^2 - 2\ell_3\ell_4\cos\gamma. \tag{26}$$

From expressions (26) can be determined the angle of transmission γ:

$$\cos\gamma = (2\ell_3\ell_4)^{-1}\{\ell_3^2 + \ell_4^2 - \ell_1^2 - \ell_2^2 - \ell_5^2 - 2\ell_1\ell_2$$
$$[\cos(\alpha_0 + \alpha)(b_{12}\cos\xi + b_{22}\sin\xi) +$$
$$+ \sin(\alpha_0 + \alpha)(b_{31}\cos\xi + b_{23}\sin\xi)] +$$
$$+ 2\ell_1\ell_5\sin\xi\cos(\beta_0 + u_{52}\alpha) +$$
$$+ 2\ell_2\ell_5[b_{22}\cos(\alpha_0 + \alpha)\cos(\beta_0 + u_{52}\alpha) +$$
$$+ b_{23}\sin(\alpha_0 + \alpha)\cos(\beta_0 + u_{52}\alpha)$$
$$+ b_{33}\sin(\alpha_0 + \alpha)\sin(\beta_0 + u_{52}\alpha)\}. \tag{27}$$

If $0<\gamma<\pi$, then the extreme values of angle γ will be determined from formula (27):

$$\cos\gamma_{max} = \frac{\ell_3^2 + \ell_4^2 - \ell_{max}^2}{2\ell_3\ell_4}, \tag{28}$$

$$\cos\gamma_{min} = \frac{\ell_3^2 + \ell_4^2 - \ell_{min}^2}{2\ell_3\ell_4}. \tag{29}$$

An optimal value $\gamma = \pi/2$ will be obtained at value of variable angle α determined from equation

$$\ell_3^2 + \ell_4^2 - \ell_1^2 - \ell_2^2 - \ell_5^2 -$$
$$-2\ell_1\ell_2[\cos(\alpha_0 + \alpha)(b_{12}\cos\xi + b_{12}\sin\xi) +$$
$$+ \sin(\alpha_0 + \alpha)(b_{13}\cos\xi + b_{23}\sin\xi)] +$$
$$+ 2\ell_1\ell_5\sin\xi\cos(\beta_0 + u_{52}\alpha) +$$
$$+ 2\ell_2\ell_5[b_{22}\cos(\alpha_0 + \alpha)\cos(\beta_0 + u_{52}\alpha) +$$
$$+ b_{23}\sin(\alpha_0 + \alpha)\cos(\beta_0 + u_{52}\alpha) +$$
$$+ b_{33}\sin(\alpha_0 + \alpha)\sin(\beta_0 + u_{52}\alpha)]=0. \tag{30}$$

The obtained results allow to determine relations between dimensions of links of the spatial five-bar mechanism for which optimal value of the angle of transmission will be arithmetical mean of its extreme values, i.e.

$$\gamma = \frac{\pi}{2} = \frac{\gamma_{max} - \gamma_{min}}{2}. \tag{31}$$

Suppose

$$\gamma_{max} = \frac{\pi}{2} + \delta \quad and \quad \gamma_{min} = \frac{\pi}{2} - \delta, \tag{32}$$

where δ is a certain angle.

Substituting of these values in (28) and (29) we will have:

$$-2\ell_3\ell_4\sin\delta = \ell_3^2 + \ell_4^2 - \ell_{max}^2;$$
$$2\ell_3\ell_4\sin\delta = \ell_3^2 + \ell_4^2 - \ell_{min}^2. \tag{33}$$

Excluding δ we will obtain

$$2(\ell_3^2 + \ell_4^2) = \ell_{max}^2 + \ell_{min}^2. \tag{34}$$

Thus if dimensions of links of the spatial five-bar mechanism satisfy condition (34), then deviations of the angle of transmission γ from optimal value $\gamma = \pi/2$ will be identical in both sides.

2.3. Conditions of Existence of Two Cranks in Spatial Five-Bar Mechanisms

At solution of numerous problems of synthesis of spatial mechanisms with two degrees of freedom it is very important to select relations between lengths of the links allowing one or two links to realize full or partial turning around chosen points.

It is known that condition of existence of cranks in planar and spherical four-bar mechanisms was first formulated by F. Grasgof. The similar problems were studied by N.G. Bruevich, V.V. Dobrovolski, F. Freidenstein, F. Duditsa and others.

For spatial four-bar mechanisms conditions of existence of one of two cranks was first formulated by D.S. Tavkhelidze.

By numerous researches into various hinged mechanisms by N.S. Davitashvili were formulated theorems of existence of cranks in spherical five-bar, six-bar and seven-bar mechanisms.

By studies of structure and kinematics of five-bar mechanisms with two degrees of freedom is ascertained that conditions of turning of input links are significantly influenced by dimensions of frame 1 and links 2 and 5 (Fig. 1), gear ratio u_{52} between input links considering their directions of movement, initial positions of input links specified by angles α_0 and β_0 and duration of the mechanism movement cycle.

For spatial five-bar mechanism (Fig. 1) we have obtained formula (12) for calculation of variable distance ℓ between movable hinges B and D, whose extreme values ℓ_{max} and ℓ_{min} are found on the base of satisfaction of conditions (15), (16).

The found extreme values ℓ_{max} and ℓ_{min} of distance ℓ allow us to write inequalities defining conditions of existence of cranks 2 and 5according to the kinematical scheme of spatial five-bar mechanism:

$$\ell_{max} \le \ell_3 + \ell_4;$$
$$\ell_{min} \ge |\ell_3 - \ell_4|. \tag{35}$$

Thus we can formulate a theorem of existence of two cranks

in spatial five-bar hinged mechanisms as follows.

In spatial five-bar hinged mechanisms two links, adjoining to the frame, will be cranks if and only if at movement of the mechanism maximum distance between the hinges created by connection of input links and corresponding couplers is less or equal to the sum of lengths of couplers and minimum distance is more or equal to the difference of these lengths.

If the assume that $\alpha 0=\beta 0=0$ and $\ell 5=0$ the spatial five-bar mechanism ABSDE (Fig. 1) will be converted into spatial four-bar mechanism ABCD (Fig. 2).

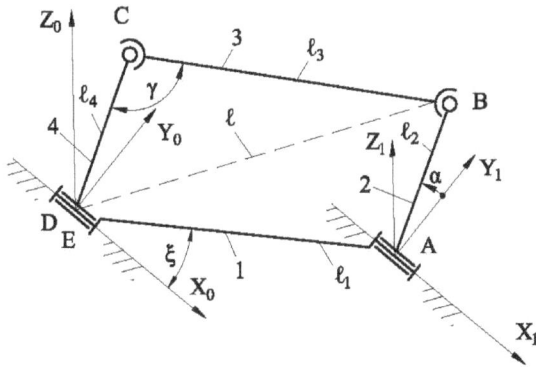

Figure 2. On the conditions of existence of cranks in a spatial four-bar mechanism.

In this case variable distance ℓ according to expression (12) will be:

$$\ell = \{\ell_1^2 + \ell_2^2 + 2\ell_1\ell_2[\cos\alpha\,(b_{12}\cos\xi + b_{22}\sin\xi) +$$

$$+\sin\alpha\,(b_{31}\cos\xi + b_{23}\sin\xi)]\}^{1/2}. \qquad (36)$$

The equation for finding α will be written as:

$$2\ell_1\ell_2[\cos\alpha\,(b_{13}\cos\xi + b_{23}\sin\xi) -$$

$$-\sin\alpha\,(b_{12}\cos\xi + b_{22}\sin\xi)] = 0. \qquad (37)$$

Accordingly value of α will be found by formula

$$\cos\alpha_{(1,2)} = \pm\frac{b_1}{\sqrt{b_1^2+b_2^2}}, \qquad (38)$$

where

$$b_1 = b_{12}\cos\xi + b_{22}\sin\xi;$$

$$b_2 = b_{13}\cos\xi + b_{23}\sin\xi.$$

Taking into account the values of α_1 and α_2 from expression (36) we will have:

$$\ell_{max} = [\ell_1^2 + \ell_2^2 + 2\ell_1\ell_2(b_{12}\cos\alpha_1 +$$

$$+b_{22}\sin\alpha_1]^{1/2}, \qquad (39)$$

$$\ell_{min} = [\ell_1^2 + \ell_2^2 + 2\ell_1\ell_2(b_{12}\cos\alpha_2 +$$

$$+b_{22}\sin\alpha_2]^{1/2}. \qquad (40)$$

Accordingly to expressions (35) conditions of existence of cranks in the spatial four-bar mechanism will be:

$$[\ell_1^2 + \ell_2^2 + 2\ell_1\ell_2(b_1\cos\alpha_1 + b_2\sin\alpha_1)]^{1/2} \le$$
$$\le \ell_3 + \ell_4;$$
$$[\ell_1^2 + \ell_2^2 + 2\ell_1\ell_2(b_1\cos\alpha_2 + b_2\sin\alpha_2)]^{1/2} \ge$$
$$\ge |\ell_3 - \ell_4|. \qquad (41)$$

3. Conclusion

By accomplished researches into five-bar mechanisms with two degrees of freedom are marked out some problems of synthesis of the mentioned mechanisms.

For the purpose of ensuring operation of the mechanism without seizure with maximum possible efficiency are determined dimensions of the links firstly considering a given angle of transmission and then depending on the dimensions of links of the spatial five-bar mechanism are calculated limits of variation of the angle of transmission. A theorem of existence of two cranks in spatial five-bar hinged mechanisms is formulated. The obtained results are suitable for designing new mechanisms and solution of practical problems.

References

[1]   G. Eason, B. Noble, and I. N. Sneddon, "On certain integrals of Lipschitz-Hankel type involving products of Bessel functions," Phil. Trans. Roy. Soc. London, vol. A247, pp. 529–551, April 1955. (references)

[2]   J. Clerk Maxwell, A Treatise on Electricity and Magnetism, 3rd ed., vol. 2. Oxford: Clarendon, 1892, pp.68–73.

[3]   A. G. Ovakimov, "Task on positions of spatial mechanisms with several degree of freedom and its solution by closed vector contour method" in Mechanics of Machines, issue 29-23, Moscow: Nauka, 1971, pp. 61–75. (In Russian).

[4]   J. Duffy, H.Y. Habib-Olahi, "A displacement analysis of spatial five-link BR-2C mechanism . 1. On the clousers of the RCRCR mechanism" in Journal of Mechanics, vol. 5, № 3, 1971, pp. 289-301.

[5]   J. Duffy, "A derivation of dual displacement equations for five-size-and seven-link spatial mechanism using spherical trigonometry. 3. Deviation of dual equations for seven-link spatial mechanism s" Rev. roum. sci. techn. Scr. mech. appl. vol. 17, 1972, № 2, pp. 291-317.

[6]   N.S. Davitashvili, "Influence of parameters of five-bar hinged mechanism on trajectory of coupler's points," Transactions of Academy of Sciences of GSSR, vol. 86, 1977, № 2, pp. 412–420, (In Russian).

[7]   N.S. Davitashvili, "Issues of kinematics and precision of five-bar hinged mechanism" Fifth World congress on Theory of Machines and Mechanisms, vol. 1, 1979, Canada, Motreal, July 8-13, pp. 557-561.

[8]   N.S. Davitashvili, "Fundamentals of the theory of synthesis, analysis and precision of linkage-lever mechanisms", Tbilisi: Technical University, 1999. –388 p. (In Russian).

[9]   N.S. Davitashvili, "Theoretical fundamentals of error and precision of hinged mechanisms with two degrees of freedom", Tbilisi: Metsniereba, 2000. –286 p. (In Russian).

[10] N.S. Davitashvili, "Dynamic investigation of plane five-link hinged mechanism with two degrees of freedom ",Tbilisi: Georgian Committee of IFToMM, 2006. – 196 p. (In Russian).

[11] N. Davitashvili, O. Gelashvili, "Synthesis of a spatial five-link mechanism with two degrees of freedom according to the given laws of Motion, in Proceedings of EUCOMES 08, the second European Congress on Mechanism Science, Springer, pp. 159-166.

19

Development of a New Internal Finishing of Tube by Magnetic Abrasive Finishing Process Combined with Electrochemical Machining

Muhamad Mohd Ridha, Zou Yanhua, Sugiyama Hitoshi

Graduate School of Engineering, Utsunomiya University, Utsunomiya-shi, Tochigi-ken, Japan

Email address:

ridha.muhamad1981@gmail.com (M. M. Ridha), yanhua@cc.utsunomiya-u.ac.jp (Z. Yanhua), sugiyama@cc.utsunomiya-u.ac.jp (S. Hitoshi)

Abstract: The research proposes a new internal surface magnetic abrasive finishing (MAF) process, which compounded with electrochemical machining (ECM) to decrease machining time. The electrochemical process changes the morphology of the aluminum tube internal surface, producing an oxidation film. Then, we removed the film by magnetic abrasive finishing, results in minimized surface roughness in a significantly reduced processing time when compared to the conventional MAF. In this research, a new experimental set up with a tool that capable of magnetic abrasive finishing and electrochemical finishing was designed and developed to study the machining feasibility. The newly developed finishing method demonstrated simultaneous process of aluminum oxide film formation by ECM and its removal by MAF. This process plays a significant role in preventing the deepening of the pit during ECM and speed up the planarization. The method was developed step by step; firstly, ECM and MAF were conducted in two separate processes. In the second experiment, we modified the finishing conditions to facilitate one-stage finishing method. An investigation of the finishing surface is focusing on the pit size that formed by ECM. The pit size indicated the residue of oxide film because it is a part of the oxidation film construction. Pits morphology changes were observed for certain finishing time to determine the minimum finishing time for its removal. Surface roughness and SEM photograph of the finishing surface were recorded and studied.

Keywords: Magnetic Abrasive Finishing, Electrochemical Machining, Internal Surface Finishing, Surface Roughness, Finishing Characteristic, Aluminum Tube Finishing

1. Introduction

The utilization of clean tubes or sanitary tubes in the field of semiconductor, chemicals, biotechnology, etc. fields is essential since the smooth surface prevents accumulation of dirt or oils in fine grooves that exists on rough metal surfaces. In a highly pressured container, dirt accumulations could cause corrosion and leading to burst and explosions [1]. In the food sector, rough surface in food tanks promotes microbial growth [2]. Due to expanding business in these fields, the recent trend shows that demand for clean tube has significantly increased. Moreover, for tube length more than 2 meters, internal finishing using machines has various mechanical constraints and finishing process is time-consuming. Method to polish tube internal surface using magnetic slurry for a thin tube was proposed in 1999 for thickness less than 5mm [3], [4]. This method works by placing a slurry that consist iron powder, white alumina and polishing agent in the tube with the present of magnetic field from the magnetic poles positioned outside of the tube. As the poles and tube rotate in the opposite direction, the movement results in finishing mechanism to the tube internal wall from the magnetic force that reacts on the slurry to the tube. The slurry moves with agitation in the tube, and this movement causes abrasives mixing while the process takes place, replacing abrasives that contacted with the surface constantly along the process. The mechanism prolonged the slurry life. However, as the tube thickness increases, the distance between poles and magnetic slurry also increase, results in the magnetic force to drop significantly. Magnetic machining jig was proposed for finishing of thick tubes (thickness 5~30mm) and has been successful in polishing thick tubes internal surface [5]–[7]. The magnetic machining jig constructed with magnet and yoke is positioned in the tube so that it creates a closed magnetic circuit that gathers magnetic flux into the circuit. This construction demonstrated more than ten times stronger machining force than without it [8]. Detailed study regarding the machining jig development and characteristics was performed for flat surface finishing and internal finishing in tubes. In this method, magnetic jig moves synchronically with the external poles and at the same time polish tube internal surface with magnetically adhered slurry at the magnet side of the jig. Compared to other metal, aluminum is softer in physical characteristic, has a lower melting point and difficult to polish. Therefore, it requires specific conditions to be polished to high surface finish. The finishing of the aluminum tube using MAF was proposed, and optimum finishing conditions was explained [9].

On the other hand, electrochemical machining is well known due to its short finishing time and high surface finishing capability for metals such as SUS304 [10], [11]. Electrochemical finishing for aluminum causes pits in most cases [12], [13]. The studies regarding pit reduction by vibration methods were reported in several studies [14], [15]. Meanwhile, El-Taweed has integrated electrochemical turning process and magnetic abrasive finishing for finishing of the aluminum rod external surface. However, the surface finishing is still relatively low [16]. Liu had studied the development of processing aluminum using electrochemical magnetic abrasive finishing for flat surface had clarified the finishing characteristic of the process for a variety of conditions [17]. Nevertheless, the study lacking focus on the pit formed by electrochemical machining that may have directly affected the surface roughness.

In this research, we proposed the magnetic abrasive finishing (MAF) combined with electrochemical machining (ECM) which is also known as electrochemical magnetic abrasive finishing (EMAF) for finishing of tube internal surface. The electrochemical process modifies the uneven surface morphology by creating oxidation film. Afterward, the MAF plays an important role in removing the film and finishing the surface. MAF quickly removes the oxidation film because of the porous aluminum oxide itself built of [18]. In the final process, with the two processes combined, results in a significant reduction in total processing time compared to the conventional MAF. The oxidation film has pits on it that is visible through SEM. We conducted an investigation on the finishing surface morphology for observations of removal of pits by MAF in the final stage to ensure achieving a minimized surface roughness. Observation of surface through SEM photographs and measurement of surface roughness was done to evaluate the morphology changes. Average surface roughness, R_a was measured using contact-type surface roughness (Surftest SV-624-3D, Mitutoyo).

2. Processing Principle

The ECM takes place in the presence of an electrolyte and current supply between workpiece (anode) and combination machining tool (cathode), which develops aluminum oxide on the workpiece internal surface. The electrochemical equation for both anode and cathode are shown as the following: [19].

$$\text{Anode: } 2OH^- \rightarrow H_2O\ 1/2O_2 + 2e^- \tag{1}$$

$$\text{Cathode: } H^+ + e^- \rightarrow 1/2H_2 \tag{2}$$

From (1), we understand that the oxygen gas is released at the anode. The oxygen reacts directly with aluminum on the surface to form aluminum oxide film. This reaction creates pit on the surface. At the same time of the ECM, MAF takes place to remove the aluminum oxide film. The slurry consists of iron powder, white alumina and polishing agent, become magnetized on the magnet side of the combination machining tool to form magnetized particles. The machining effects caused by white alumina abrasives that move with magnetic particles finishes the tube internal surface through this process.

The formation of oxidation film by ECM and removal by MAF is on-going at the same time. After a particular period, the ECM process is stopped, but the continuity of MAF is designed to remove the residual of the oxidation film. The porous and soft structure of the oxidation film accelerates the process when compared to conventional MAF process that works on the hard aluminum surface.

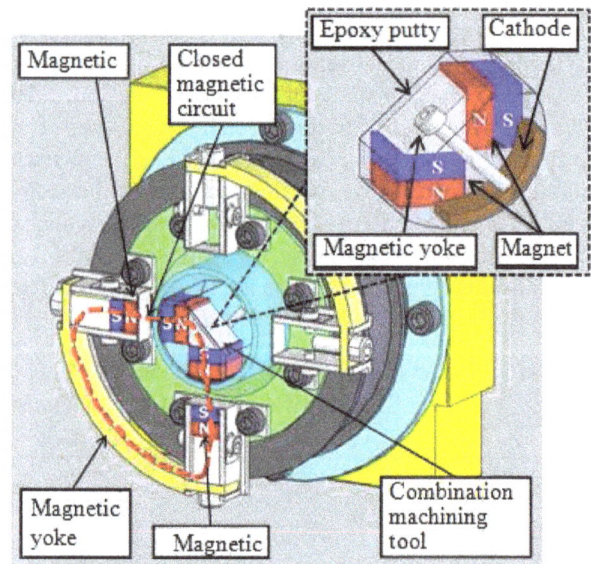

Figure 1. *Schematic of processing principle.*

Figure 1 shows the schematic of processing principle of magnetic abrasive finishing with combined electrochemical finishing in 3D model and machining tool structure. The machining tool has two functions; magnetic abrasive finishing and electrochemical finishing. Hence, we called it combination machining tool. The tool`s N-S poles, external magnetic poles and magnetic yoke are positioned inside the tube with the N-S-N-S sequence to create a closed magnetic circuit as shown in the figure. This construction creates a strong magnetic force, which push towards tube internal surface direction. When the external magnetic poles are rotated, the combination machining tool synchronically rotates with the poles. At the same time, it pushes toward the internal surface. The strong magnetic force is essential for the finishing force, which works in the tangential direction to the tube surface. Meanwhile, the workpiece is also rotated in the opposite rotation of the external poles direction, and stroke movement is applied by a crank mechanism that connected to the chuck.

3. Experimental Method and Conditions

Figure 2 shows the photograph of the experimental setup. The workpiece is fixed to the chuck and the other end to a supporter. Rotation is applied to the workpiece by turning the chuck that is connected with a spindle and pulley system to motor A. Motor B applies another rotation of opposite direction to the external magnetic poles. The stroke movement at the workpiece length direction is applied by using a crank mechanism controlled by motor C. The combination machining tool is wrapped with a polyester cloth to prevent scratches to the workpiece. The tool is magnetically adhered with the slurry and positioned in the tube accordingly.

Electrolyte sodium nitrate was pumped from the tank and flows to the tool so that the tool-tube gap fills with it. The ECM starts when the current is supplied to the anode (workpiece) and cathode (tool) by a direct current power supply. The workpiece was ultrasonically washed before and after the process with ethanol, air-dried and measured its weight. Surface contact-type surface roughness was used to measure the average roughness at three places distanced 120 degrees each.

Figure 3 shows the SEM images of the aluminum oxide structure from a study of aluminum oxide growth and characterization. In order to view aluminum oxide film thickness by the SEM photograph, it was done using aluminum purity 99.99% that was produced using magnetron sputtering [20]. However, for the current research, we are using industrial standard aluminum tube A6063 that has the purity of 97.5%. Therefore, the method to observe the film thickness is not suitable because the significant line between aluminum and aluminum oxide film could not be seen. We have decided to observe the pit structures as a parameter to evaluate the surface morphology, additional to the surface roughness measurement using contact-type surface roughness machine.

3.1. Magnetic Abrasive Finishing (MAF)

The conventional MAF process was conducted for comparison. Table 1 shows the detailed experimental conditions with finishing time 5,10,15,20,25,30,35 and 40 minutes. For this experiment, no direct current power supply or electrolyte was used as it only involves MAF. The conditions were similar to previous research conducted for aluminum tube finishing using MAF. The use of bigger size iron particles such as 330 μm causes scratches on the finishing surface. Therefore, the size 149 μm was used [9].

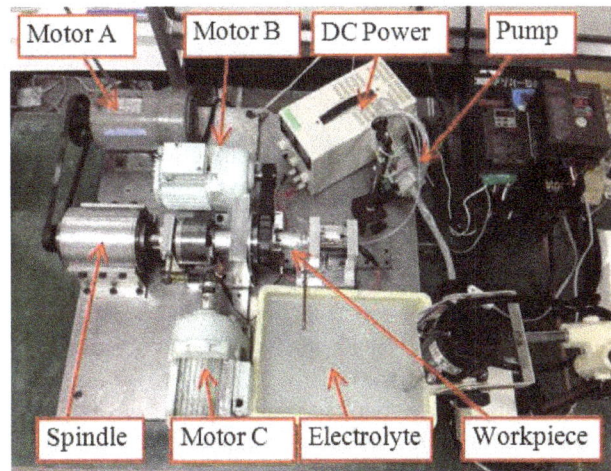

Figure 2. *Photograph of experimental setup.*

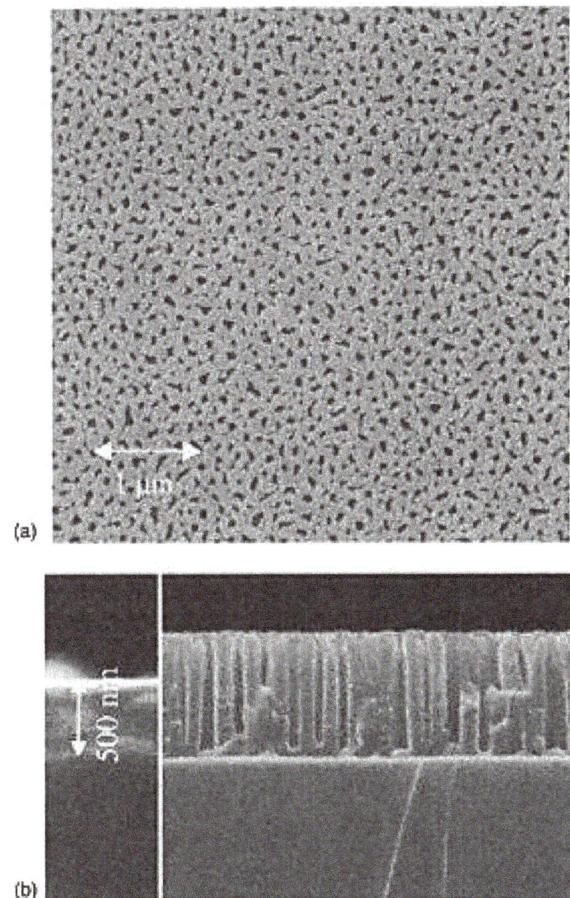

Figure 3. *SEM images (a) looking down on the surface of the porous aluminum oxide and (b) side profiles of the aluminum and the porous aluminum oxide (the expansion upon conversion from the aluminum to the oxide can be seen) [20].*

3.2. Two-Stages Finishing

The process was conducted in two-stages; the ECM followed by the MAF separately. The experiment was conducted in particular processing time combination for both processes to determine the suitable amount of time needed for each process by observing the conditions of oxide film accumulated. Table 2 shows the details experimental conditions for electrochemical finishing. The electrolyte used was sodium nitrate 20% flowed at 30 ml/min by adjusting a flow meter that connected to the pump. Electrical current is set to fix at 0.5 A from the direct current power supply. The machining time for electrochemical finishing was set at 3, 4 and 5 minutes. The combination machining tool was wrapped in a polyester cloth to prevent from scratching the tube. In this stage, no abrasive slurry was used as it involves only ECM. External poles rotation speed was set at 50 rpm with no workpiece rotation. After the process ended, the workpiece is ultrasonically washed in ethanol and measured its weight.

In the second stage, the magnetic abrasive finishing was performed with finishing conditions similar as shown in Table 1 except for the finishing time. It was conducted for 3,4,5, and 6 minutes for certain conditions. The slurry mixture was magnetically adhered on the magnet side on the combination machining tool and positioned in the tube. After the process, the workpiece was ultrasonically washed in ethanol, air-dried and measured its weight.

The finishing time combination was fixed 5 minutes for ECM followed with 5, 4 and 3 minutes of MAF. Next, it was changed 4, 3 and 2 minutes for ECM followed by a fixed 5 minutes of MAF. Surface roughness was measured, and observation on the surface finishing was made under SEM to study the pit size and morphology. This method allow us to know how long the processing time needed for ECM to reduce initial hairlines efficiently and produce the aluminum oxide film, and how long does it takes for MAF to remove the pit morphology and achieve a high finish surface.

Table 1. *Conventional MAF finishing conditions.*

Workpiece	Aluminum tube A6063 (Ø40xØ36x150 mm)
Machining tool	Ne-Fe-B rare earth permanent magnet 10x12x18 mm 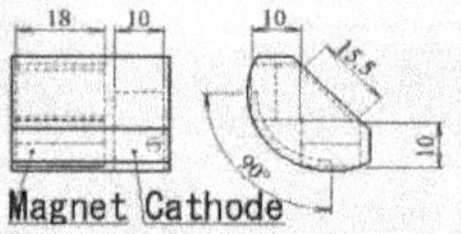
Magnetic abrasive mixture	Iron particle 3.5 g (mean diameter 149 µm); WA #10000 0.5 g; Water soluble polishing liquid 2.5 ml
Finishing time	5,10,15,20,25,30,35,40 min
Pole-tube gap	8 mm
Workpiece rotation speed	200 rpm
Poles rotation speed	50 rpm
Stroke	5 mm/s

Table 2. *ECM finishing conditions.*

Electrode-tube gap	1 mm
Electrolyte	$NaNO_3$ 20% aqueous
Electrolyte amount	30 ml/min
Current density	0.0025 A/mm^2
Poles rotation speed	50 rpm
Finishing time	3,4,5 min

3.3. One-Stage Finishing

For the one-stage finishing method, the process was conducted in one step simultaneously for ECM and MAF. Thus, additional to the quick removal of porous oxidation film, it further cuts processing time. In this method, the combination machining tool was wetted with the 2.0 ml of electrolyte onto the polyester cloth that is used to wrap it. Then, the slurry was magnetically adhered on the magnet side of the tool and positioned in the tube accordingly. The slurry composition is same for MAF as shown in Table 1. Finishing processes were conducted for 8, 9, 10, 11 and 12 minutes of which during that period, the first 2 minutes were allocated for ECM and MAF simultaneously. After 2 minutes, the current supply was shut off to stop the ECM. However, the process continues for MAF for the purpose of resurfacing the aluminum oxide film and removes the pit structures. Similarly, the workpiece is ultrasonically washed, air-dried and measure roughness and weight.

4. Experimental Results and Discussion

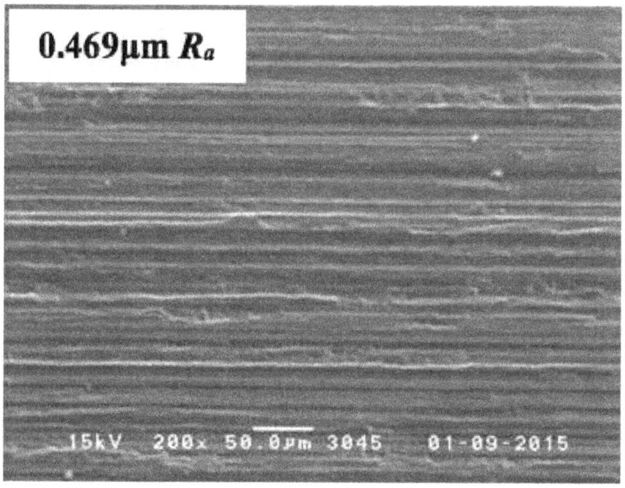

(a) Before finishing.

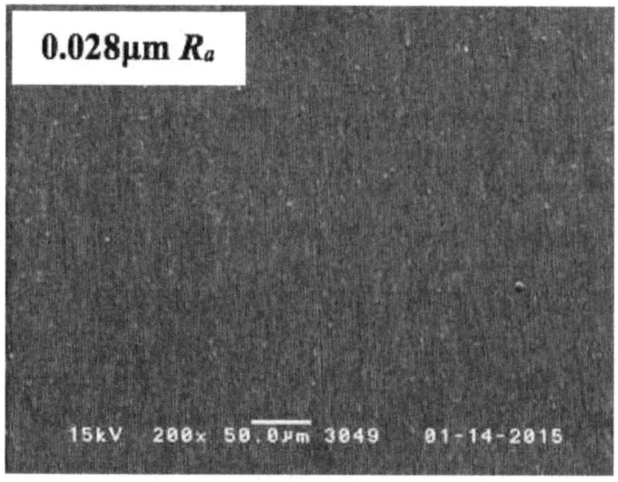

(b) After finishing 40 minutes.

Figure 4. *Surface photograph before and after finishing observed under Scanning Electron Microscope (SEM) (a) before and(b) after finishing.*

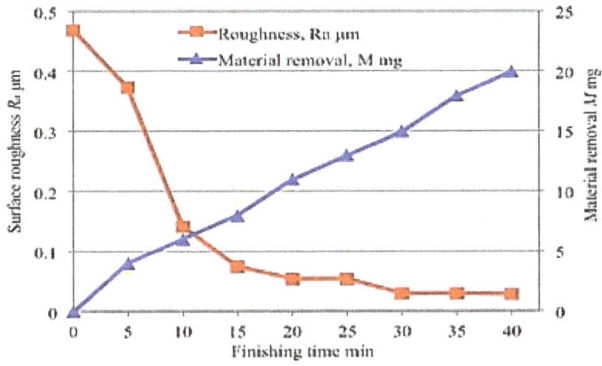

Figure 5. *The change of surface roughness and material removal weight against the finishing time.*

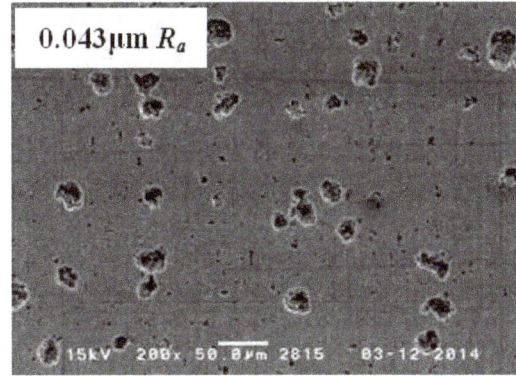

(a) Before finishing.

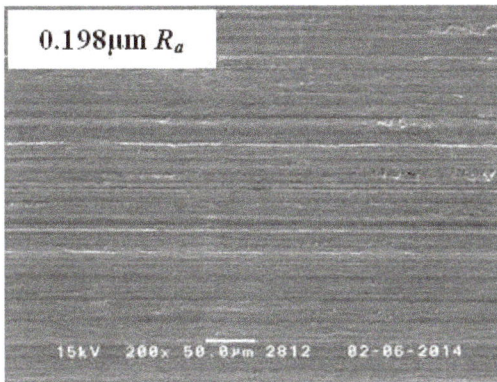

(b) After finishing (5 min ECM+ 5 min MAF).

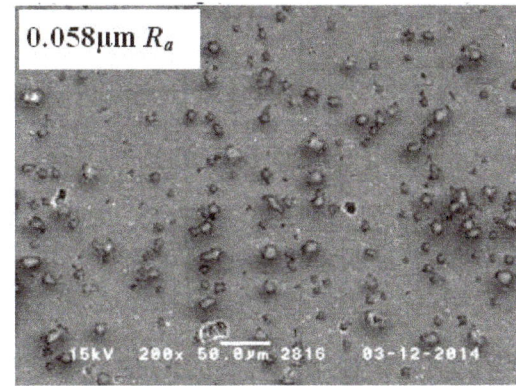

(c) After finishing (5 min ECM+ 4 min MAF).

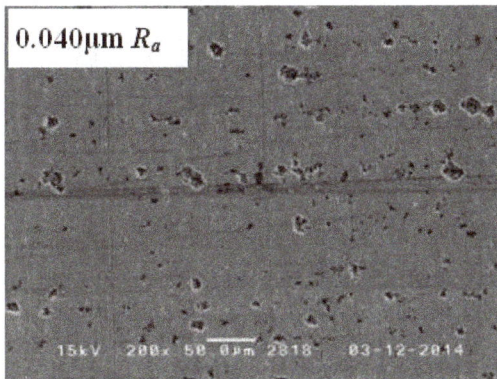

(d) After finishing (5 min ECM+ 3 min MAF).

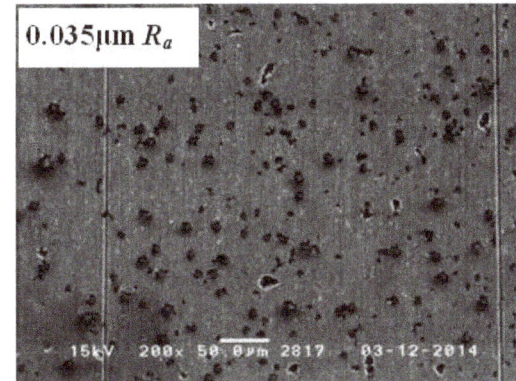

(e) After finishing (4 min ECM+ 5 min MAF).

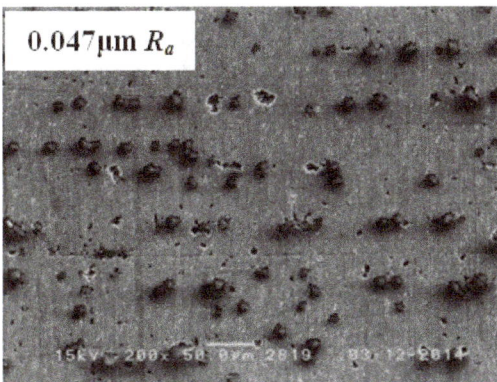

(f) After finishing (3 min ECM+ 5 min MAF).

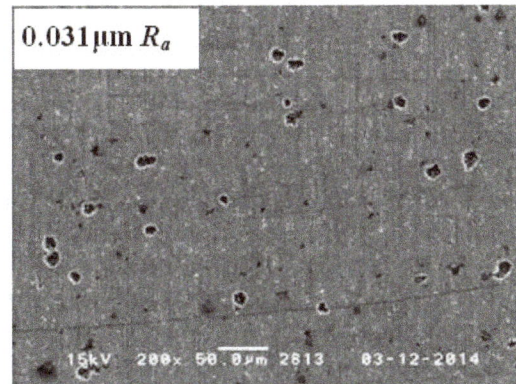

(g) After finishing (2 min ECM+ 5 min MAF)

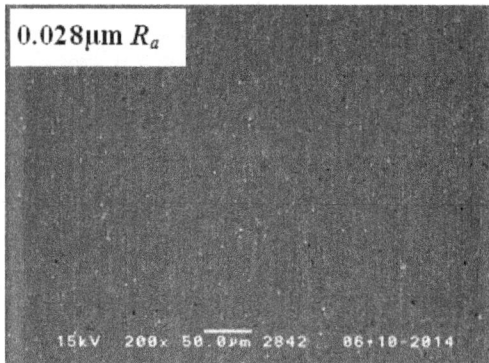

(h) After finishing (2 min ECM+ 6 min MAF).

Figure 6. *Surface photograph before and after finishing observed under Scanning Electron Microscope (SEM) with different finishing time combinations for ECM and MAF separately.*

Aluminum A6063 is softer compared to SUS304. It has Vickers hardness of 70 HV compared to the SUS304, which has 200 HV. The aluminum tube is made by extrusion process that results in hairlines formation on its internal and external surfaces. This internal surface has an initial average roughness that varies ranges from 0.2 to 0.7 μm R_a.

4.1. The conventional MAF

Figure 4(a) shows SEM photograph of the surface before finishing. The initial hairline was seen clearly before the process. Figure 4(b) shows the photograph after processing; the hairlines were removed, and surface roughness measured 0.028 μm R_a. Figure 5 shows the change of surface roughness and material removal against finishing time. The material removal shows constant removal pattern due to the usage of one size iron particle 149 μm through the process for comparison purpose. The surface was gradually leveled and finally it took 30 minutes to achieve surface roughness 0.028 μm R_a.

4.2. Two-Stages Finishing Method

In the first process which is the electrochemical process, the uneven surface undergone a planarization that creates oxidation film on the finishing surface. Figure 6 shows the surface photograph under SEM before and after finishing for different finishing time combination of ECM and MAF. From the observation through SEM, small holes or pit that resulted from etching during the ECM process could be seen. These small holes are part of the oxidation film. The holes diameter is approximately 10 to 20 μm varies depend on the depth. As the removal of oxidation film progress, the size of the pit reduced and gradually diminished as more oxidation film were removed. Figure 6(b) shows that pit still exist with equal processing time for ECM and MAF and even bigger size for shorter MAF time in Figure 6(c) and 6(d). This evident means more MAF time is needed to remove the pit structures. In Figure 6(e) ~ 6(g) we could observed that pit size gradually reduce in the reduction of ECM finishing time. Finally, in Figure 6(h) the pit was completely unseen for the processing time combination. Comparison with MAF process

shows that the current method reached a similar level of average surface roughness 0.028 μm R_a within 8 minutes finishing time, compared to the conventional method that took 30 minutes for similar level of surface roughness.

4.3. One-Stage Finishing Method

In the one-stage finishing method, the two finishing process were conducted simultaneously in order to reduce the total finishing time. Oxidation film was formed and at the same time being removed for the first two minutes of processing time. However, oxidation film was produced at a faster pace than removal by MAF. As a result, at the point the ECM process ends, an extra processing time of MAF is needed for complete removal of the pit created in the ending processing time. This is confirmed in Figure 6(b) and 6(h).

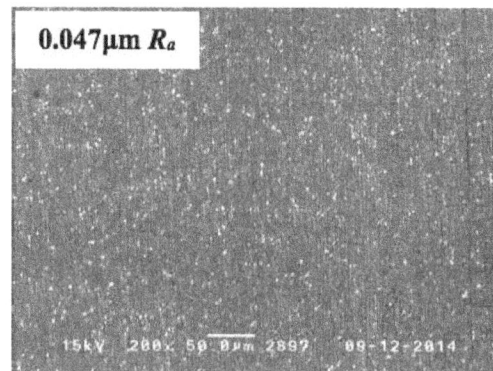

(a) After finishing (2 min EMAF+6 min MAF)

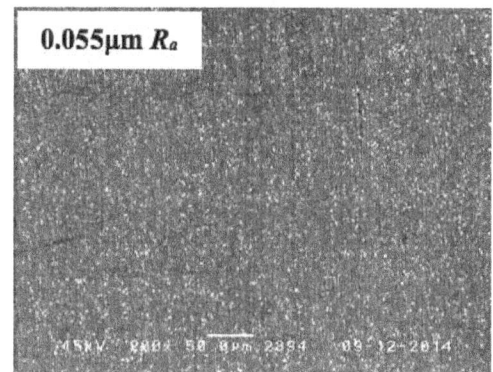

(b) After finishing (2 min EMAF+7 min MAF)

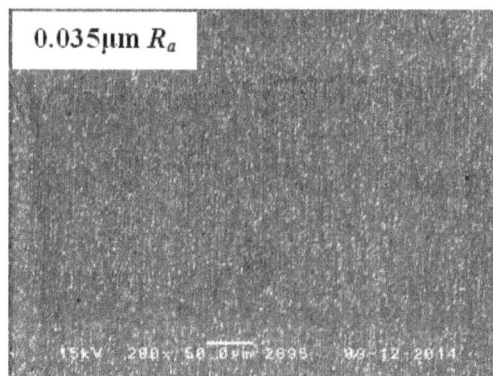

(c) After finishing (2 min EMAF+8 min MAF)

(d) After finishing (2 min EMAF+9 min MAF)

Figure 7. *Surface photograph before and after finishing observed under Scanning Electron Microscope (SEM) with different finishing time combinations for ECM and MAF separately.*

Figure 7 shows the finishing surface photograph of one-stage finishing observed under SEM. In all finishing conditions, the photograph result had showed that initial hairlines have become invisible. This swift process was due to oxidation film formation being constantly removed and created during the initial two minutes of the processing time. Surface improves to 0.029μm R_a for MAF 9 minutes as shown in Figure 7(d).

The advantage of the one-stage process is that simultaneous oxidization of aluminum and removal of oxidation film speed up the planarization process thus shortened the processing time. Since the simultaneous processing of ECM and MAF that have different finishing environment, it is a disadvantage for MAF due to change in viscosity by electrolyte as an additive. In the two-stage processes, it took a total of 8 minutes to achieve 0.028μm R_a, so we predicted it would take 6 minutes for the one-stage process based on time combination. However, the experimental result revealed that it took 11 minutes to achieve similar surface roughness.

The important issue in the combination of the two processes was the finishing environment that affected MAF due to the existence of electrolyte in the slurry, which changes the viscosity. Electrolyte mixed with slurry may also affect the electrochemical characteristic performance of the electrochemical reactions.

5. Conclusions

The results can be summarized as follows.

1) The study revealed that the newly developed finishing method referred as magnetic abrasive finishing with combined electrochemical finishing for finishing of the aluminum tube internal surface using the combination machining jig was successfully performed.

2) The finishing method for one-stage and two-stage finishing method on aluminum tube raw material internal surface had shown finishing time reduction of 60-70%, compared with conventional MAF method that took 30 minutes to achieve similar surface finish.

3) The combination of finishing time of ECM and MAF is critical in producing the improved finishing of the aluminum tube internal surface. The experiment results suggested that finishing time ratio for ECM and MAF is 2:9 to achieve mirror-finished surface or 0.030μm R_a level.

4) The novelty of this paper is in regards to the proposal of the finishing method for aluminum tube internal surface that required a specially designed combination machining tool. The paper studied the surface finishing by the removal of the pits and finishing time reduction was confirmed.

Acknowledgments

Mohd Ridha Muhamad would like to acknowledge the support from Majlis Amanah Rakyat (MARA) under Ministry of Rural and Regional Development, Malaysia Government through scholarship received. Gratitude is also extended to the Utsunomiya University Creative Department for Innovation (CDI) for their support in this research.

References

[1] J. Fisher, E. Kaufmann, and A. Pense, "Effect of Corrosion on Crack Development and Fatigue Life," *Transp. Res. Rec.*, vol. 1624, no. 98, pp. 110–117, 1998.

[2] C. G. Kumar and S. K. Anand, "Significance of microbial biofilms in food industry : a review," vol. 42, pp. 9–27, 1998.

[3] H. Yamaguchi and T. Shinmura, "Study of the surface modification resulting from an internal magnetic abrasive finishing process," *Wear*, vol. 225–229, pp. 246–255, Apr. 1999.

[4] H. Yamaguchi and T. Shinmura, "Study of an internal magnetic abrasive finishing using a pole rotation system. Discussion of the characteristic abrasive behavior," *Precis. Eng.*, vol. 24, pp. 237–244, 2000.

[5] Y. H. Zou and T. Shinmura, "Study on Internal Magnetic Field Assisted Finishing Process Using a Magnetic Machining Jig," in *Key Engineering Materials*, 2005, vol. 291–292, pp. 281–286.

[6] Y. H. Zou and T. Shinmura, "Development of ultra-precision magnetic abrasive finishing process," *JSME Annu. Meet.*, vol. 2009, no. 2, pp. 157–158, 2009.

[7] Y. H. Zou, J. N. Liu, and T. Shinmura, "Study on Internal Magnetic Field Assisted Finishing Process Using a Magnetic Machining Jig for Thick Non-Ferromagnetic Tube," in *Advanced Materials Research*, 2011, vol. 325, pp. 530–535.

[8] Y. Zou and T. Shinmura, "Development of Magnetic Field Assisted Machining Process Using Magnetic Machining Jig," *Trans. Japan Soc. Mech. Eng. Ser. C*, vol. 68, no. 669, pp. 1575–1581, Feb. 2002.

[9] M. R. Muhamad and Y. H. Zou, "A study of electrolytic combined magnetic abrasive finishing for pipe internal surface," in *Japan Society for Precision Engineering Spring Meeting 2014*, 2014, pp. 693–694.

[10] S. Lee, Y. Chen, C. P. Liu, and T. J. Fan, "Electrochemical Mechanical Polishing of Flexible Stainless Steel Substrate for Thin-Film Solar Cells," vol. 8, pp. 6878–6888, 2013.

[11] B.-H. Yan, G.-W. Chang, T.-J. Cheng, and R.-T. Hsu, "Electrolytic magnetic abrasive finishing," *Int. J. Mach. Tools Manuf.*, vol. 43, no. 13, pp. 1355–1366, Oct. 2003.

[12] K. Okubo and H. Ito, "Electropolishing for Aluminum by Periodic Reversing Current," *J. Surf. Finish. Soc. Japan*, 1986.

[13] T. Sasaki and M. Mushiro, "Influence of Electropolishing Conditions on the Occurrence of Irregular Patterns on the Surface of Anodized Aluminum," 2005.

[14] K. Tajiri and K. Tsujimoto, "Electrolytic Polishing Method of Aluminum for Non-Pitting," *Kinki Res. Surf. Treat. Alum.*, 1997.

[15] T. Nakayama, "Vibrating electropolishing process of aluminium alloys in phosphoric acid solution (8th Report)," *J. Japan Inst. Light Met.*, vol. 9, pp. 56–66, 1959.

[16] T. a. El-Taweel, "Modelling and analysis of hybrid electrochemical turning-magnetic abrasive finishing of 6061 Al/Al2O3 composite," *Int. J. Adv. Manuf. Technol.*, vol. 37, pp. 705–714, 2008.

[17] G. Y. Liu, Z. N. Guo, S. Z. Jiang, N. S. Qu, and Y. B. Li, "A study of processing Al 6061 with electrochemical magnetic abrasive finishing," vol. 14, pp. 234–238, 2014.

[18] Y. W. Jung, J. S. Byun, D. H. Woo, and Y. D. Kim, "Ellipsometric analysis of porous anodized aluminum oxide films," *Thin Solid Films*, vol. 517, no. 13, pp. 3726–3730, May 2009.

[19] Y. Kimoto, A. Yano, T. Sugita, T. Kurobe, and M. Yamamoto, *Application of micromachining*, 5th ed. Tokyo, Japan, 2010.

[20] P. G. Miney, P. E. Colavita, M. V. Schiza, R. J. Priore, F. G. Haibach, and M. L. Myrick, "Growth and Characterization of a Porous Aluminum Oxide Film Formed on an Electrically Insulating Support," *Electrochem. Solid-State Lett.*, vol. 6, no. 10, p. B42, 2003.

20

Experimental study on thermal conductivity of teak wood dust reinforced epoxy composite using Lee's apparatus method

Ramesh Chandra Mohapatra[1, *], Antaryami Mishra[2], Bibhuti Bhushan Choudhury[2]

[1]Mechanical Engineering Department, Government College of Engineering, Keonjhar, India
[2]Mechanical Engineering Departmet, Indira Gandhi Institute of Technology, Sarang, India

Email address:

rameshmohapatra75@gmail.com (R. C. Mohapatra)

Abstract: In the present work, the effective thermal conductivity of teak wood dust (TWD) filled epoxy composites at different volume fractions (6.5, 11.3, 26.8 and 35.9%) have been determined experimentally by using Lee's Apparatus. Composites of teak wood dust particles of 150μ, 200μ and 250μ sizes with varying volume fractions (6.5, 11.3, 26.8 and 35.9%) have been developed by hand lay up technique. From the tests it is observed that for each size of TWD, the thermal conductivity values of composites decreases with increase of filler content which indicates that the TWD reinforced epoxy composites have good insulation properties. It has also been found that the composite with 150μ particle size of teak wood dust at same volume fractions exhibited lowest thermal conductivity compared to composites with 200μ and 250μ of teak wood dust. Therefore the composite with particle size 150μ at 35.9% volume fraction teak wood dust may be more suitable for insulation applications. Experimental results (TWD,150μ) are also compared with the theoretical models (such as Rule of Mixture model, Russel model, Maxwell model Baschirow & Selenew model) and found that the errors associated with all the above four models with respect to experimental ones lie in the range of 20.14 to 84%, 74 to111.84%, 79.13 to 115.79% and 60.13 to 102% respectively.

Keywords: Lee's Apparatus, Epoxy, Teak Wood Dust Composite, Thermal Conductivity, Error Analysis

1. Introduction

Now a days the synthetic fibre composites are replaced by environment friendly materials such as natural fibre like wood,banana, cotton, coir, sisal, jute etc. because natural fibre composites posses better electrical resistance, good thermal and acoustic insulating properties and higher resistance to fracture. In addition, natural fibres have many advantages over synthetic fibres, for example low weight, low density, low cost, acceptable specific properties, renewable and have relatively high strength and stiffness and cause no skin irritations. A better understanding of their physical properties, mechanical and thermal behaviors will enable engineers to produce optimum design for a structure. The thermal conductivity of a composite depends upon the thermal conductive nature of the fibre, matrix properties as well as their volume fractions, sizes, shapes, thickness, orientations and perfect bonding between the constituents. In recent days natural fibre composite materials are widely used in building components to reduce heat transfer in air conditioned buildings in order to decrease energy consumption, in automotives such as car door panels, car roofs, covers etc. These materials are also familiar in interior applications such as furniture and packaging for electrical appliances. The present work relates to investigations on thermal conductivity of epoxy matrix composites filled with teak wood dust due to the following reasons (a) most of the investigations are aimed at enhancing the thermal conductivity of the polymer rather than attempting to improve its insulation capability. (b) although a large number of particulates have been used as fillers in the past, there is no report available on bio-based materials like any kind of wood dust being used for composite making. (c) investigations on thermal conductivity of particulate filled composites are rare. (d) the understanding of the relationship between the effective thermal conductivity of a composite material and the micro-structural properties (volume fractions, distribution of particles, aggregation of particles, properties of individual components, etc.) is far from satisfactory.

2. Review of Literature

Thermal conductivity is an important thermal property for selecting materials for building construction and other applications. Some studies have investigated the thermal conductivity of thermo plastic composites, but few have explored the thermal conductivity of natural fibre reinforced (wood based) composites. Russell [1] developed a model using the electrical analogy. He derived an equation for the thermal conductivity of the composite using a series parallel network. Maxwell [2] studied the effective thermal conductivity of heterogeneous materials and developed first theoretical model for two phase system. Baschirow and Selenew [3] developed a model to calculate thermal conductivities of the real media. Ten wolde et al. [4] reported thermal conductivity of composites in longitudinal direction to be $1.5 - 2.8$ times greater than to the transverse direction. Suleiman et al. [5] investigated the thermal conductivity of wood in both longitudinal and transverse directions in the temperature range of 20^0C to 100^0C. Their results showed that thermal conductivity is about 1.5 times more in the longitudinal direction than in the transverse direction due to non-homogenous nature of wood. Alsina et al. [6] presented the thermal properties of jute-cotton; sisal-cotton and ramie-cotton hybrid reinforced unsaturated polyester composites. The results showed that sisal-cotton hybrid polyester composites have the thermal conductivity 0.213-0.25W/m-^{0}k, jute-cotton hybrid polyester composites have the thermal conductivity 0.10-0.237W/m-^{0}k, and Ramie-cotton hybrid polyester composites have thermal conductivity 0.19-0.22W/m- 0 k. Abdullah [7] presented theoretical and experimental investigations of composite material as thermal insulation consists of natural fibres (white feather, Jute, Egg shell and Black feather). The results showed that the Jute composite material gives good results as composite thermal insulation compared with other natural composite materials. Haddadi et al. [8] investigated the thermal behavior of hollow conductive particle filled in epoxy resin using 3-D finite element computation. The computational results showed an increase of the effective thermal conductivity with increasing wall thickness of the hollow particle. However, for a large contact resistance and/or for a high effective thermal conductivity, it was shown that the contact resistance has a dominant influence on the effective thermal conductivity of the composite. Prisco [9] investigated experimentally the thermal conductivity of wood flour (WF) filled high density polyethylene composite (Wood plastic composite, WPC). Experimental results showed that the WPC thermal conductivity decreases with the filler content and WF content. Mohapatra et al [10] determined the thermal conductivity of the palm fibre reinforced polyester composites at different volume fractions of the fibre experimentally by using Lee's apparatus. The experimental results showed that the thermal conductivity of the composite increases with increase in fibre percentage. Mohapatra et al [11] also investigated the thermal conductivity of composites of epoxy reinforced with pine wood dust (PWD) at different volume fractions experimentally by using Lee's apparatus. The experimental results show that the incorporation of pine wood dust results in reduction of thermal conductivity of epoxy resin and there by improves its thermal insulation capability.

3. Thermal Conductivity Models

Many theoretical and empirical models have been proposed to predict the effective thermal conductivity of two phase mixtures. Some of them are cited below.

Series Model (Rule of Mixture):

$$\frac{1}{K_c} = \frac{1-\varphi}{K_m} + \frac{\varphi}{K_f} \tag{1}$$

Where c- composite, m- matrix, f-filler, Φ- volume fraction of filler and K-Thermal conductivity

Parallel model:

$$K_c = (1-\phi)K_m + \phi K_f \tag{2}$$

Where K_c - Thermal conductivity of composite, K_m-Thermal conductivity of matrix, K_f- Thermal conductivity of filler and Φ- is the volume fraction of the filler.

Maxwell model:

The derived equation is:

$$K_c = K_m[\frac{K_f + 2K_m + 2\varphi(K_f - K_m)}{K_f + 2K_m - \varphi(K_f - K_m)}] \tag{3}$$

Russel model:

$$K_c = K_m[\frac{\phi^{\frac{2}{3}} + \frac{K_m}{K_f}(1-\phi^{\frac{2}{3}})}{\phi^{\frac{2}{3}} - \phi + \frac{K_m}{K_f}(1+\phi-\phi^{\frac{2}{3}})}] \tag{4}$$

Baschirow and Selenew model:

$$K_c = K_m[1 - \frac{a^2\pi}{4} + \frac{a.\pi.p}{2}\{1-\frac{p}{a}\ln(1+\frac{a}{p})\}] \tag{5}$$

Where

$$p = \frac{K_f}{K_m - K_f}, \qquad a = (\frac{6\phi}{\pi})^{\frac{1}{3}}$$

4. Experimental Details

4.1. Materials

Teak wood dust collected from a local vendor. has been chosen as the reinforced material mostly for its light weight, low density (0.8g/cc) and   low thermal conductivity

(0.085W/m-^{0}K) . Epoxy (LY 556 and the corresponding hardener HY 951 supplied by Hindustan Ciba Geigy Ltd, India) has been used as matrix material.

Table 1. Properties of Materials

Properties	Material's value	
	Teak wood dust	**Epoxy**
Density(g/cc)	0.800	1.200
Thermal conductivity(W/m-K)	0.085	0.363
Young's modulus of elasticity(Gpa)	10.500	20.000
Tensile strength(Mpa)	95.000	75.000

4.2. Composite Preparation

The low temperature curing epoxy resin and corresponding hardener were mixed in a ratio of 10:1 by weight as recommended. The composite samples have been prepared by using hand-lay-up technique to measure the thermal conductivity (using Lee's apparatus) of teak wood dust (with average size 150μ, 200μ and 250μ respectively) reinforced in epoxy resin. A mould of 110 mm diameter and 5mm thickness was made from a stainless steel sheet (Fig.1). It was coated with wax and silicon spray was used as releasing agent for easy removal of the sample. The cast composite was cured under a load of about 50kg for 24 hours before it was removed from the mould. Then this cast was post cured in air for another 24 hours. The specimens were prepared having dimension of 110mm diameter with thickness of 5mm.

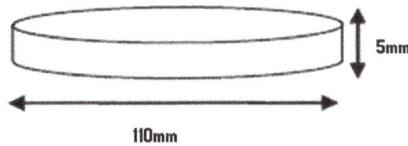

Fig. 1. Standard dimensions of specimen

4.3. Experimental Set up

The thermal conductivity test was carried out with Lee's disc apparatus as shown in Fig.2. The Nickel disc (N) is hung from the stand with the help of three strings. A heating chamber (H) with facility of passage of steam in and out is created. Metallic disc (M) is placed on the top of a heating chamber (H). Sample disc (S) is placed in between metal disc and nickel disc. Two holes are made in the nickel disc (N) and metallic disc (M) for the insertion of thermometers to measure the temperature.

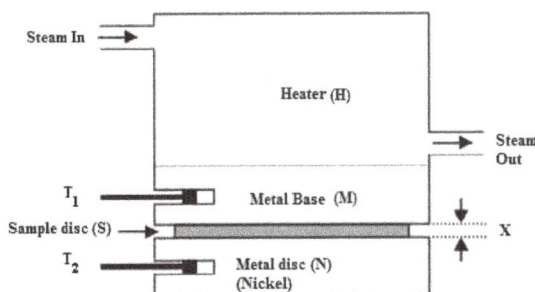

Fig. 2. Schematic diagram of Lee's disc Apparatus.

4.4. Working Procedure

Initially mass of the nickel disc (N) was measured using a balance. Diameter of the specimen was found out using Vernier Caliper and the thickness was measured using a screw gauge. After this

- The heater (H) was started by sending steam through the heating chamber. The temperatures T_1 and T_2 were recorded at a regular interval of 5 minutes till they reached the steady state.
- Then, the supply of steam was cut off and nickel disc (N) and specimen or sample disc (S) were removed. Heat was supplied to the nickel disc (N) along with the sample(S) with the help of Bunsen burner so that nickel disc along with sample is heated to a temperature 10^0C above the steady state temperature T_2. After that the Bunsen burner was removed and allowed the nickel disc (N) to cool. Temperatures were noted in every half a minute until the temperature falls about 10^0C from steady state temperature T_2.
- Variation of temperature of Nickel disc with time of cooling was plotted as shown in Fig. 3. A tangent is drawn at the steady state temperature T_2. Thus the slope of this tangent gives the rate of cooling $\partial T/\partial t$ at steady state temperature T_2.

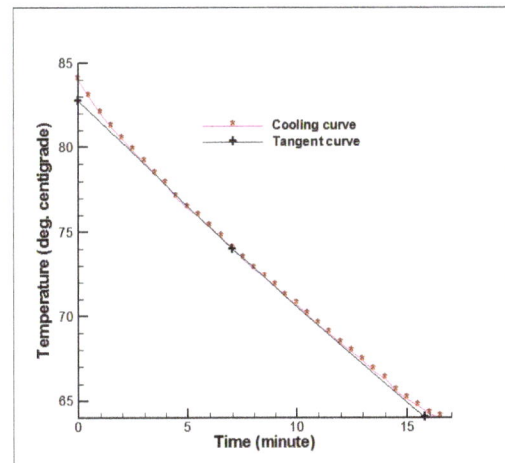

Fig. 3. Cooling rate of composite (Temperature – Time curve)

5. Thermal Conductivity Measurement

The rate of heat conducted through the specimen or sample is

$$Q = KA(T_1 - T_2)/L \qquad (6)$$

Where, L is the thickness of the sample, A is the area of cross section of the sample, K is the thermal conductivity, Q is the rate of heat transfer and $(T_1 - T_2)$ is the temperature difference.

The rate of heat lost by the nickel disc (N) to the surrounding under steady state is

$$Q = mc(\partial T/\partial t)_{T_2} \qquad (7)$$

Where, m is the mass of nickel disc (N), c is the specific heat of the brass disc (B) and $(\partial T/\partial t)$ is it's rate of cooling at T_2

Comparing equations 6 and 7

$$K = mc(\partial T / \partial t)_{T_2} / A(T_1 - T_2)/L \qquad (8)$$

$(\partial T/\partial t)_{T_2}$ and $(T_1\text{-}T_2)$ is calculated using Lee's disc apparatus. Giving the input value of mass of nickel disc (N), specific heat of the nickel disc, thickness of the sample and area of cross section of the of the sample, the thermal conductivity is calculated.

6. Results and Discussion

The thermal conductivity of the filler (TWD) and the epoxy resin have been evaluated by extrapolating the linear regression of the thermal conductivity values of the composite to 100% filler content and zero% filler content. It is found that the thermal conductivity of filler (TWD) and epoxy resin are 0.085W/m-^{0}K and 0.363W/m-^{0}K respectively. The behavior of the thermal conductivity values of different composites can now be explained using the thermal conductivity values of filler and epoxy resin. In this study, the thermal conductivity values of teak wood dust (TWD) reinforced with epoxy composites at different volume fractions of TWD is presented in Table 2. Figure 4 shows that for each size of TWD, the thermal conductivity of composite decreases with increase of filler contents which indicates that the TWD reinforced epoxy composites have good insulation properties. This is because the core of the filler is porous and air voids are created during preparation of composite. It has also been found that as the particle sizes of the TWD filler increases the thermal conductivity values with different filler content increases. In this Figure 4 the TWD filler with three different particle sizes (150μ, 200μ and250μ) are compared and found that the thermal conductivity values of 150μ (TWD filler) at same filler contents is least. From the above discussion it is concluded that choosing 150μ size is most advantageous one as it has best insulation properties on comparison to 200μ and 250μ particle sizes of TWD filler.

In this study, the effective thermal conductivity values of teak wood dust (TWD) at different volume fractions (i.e. 6.5%, 11.3%, 26.8% and 35.9% respectively) filled with epoxy composites is investigated experimentally. The obtained values are compared with four different theoretical thermal conductivity models as shown in Fig.5. From this figure, it is found that the experimental results along with all four theoretical thermal conductivity models are close to each other at low filler content. After that as the filler content increases the thermal conductivity values obtained from experiment as well as from theoretical models decreases accordingly. It has also been found that all four theoretical models shown in Fig.4 overestimate the values of thermal

conductivity obtained from the experimental one. This may be due to some of the assumptions taken for models are not practical. Further in theoretical models orientation of the fillers was assumed to be perfect, but in actual practice, when liquid matrix is powered over the fillers some of the fillers may be misaligned. However at higher volume fractions of the TWD filler the experimental values of thermal conductivity are in agreement with the predicted values. On comparison, it has been found that the errors associated with all the above four models with respect to experimental one lie in the range of 20.14 to 84%, 74 to111.84%, 79.13 to 115.79% and60.13 to 102% respectively. The values of thermal conductivity and percentage of errors associated with each method for individual composite with two components i.e. teak wood dust (TWD) filler and epoxy are given in Table 3 and Table 4 respectively

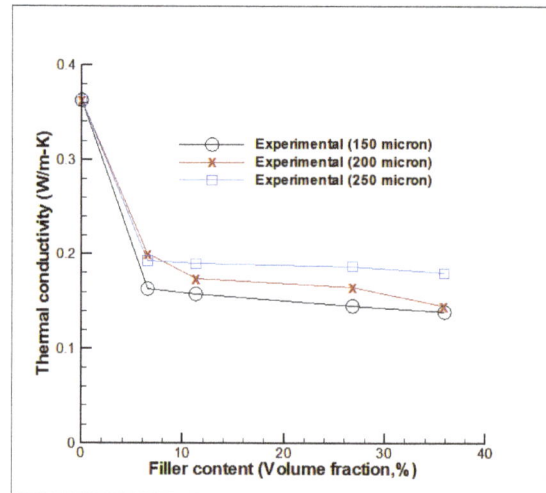

Fig. 4. *Thermal conductivity of epoxy composites as a function of filler content (TWD)*

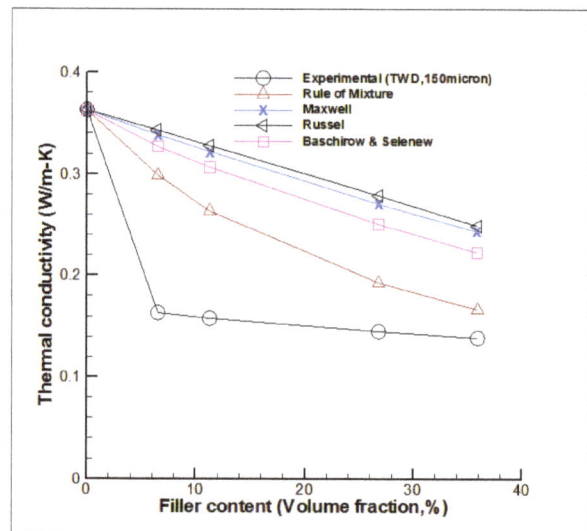

Fig. 5. *Comparison of thermal conductivity of different methods with varying filler (TWD) content*

Table 2. Measured thermal conductivity values of composites of varied composition & particle size of TWD

Sample	Particulate content (Volume fraction, %)	Thermal conductivity values of composites(W/m-K)		
		150micron	200micron	250micron
1	6.5	0.163	0.201	0.193
2	11.3	0.152	0.174	0.189
3	26.8	0.145	0.165	0.187
4	35.9	0.139	0.145	0.180

Table 3. Thermal conductivity values of composites obtained from different methods

Sample	Particulate content (Vol.%)	Effective thermal conductivities of composites (W/m-K)				
		Rule of mixture	Maxwell	Russel	Baschirow&Selenew	Experimental
1	0(Neat epoxy)	0.363	0.363	0.363	0.363	0.363
2	6.5	0.300	0.339	0.343	0.327	0.163
3	11.3	0.264	0.322	0.328	0.307	0.152
4	26.8	0.193	0.271	0.279	0.251	0.145
5	35.9	0.167	0.243	0.249	0.223	0.139

Table 4. Percentage errors with respect to the experimental value

Sample	Particulate content (Vol.%)	Percentage of errors with respect to the Experimental value			
		Rule of mixture	Maxwell	Russel	Bashirow & Selenew
1	0(Neat epoxy)	0	0	0	0
2	6.5	84.00	108.00	110.43	100.61
3	11.3	73.68	111.84	115.79	102.00
4	26.8	33.10	86.90	92.41	73.10
5	35.9	20.14	74.82	79.13	60.13

7. Conclusion

Out of the results obtained it may be concluded that the teak wood dust as reinforcing material is an eco-friendly, nontoxic, low cost and easily available material as compared to synthetic fibres. Epoxy as a matrix material has been used in a wide variety of products such as adhesives, casting compounds, body solders and encapsulates. Good chemical resistance, excellent mechanical properties and modification versatility make epoxy attractive while it's inherent low viscosity and volatility as well as moderate cure temperature allow for production ease. The measured values of effective thermal conductivity are obtained for different volume fractions of teak wood dust fillers. Incorporation of teak wood dust results in reduction of thermal conductivity of composites there by improves its thermal insulation capability From the experiments it is observed that the composite with particle size 150μ at 35.9% volume fraction teak wood dust exhibited lowest thermal conductivity i.e 0.139W/m-K. Therefore the composite with particle size 150μ at 35.9% teak wood dust may be more suitable for insulation applications and coatings. The results of this study indicate that teak wood dust reinforced composites are light in weight, economical as it is easily available and posses a good thermal insulating properties. Hence the newly developed composite materials can be used for applications in automobile interior parts, electronic packages, ceiling roofs, building constructions, sports goods and furniture etc.

References

[1] Russell HW. Principles of heat flow in porous insulation, *J Am Ceram Soc .18*(1935), 1.

[2] Maxwell JC. "A Treaties on Electricity and Magnetism," 3rd ed. New York,: Dover; (1954).

[3] Baschirow AB, Manukian AM. Thermal conductivities of polymers at various temperatures and pressures Mech. Polim (1974), 3, pp.564.

[4] TenWolde, A., McNatt, J.D., Krahn, L. "Thermal Properties of wood and wood panel products for use in buildings", Oak Ridge NationalLaboratoryhttp://www.fpl.fs.fed.us/documnts/pdf1988/tenwo88a.pdf.

[5] Suleiman B. M., Larfeldt J., Leckner B. and Gustavsson M.,"Thermal conductivity and diffusivity of wood," Wood Sci. Technol. Vol. 33 (6), 1999, pp. 465-473.

[6] Alsina, OLS, de Carvalho, LH, Ramos Filho, FG, d Almeida. JRM, "Thermal properties of hybrid lingo cellulosic fabric reinforced polyester matrix composites," Journal of Polymer Testing, Vol. 24, 2005, pp. 81-85.

[7] Abdullah Fadhel Abbas, "Theoretical and Experimental investigation of natural composite materials as thermal insulation," Al- Qadisiya Journal for engineering sciences, Vol. 4 (2), 2011.

[8] Haddadi Manel, Agoudjil Boudjemaa, Boudenne A hollow particles,"bderrahim and Garnier Bertrand, "Analytical and numerical investigation on effective thermal conductivity of polymer composites filled with conductive hollow particles,"International Journl of Thermophysics,2013.

[9] Prisco Umberto, "Thermal conductivity of flat-pressed wood plastic composites at different temperatures and filler content," Science and engineering of composite materials, Vol. 0, Issue 0, 2013, pp. 1-8.

[10] Mohapatra Ramesh Chandra, Mishra Antaryami and Choudhury Bibhuti Bhushan, "Investigations on Thermal Conductivity of Palm Fibre Reinforced Polyester Composites," IOSR Journal of Mechanical and Civil Engineering (IOSR-JMCE) Volume 11, Issue 1 Ver. I (Feb. 2014), PP 01-05.

[11] Mohapatra Ramesh Chandra, Mishra Antaryami and Choudhury Bibhuti Bhushan, "Measurement on Thermal Conductivity of Pine Wood Dust Filled Epoxy Composites" American journal of Mechanical Engineering, Vol. 2, Issue 4 (2014), pp 114-119. DOI:10.12691/ajme-2-4-3.

21

Adaptive Tracking Control of a PMSM-Toggle System with a Clamping Effect

Yi-Lung Hsu, Ming-Shyan Huang, Rong-Fong Fung[*]

Department of Mechanical & Automation Engineering, National Kaohsiung First University of Science and Technology, Kaohsiung, Taiwan

Email address:

u9615907@nkfust.edu.tw (Yi-Lung Hsu), rffung@nkfust.edu.tw (Rong-Fong Fung)

Abstract: This paper discusses an adaptive control (AC) designed to track an energy-saving point-to-point (ESPTP) trajectory for a mechatronic system, which is a toggle mechanism driven by a permanent magnet synchronous motor (PMSM) with a clamping unit. To generate the PTP trajectory, we employed an adaptive real-coded genetic algorithm (ARGA) to search for the energy-saving trajectory for a PMSM-toggle system with a clamping effect. In this study, a high-degree polynomial was used, and the initial and final conditions were taken as the constraints for the trajectory. In the ARGA, the parameters of the polynomials were determined by satisfying the desired fitness function of the input energy. The proposed AC was established by the Lyapunov stability theory in the presence of a mechatronic system with uncertainties and the impact force not being exactly known. The trajectory was tracked by the AC in experimental results so as to be compared with results produced by trapezoidal and high-degree polynomials during motion.

Keywords: Adaptive Control, ARGA, Clamping Effect, Energy-Saving, Trajectory Planning

1. Introduction

This paper discusses adaptive control (AC) designed to track an energy-saving point-to-point (ESPTP) trajectory for a PMSM-toggle system. In general, this example is referred to as point-to-point control, and it takes into account low acceleration and jerk-free motion [1]. Astrom and Wittenmark [2] presented a general methodology for the off-line tridimensional optimal trajectory planning of robot manipulators in the presence of moving obstacles. Planning robot trajectory by using energetic criteria provides several advantages. On one hand, it yields smooth trajectories and is easy to track, while reducing the stress in the actuators and manipulator structures. Moreover, the minimum amount of energy may be desirable in several applications, such as those with energy-saving control or a quantitatively limited energy source [3]. Examples of minimum-energy trajectory planning are provided in [4]. However, the selection of a suitable profile for a specific application is still a challenge since it affects overall servo performance. Thus, in this study, the authors designed the kinematics of the trajectory profiles for motion tracking control within a PTP trajectory.

The AC techniques proposed in this study are essential to providing stable, robust performance for a wide range of applications such as robot control [5-9] and process control [10]. Most such applications are inherently nonlinear. Moreover, a relatively small number of general theories exist for the AC of nonlinear systems [11]. Since the application of a mechatronic system has minimum-energy tracking control problems for elevator systems, the AC technique developed by Chen [12], who made use of conservation of energy formulation to design control laws for the fixed position control problem, was adopted to control the PMSM-toggle system in this study. In addition, an inertia-related Lyapunov function containing a quadratic form of a linear combination of position- and speed-error states was formulated.

The difference between previous studies [13-18] and this study is that this study takes the clamping unit into consideration. The main contribution of this study is that the proposed AC adapts not only to parametric uncertainties of mass variations, but also to external disturbances. The performance with external disturbances is validated through the results obtained both numerically and experimentally on the energy-saving point-to-point trajectory processes for a PMSM-toggle system with a clamping unit.

2. Modeling of the Mechatronic System

2.1. Electrical Model

The permanent magnet synchronous motor drive toggle system is shown in Fig. 1, and the electromagnetic torque developed by the PMSM is

$$\tau_e = K_t i_q, \tag{1}$$

where τ_e is the electromagnetic torque, i_q is the current, and K_t is the motor torque constant. Usually, the PMSM is controlled by voltage command v_q^*. The machine model of a PMSM can be described as a rotating rotor coordinate. The electrical equation is

$$L_q \dot{i}_q + R_s i_q + \omega_s \lambda_d = v_q^*, \tag{2}$$

where L_q is the inductance, $\dot{i}_q$ is di_q / dt, R_s is the stator resistance, and ω_s and λ_d are the inverter frequency and stator flux linkages, respectively.

The applied torque can be obtained as follows:

$$\tau = n\left(\tau_e - J_m \dot{\omega}_r - B_m \omega_r\right), \tag{3}$$

where τ_e is the electromagnetic torque, the variables ω_r and $\dot{\omega}_r$ are the angular speed and acceleration of the rotor, respectively, B_m is the damping coefficient, and J_m is the moment of inertia. It is noted that $\omega_s = p\omega_r$, where p is the number of pole pairs and n is the ratio of the geared speed-reducer. Eqs. (2) and (3) represent the mathematical model of the PMSM. They give the voltage and motor torque variation with respect to time.

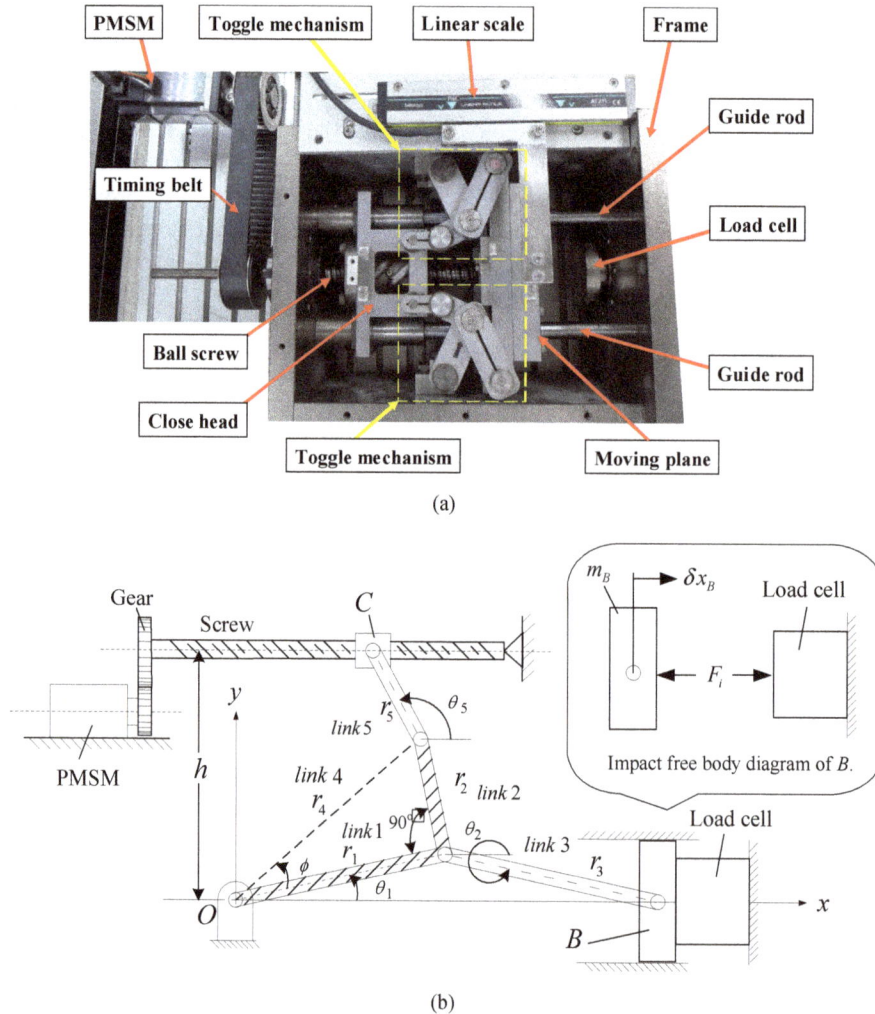

(a)

(b)

Fig. 1. *PMSM-toggle system with a clamping unit. (a) Photograph of experimental device. (b) Physical model.*

2.2. Impact Model

In this section, we consider the motion in a given stroke of the toggle mechanism undergoing impact when two objects collide over a very short period of time. The continuous force model approach [19] employs a logical spring-damper element to estimate the impact force between the two masses of the mechatronic system as,

$$F_i = K_l z + D_l \dot{z}, \qquad (4)$$

where K_l is the elastic spring coefficient, z is the relative displacement or penetration between the surfaces of the two colliding bodies, $\dot{z}$ is the relative speed, and D_l is the damping coefficient. For the time period of $t_i^- \leq t \leq t_i^+$, the differential-algebraic equation can also be rewritten in matrix form. The impact model of the toggle mechanism with a clamping unit is shown in Fig. 2.

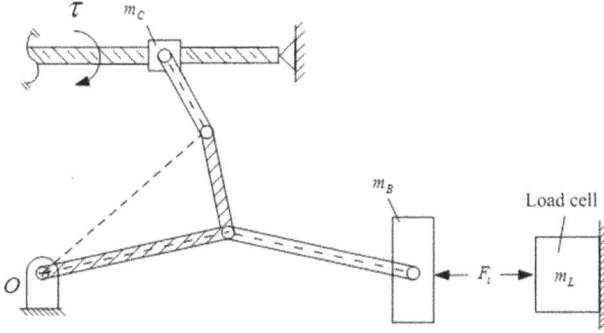

Fig. 2. *Impact model of toggle mechanism with a clamping unit.*

2.3. Mathematical Model of the Toggle Mechanism

The toggle mechanism of the electrical injection molding machine was driven by a PMSM. The experimental setup and physical model are shown in Fig. 1(a) and Fig. 1(b), respectively. The differential-algebraic equations of the toggle mechanism are summarized in matrix form [20]. The matrix form of the equations can be written as:

$$\hat{M}(v)\ddot{v} + \hat{N}(v, \dot{v}) = \hat{Q}(v)U + \hat{D}(v). \qquad (5)$$

where

$$\hat{M} = M^{vv} - \mathbf{M}^{vu}\mathbf{\Phi}_u^{-1}\mathbf{\Phi}_v - \mathbf{\Phi}_v^{\mathbf{T}}\left(\mathbf{\Phi}_u^{-1}\right)^{\mathbf{T}}\left[\mathbf{M}^{uv} - \mathbf{M}^{uu}\mathbf{\Phi}_u^{-1}\mathbf{\Phi}_v\right],$$

$$\hat{N} = \left[N^v - \mathbf{\Phi}_v^{\mathbf{T}}\left(\mathbf{\Phi}_u^{-1}\right)^{\mathbf{T}}\mathbf{N}^u\right] + \left[\mathbf{M}^{vu}\mathbf{\Phi}_u^{-1} - \mathbf{\Phi}_v^{\mathbf{T}}\left(\mathbf{\Phi}_u^{-1}\right)^{\mathbf{T}}\mathbf{M}^{uu}\mathbf{\Phi}_u^{-1}\right]\gamma,$$

$$\hat{Q} = B^v - \mathbf{\Phi}_v^{T}\left(\mathbf{\Phi}_u^{-1}\right)^{T}\mathbf{B}^u, \quad U = i_q,$$

$$\hat{D} = D^v - \mathbf{\Phi}_v^{T}\left(\mathbf{\Phi}_u^{-1}\right)^{T}\mathbf{D}^u.$$

The elements of the vectors $\mathbf{u}$, v and matrices $\mathbf{\Phi}_u$, $\mathbf{\Phi}_v$, $\mathbf{M}^{uu}$, $\mathbf{M}^{uv}$, $\mathbf{M}^{vu}$, M^{vv}, N^u and N^v are detailed in [20]. The system Eq. (5) is an initial value problem and can be integrated by using the fourth order Runge-Kutta method.

3. Energy-Saving Trajectory Planning

This section discusses how the AC was designed to track an ESPTP trajectory for a PMSM-toggle system. The degrees of the polynomial depended on the number of end-point conditions, which are desired for smoothness in the resulting motion. In the simplest case, the motion is defined during the initial time t_0 and final time T, and it satisfies the end-point conditions of position, velocity and acceleration at any time. From the mathematical point of view, the problem is then to find a function such that

$$v = v(t), \qquad t \in [t_0, T]. \qquad (6)$$

This problem can be easily solved by considering the polynomial function

$$v(t) = a_0 + a_1 t + a_2 t^2 + a_3 t^3 + ... + a_m t^m, \qquad (7)$$

where each coefficient a_i, $i = 0,...,m$, is a real number, and a_m is a non-negative real number. The $m+1$ coefficients a_i, $i = 0,...,m$, were determined such that the end-point constraints were satisfied. A high-degree polynomial was used to describe the trajectory, and it satisfied the desired constraints of position, velocity and acceleration at the end points.

For the PTP trajectory, we considered the profile with zero initial displacement, θ_1^h displacement at the final time T, a displacement and speed of 0 at t0, and $t = T$. Thus, we obtained the following end-point conditions:

$$v(t_0) = 0, \; \dot{v}(t_0) = 0, \qquad (8)$$

$$v(T) = \theta_1^h, \; \dot{v}(T) = 0. \qquad (9)$$

By using these four conditions and substituting them into equation (9) when $n = 12$, we obtain:

$$a_2 = \frac{3\theta_1^h}{T^2} + a_4 T^2 + 2a_5 T^3 + 3a_6 T^4 + 4a_7 T^5 + 5a_8 T^6 + 6a_9 T^7 + 7a_{10} T^8 + 8a_{11} T^9 + 9a_{12} T^{10}, \qquad (10)$$

$$a_3 = \frac{-2\theta_1^h}{T^3} - 2a_4 T - 3a_5 T^2 - 4a_6 T^3 - 5a_7 T^4 - 6a_8 T^5 - 7a_9 T^6 - 8a_{10} T^7 - 9a_{11} T^8 - 10a_{12} T^9. \qquad (11)$$

It is seen from Eqs. (10) and (11) that the two coefficients a_2 and a_3 can be obtained if the nine coefficients ($a_4 \sim a_{12}$) are known. In our trajectory design, the nine coefficients ($a_4 \sim a_{12}$) were determined by the adaptive real-coded genetic algorithm (ARGA) method with an energy-saving fitness function.

The PMSM is considered thermodynamically as an energy converter. It takes electrical energy from a controlled input and then outputs it as mechanical work to drive the toggle mechanism system with a clamping unit. The input absolute electrical energy (IAEE) to the system is defined as

$$E_i = \int_0^T \left| i_q v_q \right| dt \qquad (12)$$

where i_q is the electric current and v_q is the voltage command.

4. Adaptive Real-Coded Genetic Algorithm

It is important that crossover probability and mutation probability are set correctly for the genetic algorithms;

improper settings will cause algorithms to only find local optimums and will also cause premature convergence. Therefore, an efficient method that allows for fast settings is essential. To resolve this, a mechanism to adjust the crossover probability and mutation probability according to the algorithmic performance was considered. In this study, the adaptive real-coded genetic algorithm for polynomial coefficient identification of the ESPTP trajectory of the PMSM-toggle system was employed.

In equations (10) and (11), there were nine unknown coefficients $a_4 \sim a_{12}$ to be determined by the ARGA. First, we defined the decision vector as

$$\mathbf{z} = [a_4, a_5, a_6, a_7, a_8, a_9, a_{10}, a_{11}, a_{12}]. \qquad (13)$$

The procedure carried out for the ARGA is shown in Fig. 3. In this study, the procedure was reproduced through roulette wheel selection, while the crossover and uniform mutation were carried out through the methods described in [21].

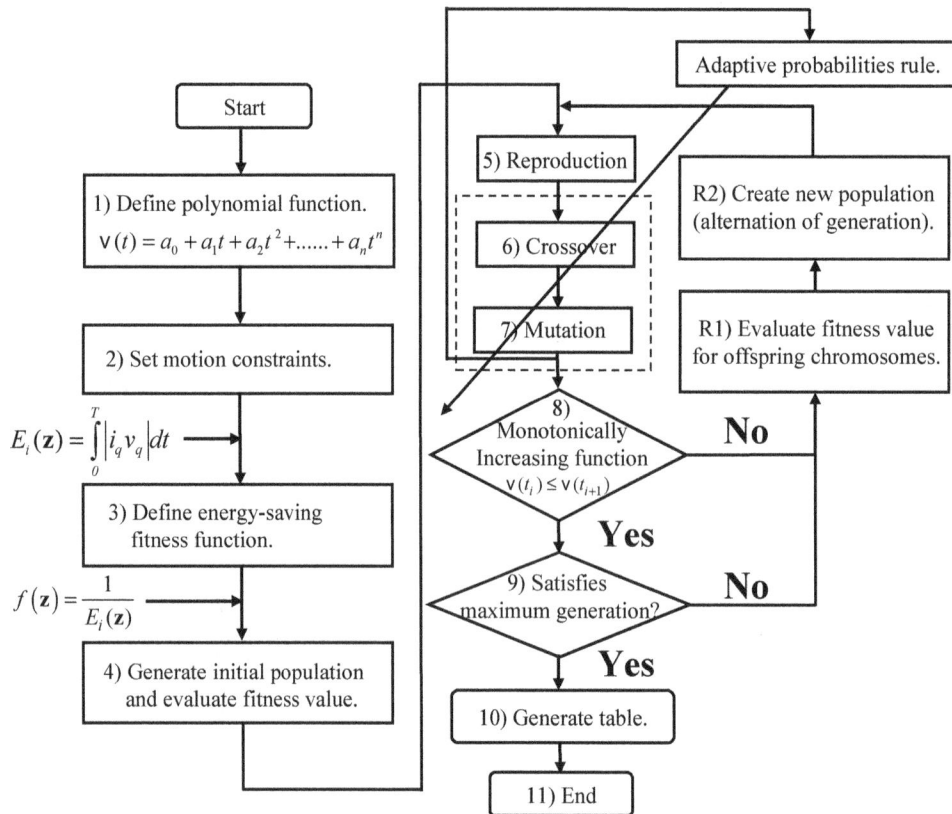

Fig. 3. *Flow chart for carrying out adaptive real-coded genetic algorithm.*

4.1. Fitness Function

How the fitness function is defined is the key to the genetic algorithm since the fitness function is a figure of merit and could be computed by using any domain knowledge. In the proposed ESPTP trajectory planning problem, the researchers defined the input energy function as the fitness function $f(\mathbf{z})$ as

$$f(\mathbf{z}) = \sum_{i=1}^m E_i(\mathbf{z}), \qquad (14)$$

where m is the total number of samples and $E_i(\mathbf{z})$ is the input energy of the ith sampling time.

4.2. Adaptive Probability Law

To reduce premature convergence and improve the convergence rate of the traditional real-coded genetic algorithm (TRGA), the adaptive probabilities of crossover and mutation were used in the ARGA. "Crossover" is the breeding of two parents to produce a single offspring which possesses features of both parents and thus may turn out better or worse than either parent according to the objective function. The primary purpose of mutation is to introduce variation, help bring back certain essential genetic traits, and avoid premature convergence of the entire feasible space caused by certain super chromosomes.

To reduce premature convergence and improve the convergence rate of the TRGA, the adaptive probabilities rule [21] of crossover and mutation were used in the ARGA. The probabilities of crossover Γ_c and mutation Γ_m are respectively given as follows:

$$\Gamma_c = \Gamma_c' \times \left(1 + \alpha \frac{\left(F_{\text{avg}}\right)^{\delta_c}}{\left(F_{\text{max}} - F_{\text{min}}\right)^{\delta_c} + \left(F_{\text{avg}}\right)^{\delta_c}}\right), \quad (15)$$

$$\Gamma_m = \Gamma_m' \times \left(1 + \beta \frac{\left(F_{\text{avg}}\right)^{\delta_c}}{\left(F_{\text{max}} - F_{\text{min}}\right)^{\delta_c} + \left(F_{\text{avg}}\right)^{\delta_c}}\right), \quad (16)$$

where F_{max}, F_{min} and F_{avg} are the maximum, minimum and average individual fitness values of (14), respectively, Γ_c' and Γ_m' are the crossover and mutation probabilities, respectively, and α, β, and δ_c are coefficient factors. In this study, the values $\alpha = 0.24$, $\beta = 0.17$, and $\delta_c = 0.22$ [21] were used. From Eqs. (15) and (16), it can be seen that the adaptive Γ_c and Γ_m vary with fitness functions. Γ_c and Γ_m increase when the population tends to get stuck at a local optimum (when attraction basins are found around locally optimal points) and decrease when the population is scattered in the solution space.

4.3. Increasing Function

For the sake of tracking the motion profile of the mechatronic system, the trajectory displacement needs to be designed as a monotonically increasing function from the start point to the end point. In this study, $v(t)$, $0 \le t \le T$ was assumed as the monotonically increasing function:

$$v(t_i) \le v(t_{i+1}), \quad t_i \le t_{i+1}, \quad (17)$$

where the subscript i represents the ith sampling time. This constraint of the monotonically increasing function had to be included in the procedure of the ARGA as shown in Fig. 3.

5. Adaptive Control Design

In this study, the researchers used the law of AC to describe what happens when two objects collide. "To adapt" means to change a behavior to conform to new circumstances. The AC law can control the two objects and balance the speed. The AC system is shown in Fig. 4, where x_B^* and x_B are the slider command position and slider position of the PMSM-toggle system, respectively. The slider position x_B is the desired control objective and can be manipulated by the relation $x_B = 2\eta \cos\theta_1$, where the angle $\theta_1 = v$ is the experimentally measured state as found by use of a linear encoder system.

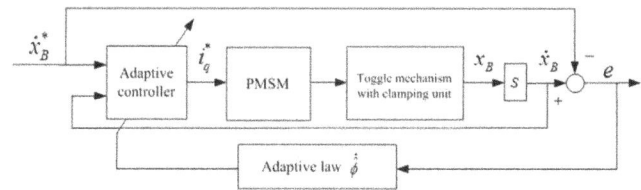

Fig. 4. Block diagram of the AC system.

In order to design an AC, the researchers rewrote Eq. (5) as a second-order nonlinear equation:

$$u(t) = f(v;t)\ddot{v}(t) + G(v;t) - d(t), \quad (18)$$

where

$$f(v;t) = \hat{Q}^{-1}\hat{M}, \quad G(v;t) = \hat{Q}^{-1}\hat{N}, \quad d(t) = \hat{Q}^{-1}\hat{D},$$

and $u(t)$ is the control input voltage. It was assumed that the exact mass of slider B and the exact impact force F_i could not be known. With these uncertainties, the first step in designing the AC was to select a Lyapunov function, which is a function used for tracking error and parameter error. An inertia-related Lyapunov function containing a quadratic form of a linear combination of position- and speed-error states was chosen as follows [15]:

$$V = \frac{1}{2}s^T f(v;t)s + \frac{1}{2}\tilde{\phi}^T \Upsilon^{-1}\tilde{\phi}, \quad (19)$$

where

$$s = \lambda_e e + \dot{e}, \quad e = v - v^*,$$

$$\Upsilon = \begin{bmatrix} \gamma_1 & 0 \\ 0 & \gamma_2 \end{bmatrix}, \quad \tilde{\phi} = \phi - \hat{\phi}, \quad \phi = \begin{bmatrix} m_B & F_i \end{bmatrix}^T, \quad \hat{\phi} = \begin{bmatrix} \hat{m}_B & \hat{F}_i \end{bmatrix}^T.$$

and in which λ_e, γ_1 and γ_2 are positive scalar constants. The auxiliary signal s may be considered as a filtered tracking error.

Differentiating Eq. (19) with respect to time gives

$$\dot{V} = s^T f(\mathrm{v};t)\dot{s} + \frac{1}{2}s^T \dot{Q}^{-1}\hat{M}s + \frac{1}{2}s^T \hat{Q}^{-1}\dot{\hat{M}}s + \tilde{\phi}^T \Upsilon^{-1}\dot{\tilde{\phi}}, \quad (20)$$

and by multiplying the variable $\dot{s}$ with Eq. (20), we obtain

$$f(\mathrm{v};t)\dot{s} = f(\mathrm{v};t)(\lambda_e \dot{e} - \ddot{x}_B^* + \ddot{x}_B) = A(\bullet) + B(\bullet)\phi - 2\eta \sin \theta_1 u, \quad (21)$$

where $A(\bullet)$ and $B(\bullet)$ are described in reference [15]. Substituting Eq. (21) into Eq. (20) gives

$$\begin{aligned}
\dot{V} &= s^T \left[A(\bullet) + B(\bullet)\phi - 2\eta \sin \theta_1 u \right] + \frac{1}{2}s^T \dot{Q}^{-1}\hat{M}s + \frac{1}{2}s^T \hat{Q}^{-1}\dot{\hat{M}}s + \tilde{\phi}^T \Upsilon^{-1}\dot{\tilde{\phi}} \\
&= s^T \left[A'(\bullet) + B'(\bullet)\phi - 2\eta \sin \theta_1 u \right] + \tilde{\phi}^T \Upsilon^{-1}\dot{\tilde{\phi}},
\end{aligned} \quad (22)$$

where $A'(\bullet)$ and $B'(\bullet)$ are described in reference [15]. If the control input is selected as

$$u = \frac{\left[A'(\bullet) + B'(\bullet)\hat{\phi} + K_V s \right]}{2\eta \sin \theta_1}, \quad (23)$$

where K_V is a positive constant, then Eq. (22) becomes

$$\dot{V} = -s^T K_V s + \tilde{\phi}^T \left[\Upsilon^{-1}\dot{\tilde{\phi}} + B'(\bullet)^T s \right]. \quad (24)$$

By selecting the adaptive update rule as

$$\dot{\tilde{\phi}} = -\dot{\hat{\phi}} = -\Upsilon B'(\bullet)^T s, \quad (25)$$

and substituting it into Eq. (24), it then becomes

$$\dot{V} = -s^T K_V s \leq 0. \quad (26)$$

Since $\dot{V}$ in Eq. (26) is negative semi-definite, then V in Eq. (19) is upper-bounded. As V is upper-bounded and $f(\mathrm{v};t)$ is a positive-definite matrix, it can be said that s and $\tilde{\phi}$ are bounded.

6. Numerical Simulations and Experiment Results

6.1. Numerical Simulations

This section discusses how the researchers simulated the ESPTP motion profile for the PMSM-toggle system. The trajectory profile $\mathrm{v}(t)$ was chosen as a monotonically increasing function. The input absolute electrical energy E_i was calculated by the fourth-order Runge-Kutta method via a Windows supported MATLAB package with a sampling time of $\Delta t = 0.01 \sec$ and the time interval being from 0 to 1 sec. In the numerical simulations, we adopted the parameters of the PMSM-toggle system obtained as follows:

$$m_2 = 1.82 \text{ (kg)}, \; m_3 = 1.61 \text{ (kg)}, \; m_5 = 0.95 \text{ (kg)}, \; m_B = 8.86 \text{ (kg)}, \; m_C = 5.58 \text{ (kg)},$$
$$r_1 = 0.06 \text{ (m)}, \; r_2 = 0.032 \text{ (m)}, \; r_3 = 0.06 \text{ (m)}, \; r_4 = 0.068 \text{ (m)}, \; r_5 = 0.03 \text{ (m)}, \; l_d = 0.01 \text{ (m)},$$
$$h = 0.068 \text{ (m)}, \; K_t = 0.565 \text{ (Nm/A)}, \; J_m = 6.72 \times 10^{-5} \text{ (Nms}^2), \; B_m = 1.21 \times 10^{-2} \text{ (Nms/rad)},$$
$$K_l = 1.056 \times 10^7 \text{ (kN/m)}, \; D_l = 970 \text{ (Ns/ m)}.$$

In the numerical simulations, the fitness value increased as the generation number increased, and almost all of the genes $(a_4, a_5, \cdots, a_{12})$ of the chromosome converged near the 30th generation for the twelfth-degree polynomial as shown in Figs. 5(a)-5(d). Figures 5(a) and 5(b) show the displacements and speeds. From the comparisons in Fig. 5(c), it is demonstrated that the ARGA is more efficient in identifying polynomial coefficients than the TRGA. The energy used was less than 9×10^{-3} J. It is thus concluded that the ARGA does not only find local optimums while preventing premature convergence, the fitness values of the ARGA are greater than those of the TRGA.

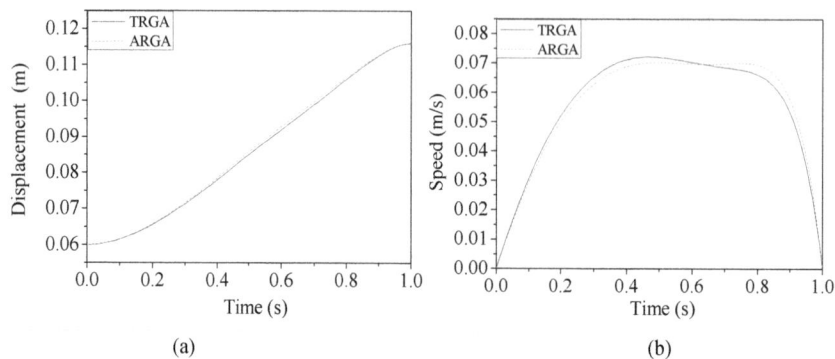

(a)   (b)

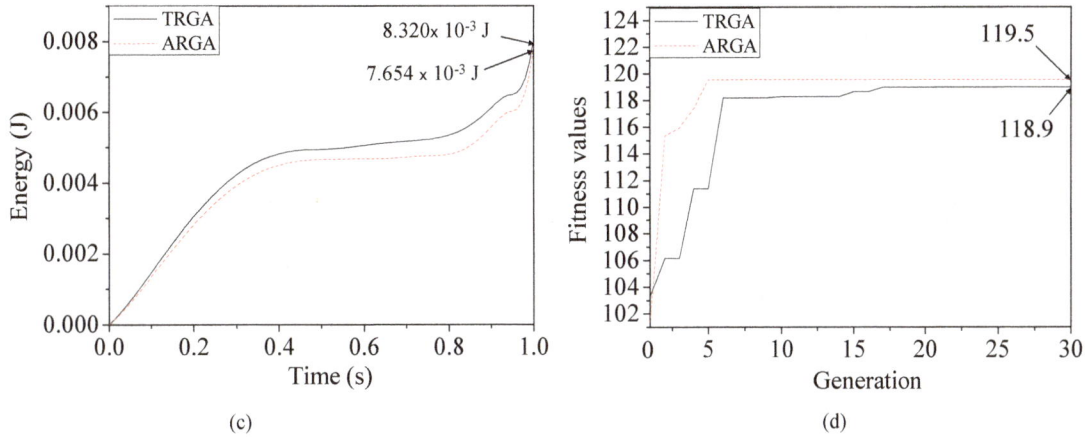

Fig. 5. *Comparisons of the TRGA and ARGA for 12th-degree polynomials in numerical simulations. (a) Displacement. (b) Speed. (c) IAEE. (d) Fitness values.*

The comparisons of dynamic responses of the PMSM-toggle system for trapezoidal, fourth-degree, and twelfth-degree polynomials are shown in Figs. 7(a)-7(d). The speeds are compared in Fig. 7(c). The displacement- and speed-error comparisons with respect to the trapezoidal, fourth-degree and twelfth-degree polynomials are shown in Figs. 7(b) and 7(d). (The fourth-degree and twelfth-degree polynomials were formulated based on ESPTP trajectories.) The final identification of the polynomial coefficients $a_4 \sim a_{12}$, the values of the fitness function of the mechatronic system, and the highest fitness value were found by using the twelfth-degree polynomial. The total energy values are also compared in Table 1, where the final values are about 8.320×10^{-3} J, 9.661×10^{-3} J, and 7.654×10^{-3} J. The lowest value is that of the twelfth-degree polynomial, and the

trapezoidal polynomial had a relative reduction of -8% in input energy.

6.2. Experimental Setup

A photo of the PMSM-toggle system with a clamping unit is shown in Fig. 1(a), and the experimental equipment used is shown in Fig. 6. The control algorithm was implemented by using a Celeron computer, and the control software used was LabVIEW. The PMSM was driven by a Mitsubishi HC-KFS13 series. The specifications were set as follows: rated torque of 1.3 Nm, rated rotation speed of 3k rpm, rated output of 0.1 kW, and rated current of 0.7 A. The servo-motor was driven by a Mitsubishi MR-J2S-10A.

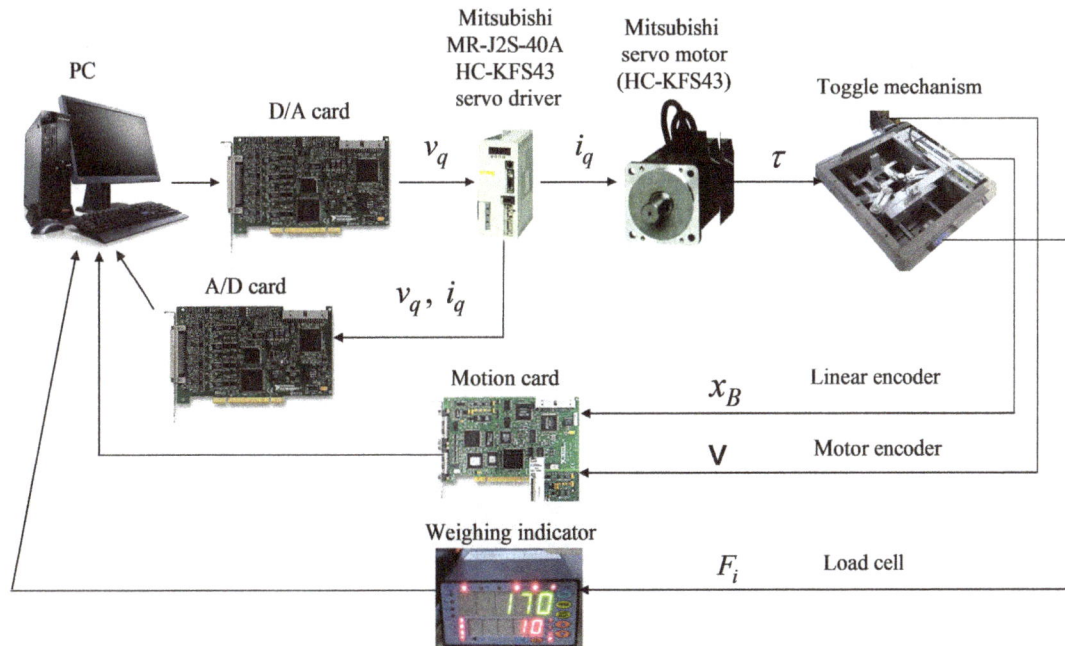

Fig. 6. *Experimental equipment for the PMSM-toggle system with a clamping unit.*

6.3. Experimentation

For the ESPTP trajectory processes of a PMSM-toggle system, the control objective was to control the position of

slider B to move from the start-position of 0 m to the end-position of 0.116 m with the clamping point at 0.1159 m. The numerical simulations and experimental results of trapezoidal, fourth-degree and twelfth-degree polynomials for the ESPTP trajectory displacement and speed tracking control by the AC are shown in Figs. 7(a)-7(h). The control gains are $\lambda_e = 0.5$, $\gamma_1 = 104$, $\gamma_2 = 209$, $K_V = 6$. Figures 7(a) and 7(b) show the displacement, and their tracking error is less than about -0.5 mm of the ESPTP trajectory of the numerical simulations and experimental results. Figures 7(c) and 7(d) show the speed of the PMSM-toggle system, and their errors are slight. Moreover, the command input, input current, impact force and input energy with the clamping effect are shown in Figs. 7(e)-7(h). As seen from the experimental results, accurate tracking control performance of the PMSM-toggle system with a clamping unit can be obtained after the clamping point of the ESPTP trajectory of the AC system, and adaptive characteristics were achieved for the AC system. The final input energy values were about 79.9 J, 99.4 J and 51.1 J, respectively. The total energy comparisons are shown in Table 1. The lowest value is that of the twelfth-degree polynomial, and the trapezoidal polynomial had a relative reduction of -36% in input energy. In conclusion, the clamping time was shorter and the speed profile was smoother. Moreover, a better energy-saving effect can be achieved for a PTP trajectory with a clamping effect by using the AC system.

Table 1. *Comparisons of input energy for the numerical simulation and experimental trajectory planning of the trapezoidal, 4th-degree, and 12th-degree polynomials of the PMSM-toggle system.*

Trajectory planning of the trapezoidal, 4th-, and 12th-degree polynomials	IAEE (J)	
	Numerical simulation by ARGA without a clamping unit	Experimental results by AC with a clamping unit
Trapezoidal	8.320×10^{-3} J	79.9 J
4th-Degree	9.661×10^{-3} J	99.4 J
12th-Degree	7.654×10^{-3} J.	51.1 J

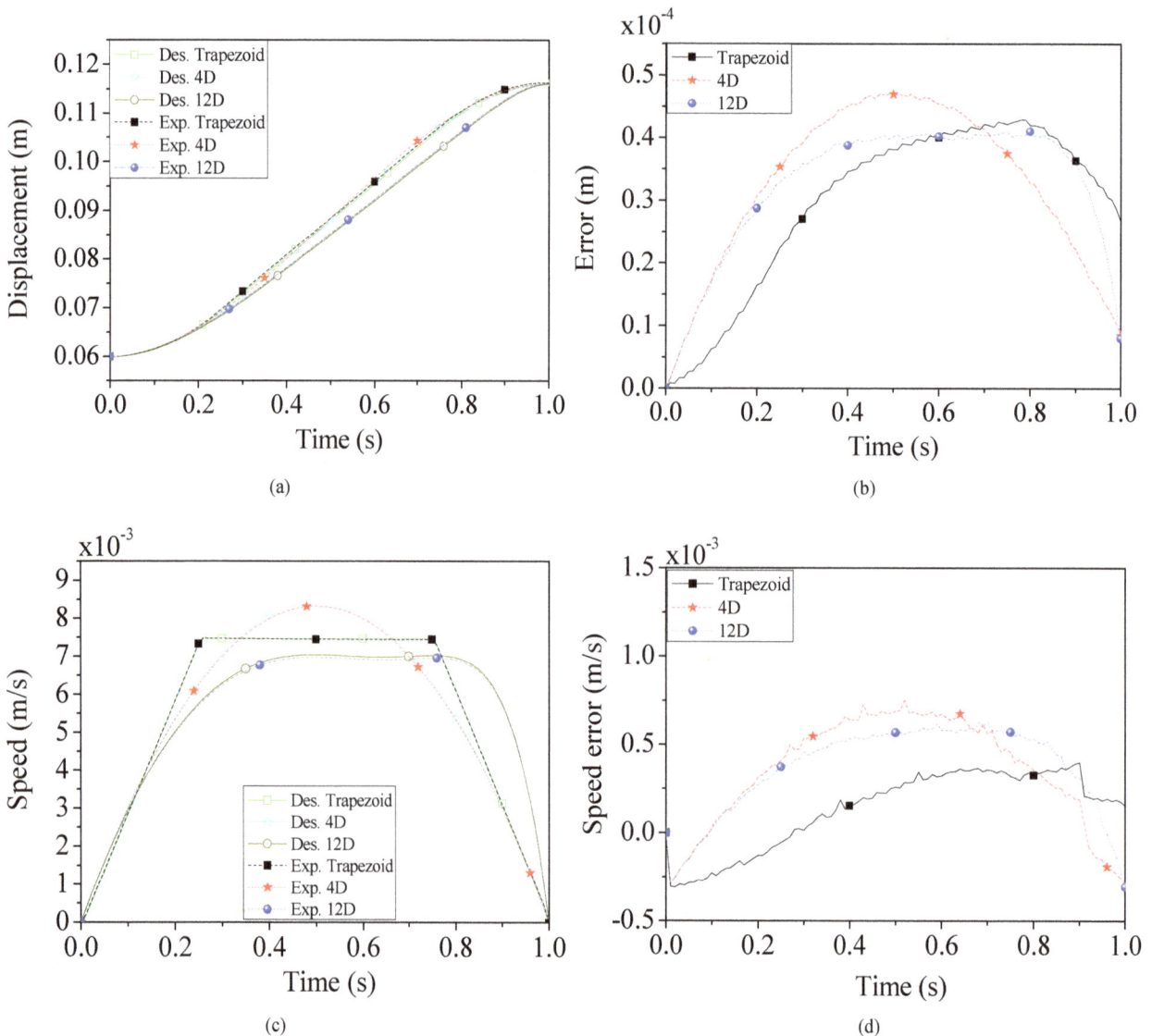

(a)

(b)

(c)

(d)

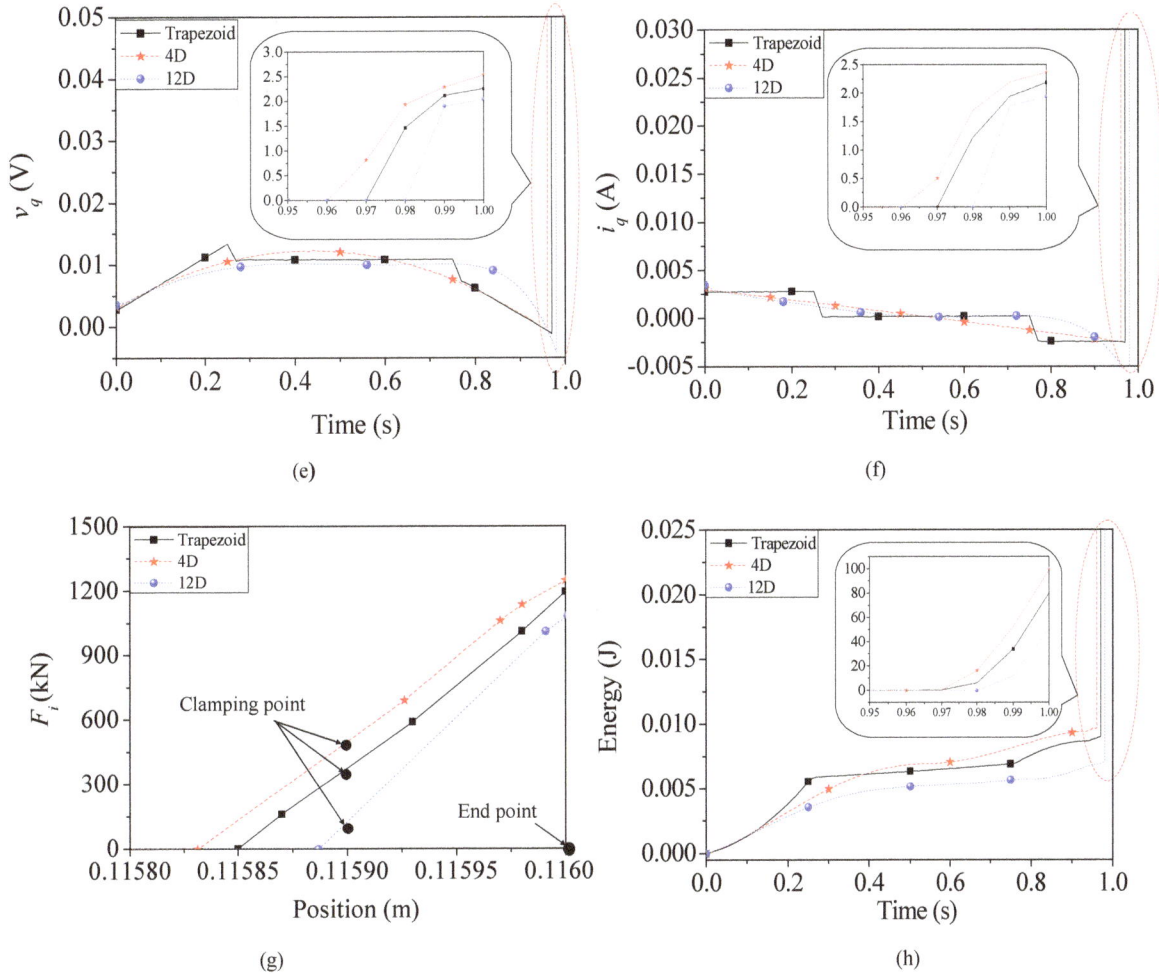

Fig. 7. *Comparisons of numerical simulations and experimental results for trapezoidal, fourth-degree and twelfth-degree polynomials by AC. (a) Displacement tracking. (b) Displacement tracking error. (c) Speed tracking. (d) Speed tracking error. (e) Input voltage. (f) Input current. (g) Impact force. (h) IAEE.*

7. Conclusion

A mathematical model was put into use for a PMSM-toggle system with a clamping unit, and the ESPTP trajectory for the mechatronic system was successfully planned by the adaptive real-coded genetic algorithm method described in this paper. The proposed AC was established by the Lyapunov stability theory for a mechatronic system with uncertainties and the impact force not being exactly known. The proposed methodology described in this paper was applied to a mechatronic system with a clamping unit. The mechatronic system required the design of an ESPTP trajectory which can be interpreted by any continuous function and which has different motion constraints at the start and end points. The results demonstrate that the adaptive control performance in the PTP trajectory with a clamping effect is successful for a mechatronic system.

Acknowledgement

The financial support from the Ministry of Science and Technology of the Republic of China with contract number MOST 103-2221-E-327 -009 -MY3 is gratefully acknowledged.

References

[1] Pu, J., Weston, R. H. and Moore, P. R., Digital Motion Control and Profile Planning for Pneumatic Servos, *ASME J. of DSMC*, Vol. 114, No. 4, pp. 634-640, 1992.

[2] Astrom, K. J. and Wittenmark, B., Adaptive Control, *Addison-Wesley, MA*, 1994.

[3] Slotine, J. J. E. and Li, W., Composite Adaptive Control of Robot Manipulators, *Automatica*, Vol. 25, No. 4, pp. 509-519, 1989.

[4] Biagiotti, L. and Melchiorri, C., Trajectory Planning for Automatic Machines and Robots, *Springer-Verlag*, 2008.

[5] Mohamed, Y. A. -R. I., Design and Implementation of a Robust Current-Control Scheme for a PMSM Vector Drive With a Simple Adaptive Disturbance Observer, *IEEE Trans. on Industrial Electronics*, Vol. 54, No. 4, pp. 1981-1988, 2007.

[6] Kim, K. H., Model Reference Adaptive Control-Based Adaptive Current Control Scheme of a PM Synchronous Motor with an Improved Servo Performance, *IET Electric Power Applications*, Vol. 3, No. 1, pp. 8-18, 2009.

[7] Shihua, L. and Zhigang, L., Adaptive Speed Control for Permanent-Magnet Synchronous Motor System with Variations of Load Inertia, *IEEE Trans. on Industrial Electronics*, Vol. 56, No. 8, pp. 3050-3059, 2009.

[8] Lee, D. H., Lee, J. H. and Ahn, J. W., Mechanical Vibration Reduction Control of Two-Mass Permanent Magnet Synchronous Motor Using Adaptive Notch Filter with Fast Fourier Transform Analysis, *IET Electric Power Applications*, Vol. 6, No. 7, pp. 455-461, 2012.

[9] Cho, S. H. and Helduser, S., Robust Motion Control of a Clamp-Cylinder for Energy-Saving Injection Moulding Machines, *Journal of Mechanical Science and Technology*, Vol. 22, No. 12, pp. 2445-2453, 2008.

[10] Kendra, S. J., Basila, M. R. and Cinar, A., Intelligent Process Control with Supervisory Knowledge-Based Systems, *IEEE Control Systems*, Vol. 14, No. 3, pp. 37-47, 1994.

[11] Wai, R. J., Adaptive Sliding-Mode Control for Induction Servomotor Drive, *IEE Proceedings on Electric Power Applications*, Vol. 147, No. 6, pp. 553-562, 2000.

[12] Chen, K. Y., Huang, M. S. and Fung, R. F., Adaptive Minimum-Energy Tracking Control for the Mechatronic Elevator System, *IEEE Trans. on Control Systems Technology*, Vol. PP, No. 99, 2013.

[13] Fung, R. F. and Chen, K. W., Dynamic Analysis and Vibration Control of a Flexible Slider-Crank Mechanism Using PM Synchronous Servo Motor Drive, *Journal of Sound and Vibration*, Vol. 214, No. 4, pp. 605-637, 1998.

[14] Lin, F. J., Fung, R. F. and Wai, R. J., Comparison of Sliding-Mode and Fuzzy Neural Network Control for Motor-Toggle Servomechanism, *IEEE/ASME Transactions on Mechatronics*, Vol. 3, No. 4, pp. 302-318, 1998.

[15] Chuang, C. W., Huang, M. S., Chen, K. Y. and Fung, R. F., Adaptive Vision-Based Control of a Motor-Toggle Mechanism: Simulations and Experiments, *Journal of Sound and Vibration*, Vol. 312, No. 4-5, pp. 848-861, 2008.

[16] Fung, R. F. and Chang, C. F., Force/Motion Sliding Mode Control of Three Typical Mechanisms, *Asian Journal of Control*, Vol. 11, No. 2, pp. 196-210, 2009.

[17] Cerman, O. and Hušek, P., Adaptive Fuzzy Sliding Mode Control for Electro-Hydraulic Servo Mechanism, *Expert Systems with Applications*, Vol. 39, No. 11, pp. 10269-10277, 2012.

[18] Cho, S. H. and Fung, R. F., Virtual Design of a Motor-Toggle Servomechanism with Sliding Mode-Combined PID Control, *Proceedings of the Institution of Mechanical Engineers, Part C: Journal of Mechanical Engineering Science*, pp. 1-10, 2014. DOI: 10.1177/0954406214531944

[19] Khulief, Y. A. and Shabana, A. A., A Continuous Force Model for the Impact Analysis of Flexible Multibody Systems, *Mechanism and Machine Theory*, Vol. 22, No. 3, pp. 213-224, 1987.

[20] Hsu, Y. L., Huang, M. S. and Fung, R. F., Convergent Analysis of an Energy-Saving Trajectory for a Motor-Toggle System, *Journal of Vibration Engineering and Technologies*, Vol. 3, No. 1, pp. 95-112, 2015.

[21] Du, Y., Fang, J. and Miao, C., Frequency-Domain System Identification of an Unmanned Helicopter Based on an Adaptive Genetic Algorithm, *IEEE Trans. on Industrial Electronics*, Vol. 61, No. 2, pp. 870-881, 2014.

22

Bibliometric Study of Welding Scientific Publications by Big Data Analysis

Pavel Layus, Paul Kah

Lappeenranta University of Technology, Skinnarilankatu, Lappeenranta, Finland

Email address:
pavel.layus@lut.fi (P. Layus), paul.kah@lut.fi (P. Kah)

Abstract: Researchers are nowadays overloaded with scientific information, and it is often difficult to obtain a clear overview of existing topical research in some particular field. Big data tools and instruments can be utilized to define trending research topics by analyzing recent publications. This paper analyses 12000 articles related to arc welding from the Scopus database for the period 2001-2012 using VOS viewer and Microsoft Excel. The most commonly occurring keywords are presented statically and as a time series. The results of this paper provide an overall landscape of scientific research in the field of arc welding and help indicate trends of emerging topics in welding research. This work is of value to both industry and academia as an indicator of changes in the field and areas of current interest. Some guidelines for potential future research on the subject are provided.

Keywords: Bibliometrics, Scopus, Keywords, VOS Viewer, Big Data, Research Trends, Welding

1. Introduction

The science of welding includes a great variety of research fields, such as metallurgy, mathematical modelling, physics, thermodynamics, heat transfer and many others. While the joining of metals has been used for centuries, welding technology has started its development from the invention of electricity in 19th century. Since that time, the science of welding has been in the core of numerous outstanding engineering achievements. Currently, welding is the primary way of joining metallic materials, and it plays a major role in automotive, steel structures, shipbuilding, agricultural and many other manufacturing industries. The global welding products market was valued at USD 17.47 billion in 2013 and is expected to reach USD 23.78 billion by 2020 and expand at a compound annual growth rate (CAGR) of 4.5% between 2014 and 2020 [1]. A wide range of review articles and books (for instance [2-9]) were written on welding science, reading which could help gaining deeper understanding of scientific research in the field of welding.

Nowadays research data on various topics, including welding, are becoming more accessible than ever, thanks to the development of extensive online bibliographic databases containing abstracts and citations of academic journal articles. Notable examples of such databases are Scopus and Web of Science, and most recently published article abstracts

and full texts can be accessed via such databases. The vast amount of available information, which is growing from year-to-year [10], is challenging to use and analyze efficiently. This challenge calls for the development of suitable approaches and tools to convert big data into understandable, usable and practical information. Research and development institutions and industry have become aware of the potential of big data to provide competitive advantage, and bibliometric analysis has been used in a number of fields as a way of highlighting and delineating trends [11-20].

Despite the huge amount of publications on welding, no attempt has yet been made to conduct bibliographic analysis of scientific publications in the field of welding. Moreover, there has been no efforts to evaluate trends in the field of welding. Research study that would address these topics is essential to reveal and explain the developments in the field, bring a deeper understanding of the impact of research on the literature and comprehensive advices for the future research in the field of welding. The motivation of this research is to find out which aspects of the science of welding were in the research spotlight during the last decade.

Current paper persuades an attempt to present most prominent trends and highlights in the field of arc welding research conducted over the last decade (2001-2012) by performing in-depth bibliographic analysis of scientific publications. This research work applies the approaches and tools of big data to quantitatively analyze about 10000 scientific journal publications related to the topic of arc welding. The hypothesis of this study states that using numerical analysis of a quantitatively collected dataset, it is possible to spot current trends and important topics in scientific research in the field of welding, and to find out which topics are of increasing interest and which of decreasing importance.

This broad bibliographic study of scientific publications on welding could be valuable to several beneficiary groups, such as researchers, educators and industrial professionals. Researchers would receive a special instrument that lets them determine previous and current research highlights as well as formulate future possible topics for the science of welding. Educators would get an information regarding overall landscape of the science of welding, which will allow them to embrace it in the teaching materials for modern welding technology courses and trainings. Industrial professionals would obtain in-depth information and use concepts that would support them in planning and carrying out their research and development as well as commercial projects in the field of welding technology. The value of this paper is that it presents an overall landscape of scientific research in arc welding, based on quantitative data, revealing various trends in the field. In addition, it further demonstrates the worth of bibliometric analysis as a basis for research of trends in engineering.

1.1. Big Data in the Context of Bibliometric Research

Big data analysis of scientific publications allows the measuring of trends based on bibliometric information such as keywords, date of publication, references and other records. Bibliometric data analysis has been applied in many research areas, including environmental assessment [16], sustainable hydropower development [17], nucleation techniques [18], risks of engineering nanomaterials [19] and even intercultural relations [20]. Datasets of scientific publications for analysis can be constructed from various databases, of which the most important and comprehensive in the context of this study are the Scopus and Web of Science databases. These databases offer some tools to analyze the scientific datasets, their functionality is, however, rather limited. Specialist software products for bibliometric research are available that offer more tools and functions, such as the software tool used in this work, VOS viewer [21].

1.2. Comparison of Scopus and Web of Science

Several scientific document databases are available, of which Scopus and Web of Science are among the leading database services. Although the two databases differ in coverage, they can be considered to be complementary, and which database is most suitable depends on the discipline, topic and period of analysis [22]. This study began with a comparison of the Scopus and Web of Science databases and selection of the most suitable database.

The Scopus database contains approximately 34278 journal entries, whereas Web of Science contains about 16957 journals. Scopus and Web of Science might include journals with an extremely low impact factor, indicating potentially low-quality publications. In an attempt to select higher quality journals it was decided to match records from Scopus and Web of Science with the Ulrich database [23], which contains a large number of highly rated journals. 20464 Scopus journals matched the Ulrich database and 13607 journals from Web of Science.

It can be seen that Scopus contains almost twice as many journals as Web of Science. However, as study [24] shows, the distribution of journal topics is different in Scopus and Web of Science. Since the primary focus of this research is welding, journals on Engineering and Technology are of interest. 33% of journals indexed in Scopus are in the Natural Sciences and Engineering domain. Web of Science shows a higher percentage of journals in the same domain – 43%. Nevertheless the number of indexed journals is significantly larger in Scopus, and therefore Scopus still represents the larger dataset (6730>5810). The journal topics distribution in Scopus and in Web of Science databases are shown in Fig. 1.

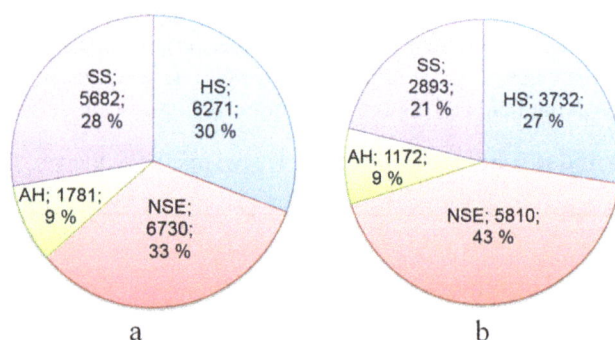

Fig. 1. a - SCOPUS database journals domain distribution, b – Web of Science database journals domain distribution.

Where HS is Health Sciences journals; NSE is Natural Sciences and Engineering journals; AH is Arts and Humanities journals and SS is Social Sciences journals.

Although Scopus contains more journals in the natural sciences and engineering domain, this does not mean that Scopus contains more articles on welding. Therefore, the next step of the database comparison was to evaluate the number of scientific articles on welding. This evaluation can be made by searching for article titles that contain the word "weld" with all possible endings, e.g. welding, welded. The selected timeframe was the last 14 years, i.e. 2000-2014. As can be seen from Fig. 2, Scopus contains more entries with the word "weld" or one of its variants in the title for each year during 2000-2014.

In view of the larger journal coverage and greater number of scientific articles on welding, the Scopus database was selected as the source of the dataset for this study.

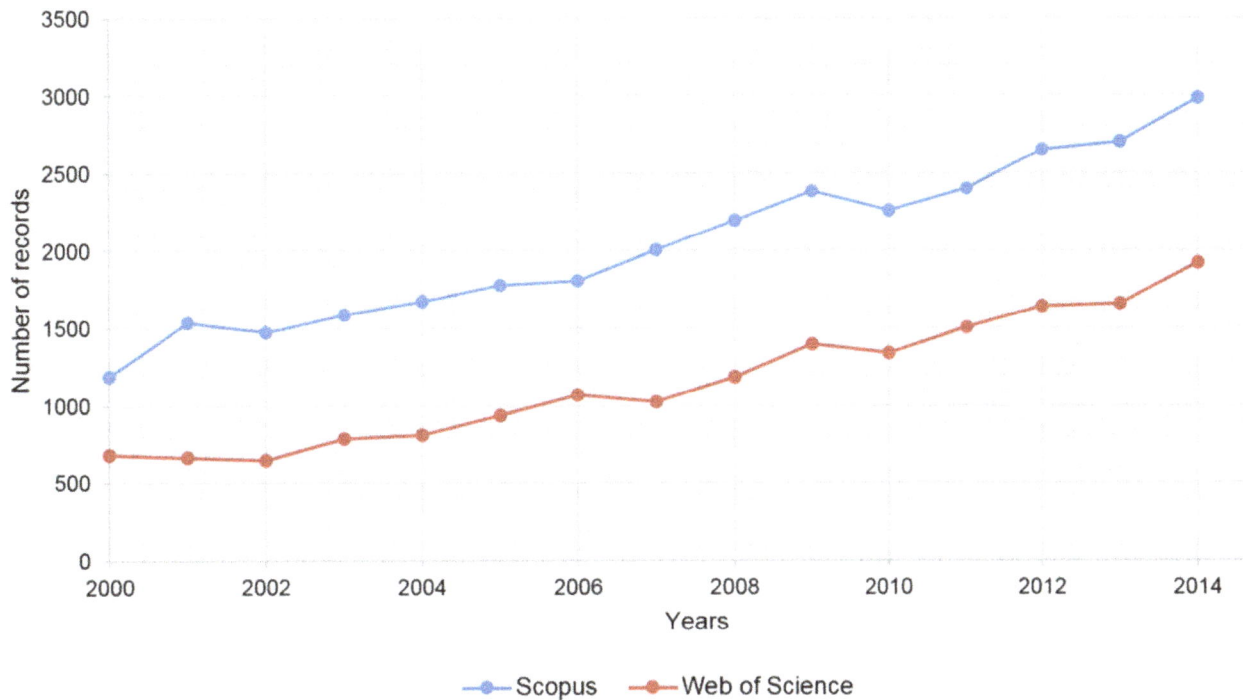

Fig. 2. *Number of scientific articles related to welding in the Scopus and in Web of Science databases in 2000-2014.*

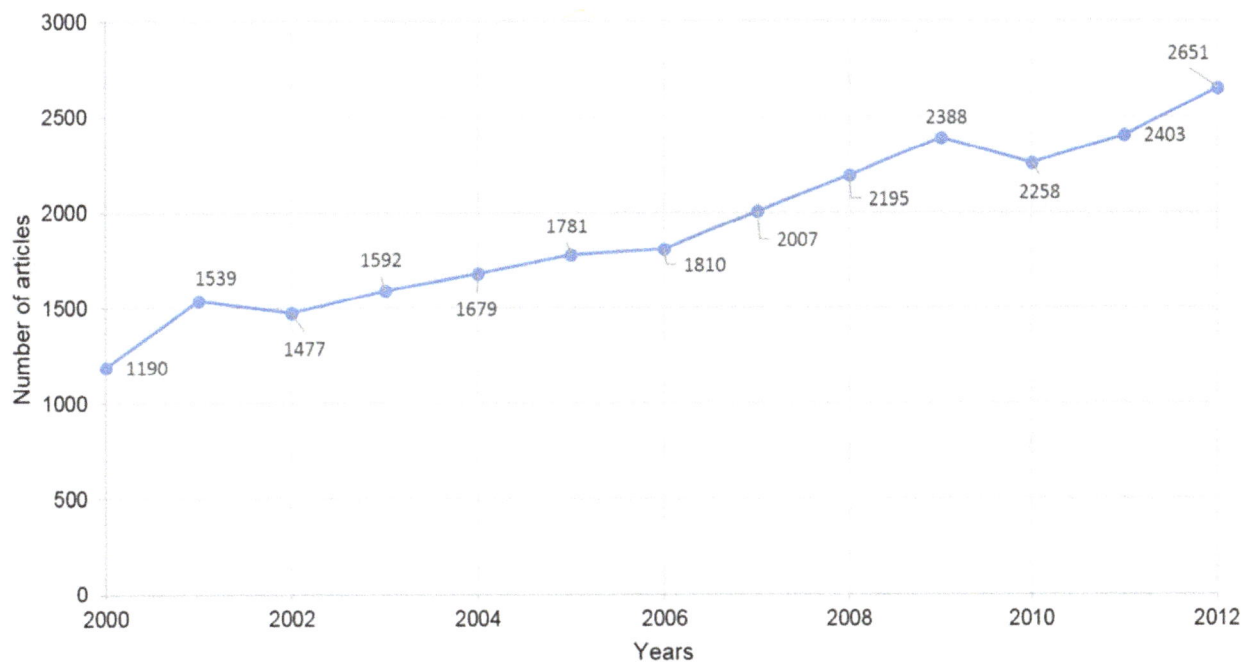

Fig. 3. *Number of articles published in the field of welding presented in Scopus over 2001-2012 period.*

2. Analysis Process and Setup

The data for the analysis were acquired first by a search query "weld*". After manual cleaning of the dataset, about 14000 indexed scientific published articles in English listed on Scopus for 2001-2012 in the field of arc welding processes were selected. After the dataset was finalized, the article entries were analyzed using Scopus tools, VOS viewer 1.5.7 and Microsoft Excel.

2.1. Scopus Analysis

Scopus has a set of tools for analyzing the articles dataset. In this research, the total share of arc welding articles and its yearly distribution is given based on Scopus tools results. The size of the Scopus database increases on average 8.7% yearly. The number of journal article publications on welding is also increasing yearly, but at a slower pace – 5% on average. Therefore, the number of publications on welding as a share of total publications is decreasing. A possible explanation is an increase in the range of welding-related keywords introduced each year, since science becomes more diverse. However, it is also possible that publication on welding or related sciences are decreasing. Fig. 3 illustrates the number of papers in the field of welding in Scopus database.

Fig. 3 illustrates the number of papers published in the field of welding over 2001-2012. It can be seen that the number of articles follows a generally increasing trend; however, the jump in 2008-2009 is an interesting phenomena. The authors have not found any obvious facts or trends to explain this sudden change. The decrease in the number of publications during 2010 might be due to not all publications having been indexed and uploaded to Scopus.

2.2. VOS Viewer 1.5.7 Analysis

VOS viewer 1.5.7 software allows graphical representation of keywords clustering. The keywords are clustered to show their connection to each other and indicate the degree of similarity in meaning. The software analyses which keywords more often occur together in one article and puts them close to each other on the plot. The results of the clusterization process are presented in Fig. 4. It can be seen that keywords are divided into five clusters, which were manually labeled based on knowledge of welding research topics as:

- Materials and microstructure
- Mechanical properties and mechanical fracture modes
- Welding parameters and chemical properties and processes
- Computer tools and welding processes
- Applications

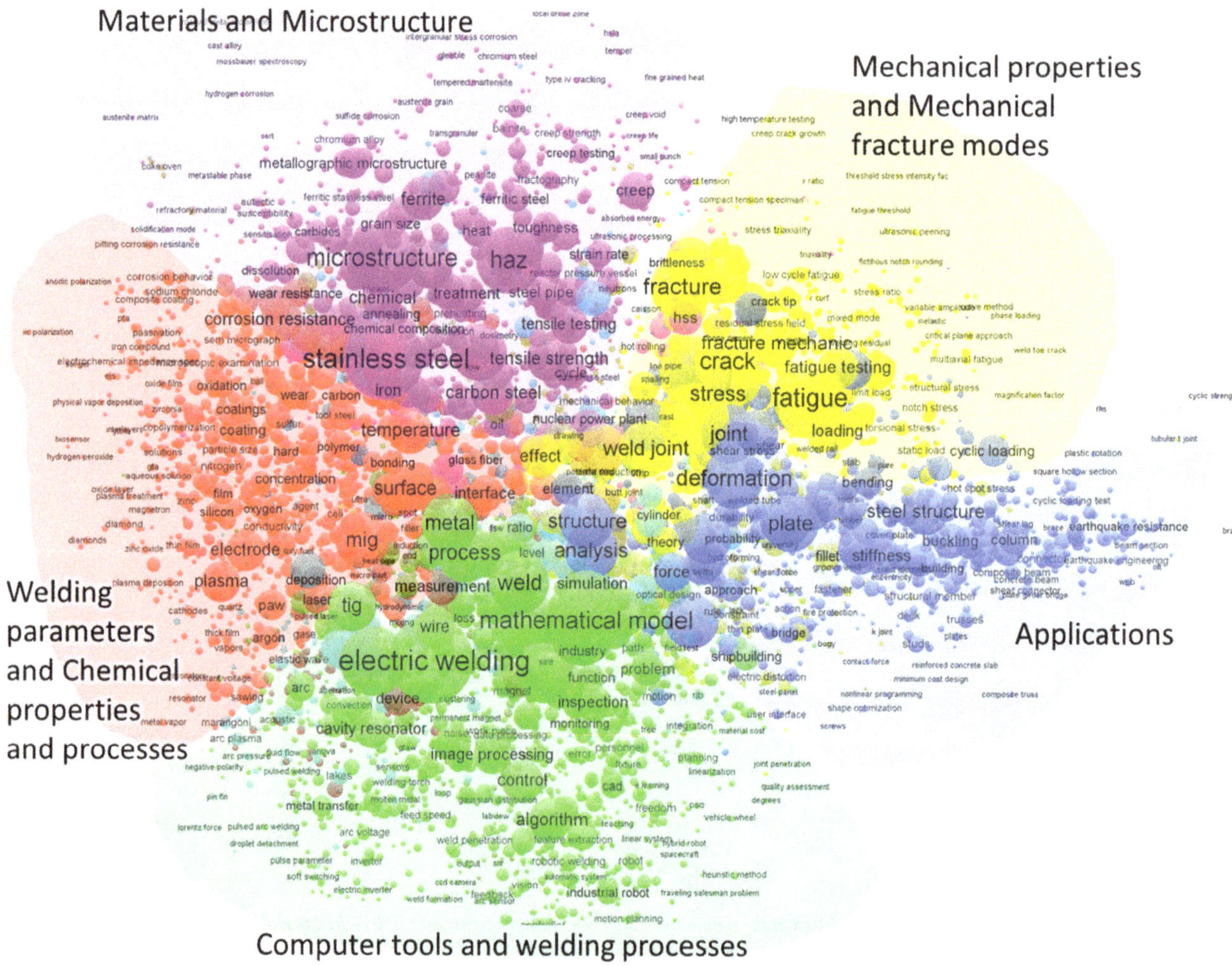

Fig. 4. Cluster distribution of keywords.

As can be seen in Fig. 4, some keywords do not perfectly fit into the clusters; however, the vast majority of keywords allow cluster borders to be defined. Fig. 4 illustrates another problematic area of this approach, as some keywords are non-descriptive, such as "effect", "process" or "weld". In the current analysis, these non-descriptive keywords do not significantly influence the overall picture, as the clusters are clearly defined.

2.3. Microsoft Excel Analysis

Microsoft Excel analysis was performed to present the overall research topics and to illustrate trends in scientific research in the field of welding. In order to present trends, the twenty most commonly occurring keywords were selected from each year. The next step was to put them in Excel and fill the number of occurrences of each keyword in the selected period. Related keywords were manually combined into ten groups based on the authors' knowledge of scientific research in the field of welding. Resulting groups correlated closely with VOSviewer clusters (Fig. 4). The close correlation suggests that the manual choice of keyword groups was correct. Some keywords, such as, for instance, the keywords "weld" and "welding", were omitted, as they are non-descriptive. Fig. 5 shows the percentage of each keyword group. It can be seen that materials, mechanical properties, mechanical failure modes and computer tools have the largest shares of 18%, 13%, 13% and 12%, respectively.

The largest keyword groups (i.e. materials, computer tools, mechanical failure modes, and mechanical properties) consisted of the keywords listed in Table 1.

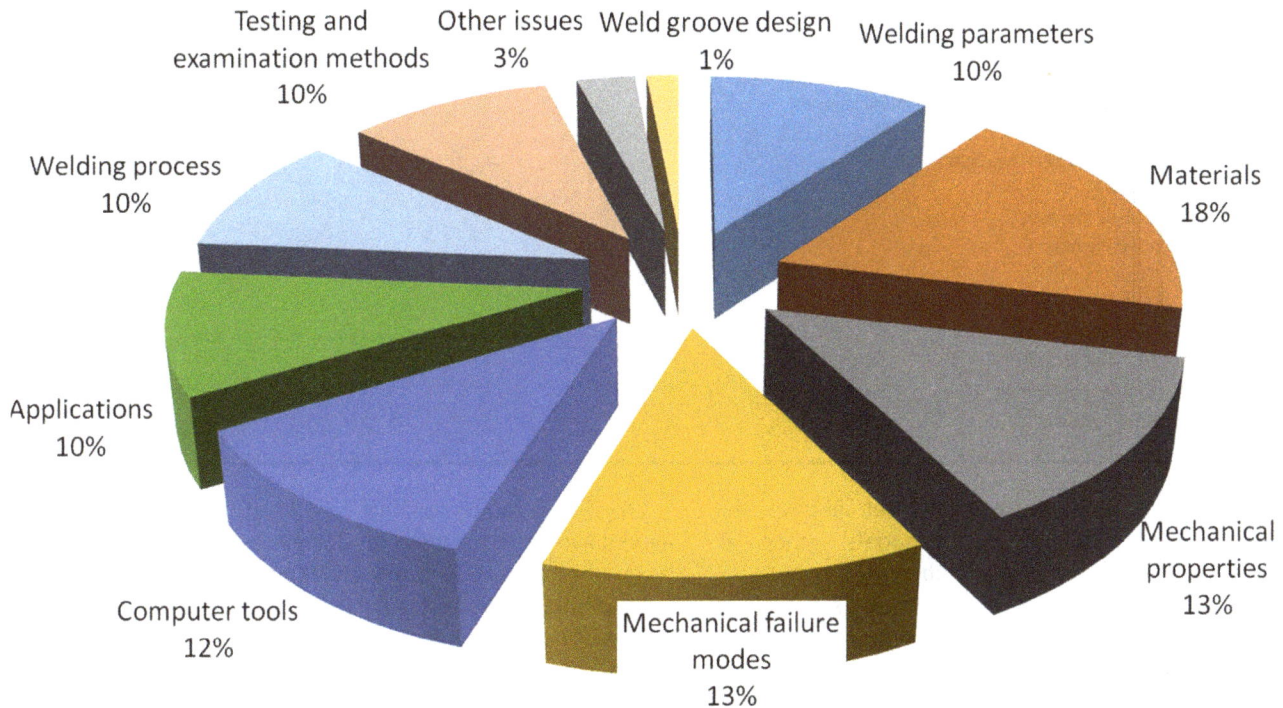

Fig. 5. *Keyword groups of welding articles.*

Table 1. Most commonly occurring keywords and groups to which they were allocated.

Keywords group	Keyword(s)	Number of occurrences	Total
Materials	Microstructure (Base metal, Ferrite, Austenite, HAZ, Martensite, Solidification, Weld metal)	3391	8531
	Stainless steel (304 stainless steel, Austenitic stainless steel, Duplex stainless steel, Martensitic stainless steel)	1697	
	Chemical composition (Chlorine compounds, Hydrogen, Carbides, Carbon, Iron, Silicon, Tungsten)	1182	
	Steel (Carbon steel, Ferritic steel, High strength steels, Low carbon steel)	822	
	Steel metallurgy	382	
	Metal	356	
	Weldability	201	
	Interfaces (materials)	116	
	Dissimilar metals	113	
	Alloy	112	
	Plastic	95	
	Nanostructured materials	64	
Computer tools	FEM	1453	5446
	Computer simulation	833	
	Numerical methods	750	
	Mathematical model	681	
	Optimization	313	
	3D	309	
	Image processing	245	
	Algorithm	185	
	Neural networks	146	
	Parameter estimation	144	

Keywords group	Keyword(s)	Number of occurrences	Total
Mechanical failure modes	Forecasting	140	5877
	Modeling	134	
	Analytical model	113	
	Corrosion (Pitting Corrosion, Coatings, Painting)	1528	
	Crack	1516	
	Fatigue (Fatigue crack propagation, Fatigue damage, Fatigue life)	1339	
	Fracture (Fracture mechanics, Brittle fracture)	929	
	Creep	233	
	Buckling	183	
	Wear resistance	149	
Mechanical properties	Stress (Yield stress, Stress analysis, Stress concentration, Stress intensity factors)	1488	6319
	Strength of materials	575	
	Loading (Cyclic loading, Structural analysis)	528	
	Deformation	500	
	Hardness (Microhardness)	363	
	Strain	361	
	Tensile strength	333	
	Fracture toughness	312	
	Ductility	249	
	Tribology (Friction coefficient)	203	
	Stiffness	188	
	Material property	156	
	Plastic deformation	143	
	Fatigue strength	136	
	Toughness	126	

Table 1 shows rather expected results; for instance, the most trending keyword in the materials group is microstructure (including its subtopic keywords). Microstructure is one of the most important topics in welding, as the mechanical properties of the weld are defined by the type of microstructure [25-27].

Articles dedicated to microstructure are mostly written about the heat-affected zone (31%), which was expected since it is closely connected to arc welding processes. The keywords "weld metal" and "base metal" are represented by significant percentages – 15% and 12% respectively. Other important topics are iron-phases (austenite (15%), ferrite (13%) and martensite (8%)). The distribution is presented in Fig. 6a.

Likewise, it can be seen that welding of stainless steels is an important topic in the materials domain, as stainless steel is the second most popular keyword, including the subtopic keywords (304 stainless steel, austenitic stainless steel, duplex stainless steel, and martensitic stainless steel). Over half of the journal articles on welding of stainless steel are dedicated to austenitic stainless steel (61%). Other important research areas include martensitic (17%) and duplex stainless steels (13%). The distribution is presented in Fig. 6b.

The next step of the study was to perform time series analysis to determine research trends. The time series analysis results for selected keyword groups and the distribution of keywords in the group of keywords is shown in Fig. 7.

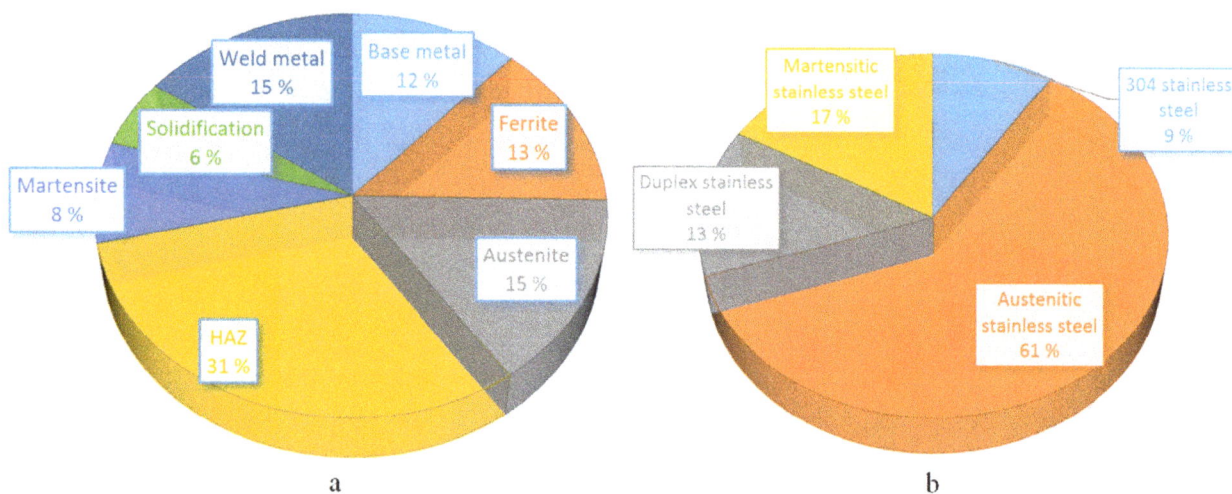

Fig. 6. *Keyword distribution of scientific articles on welding, a –Microstructure group, b - Stainless steels group.*

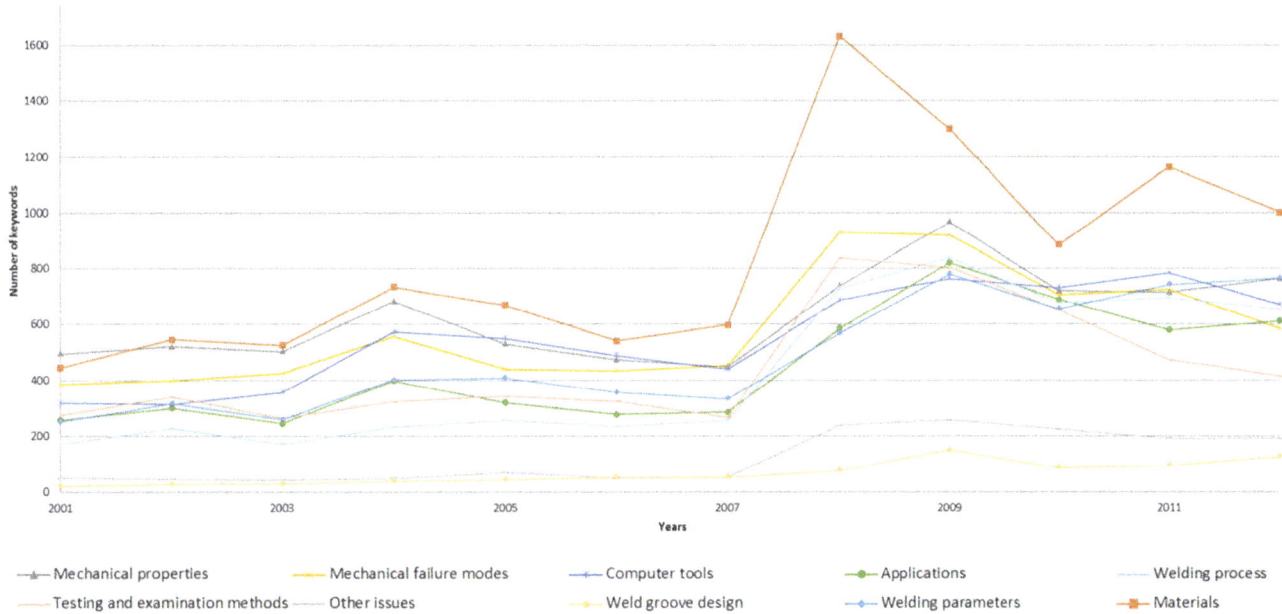

Fig. 7. *Time series analysis of major keyword groups.*

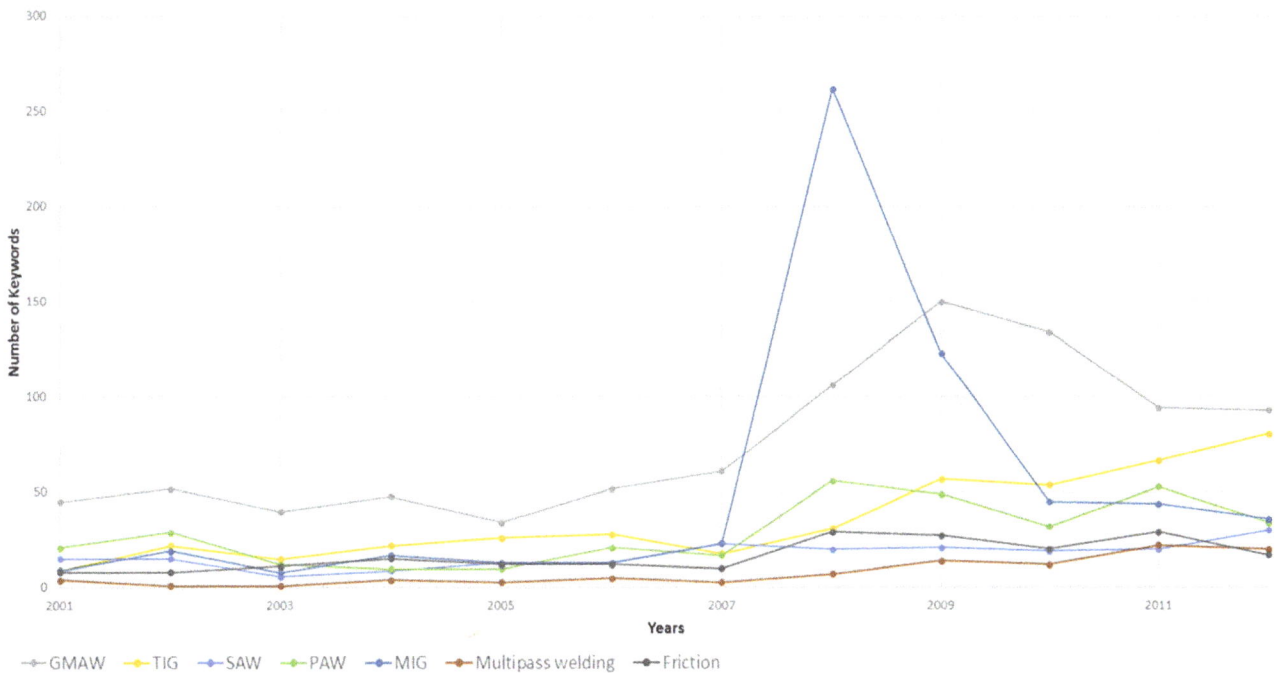

Fig. 8. *Welding processes keyword group time series analysis, where GMAW is Gas Metal Arc Welding, TIG is Tungsten Inert Gas Welding, SAW is Submerged Arc Welding, PAW is Plasma Arc Welding and MIG is Metal Inert Gas Welding.*

The decline in 2012 most likely occurred due to non-complete Scopus records for recent years, and, the decline is expected to disappear in the next couple of years, when all papers from 2012 are indexed. The keyword group materials shows an interesting increase in 2008. One possible explanation is the increase of interest in dissimilar materials welding. The decline in testing and examination methods in the last years can be connected with the fact that this field of science is well established and does not attract many researchers.

One other group of keywords, which deserves attention, is arc welding processes, which is presented in Fig. 8. The

number of articles related to TIG welding steadily increased from 2008 to 2012, which may be connected with intensive development of robotic welding systems, which often utilize TIG welding processes. The number of GMAW related articles reached a maximum in 2009, after which the prevalence of this topic has decreased. The drop in the number of articles may be due to GMAW being a rather well-known welding procedure, and therefore suffering from a decreasing amount of research interest in recent years. The number of studies on multipass welding is increasing, from 3 in 2007 to 20 in 2012, mostly due to an increase in publications related to simulation of multipass welding and predictions of the welding process outcome.

3. Limitations

The major limitations of the study are primarily related to the selected dataset: the particular types of academic publications selected, the selection of English language publications only, incomplete records for some academic publications and the relatively short time period covered.

The Scopus database presents various types of scientific works, including journal articles, conference proceedings, book series, trade publications, books and reports. However, the research presented here only analyzes journal articles, which represent 53.7% of the total number of publications. Journal articles usually demonstrate research credibility, as in most cases journal articles are peer-reviewed, which indicates relatively high scientific value of the research. Journal articles form the largest part of publications on Scopus comparing to other publication groups (e.g. conferences, books), thus ensuring a sufficiently large dataset for statistically valid analysis.

During the current research work it was decided to focus solely on the period from 2001 to 2012. The end point of 2012 was chosen because some of the latest research publications from 2013-2015 may not have been uploaded to the Scopus database. In preliminary study, it was noted by the authors that a number of articles were only added to the Scopus database a couple of years after the publication date. Therefore, in order to obtain a more accurate dataset it was decided to utilize data only until 2013. Publications prior to 2001 are also not included, because a significant part of these papers might not be included in the Scopus database.

Another limitation of this research comes from possible inaccuracy in keyword analysis of the publications. The Scopus database contains many keywords with identical meaning but different grammatical form or spelling, and despite manual keyword cleaning, the dataset keywords may not be completely accurate. The error can be of the order 1-3%. Since this research considers only publications on arc welding, papers on other welding processes, such as laser or friction welding were excluded from the dataset. The exclusion process was performed by excluding articles with corresponding keywords, such as laser or friction. This process also might influence the accuracy of results, since articles that contain both arc and friction welding keywords are excluded.

Further limitations of this study are related to the accuracy and quality of the records obtained from the Scopus database. The obtained dataset of 14882 entries was manually examined, and entries with incomplete records were discarded. About 13960 articles (94%) had complete records and were used for further analysis in this study. According to [28], the Scopus database might contain duplicate entries of up to about 13%. Some duplicate entries might be left in the dataset after manual cleaning and might have an influence on the accuracy of the results.

4. Conclusion

The current research presents first bibliometrics work on scientific research in the field of arc welding. The study provides an overview of welding research articles for the period 2001-2012. The analysis is based on bibliometric information, such as keywords, date of publication, references and other records, which were processed with Scopus tools, VOS viewer and Microsoft Excel.

It was found that the number of publication in the topic of welding increased by 5% year-on-year. It was seen that keywords related to materials, mechanical properties, mechanical failure modes and computer tools had the largest shares of 18%, 13%, 13% and 12% respectively. Articles about microstructure were mostly related to the heat-affected zone (31%), which is a major concern for microstructural changes in welding. The keywords "weld metal" and "base metal" were represented by significant percentages – 15% and 12% respectively. Likewise, welding of stainless steel is an important topic in the materials domain, as stainless steel is the second most popular keyword, including its subtopic keywords: 304 stainless steel, austenitic stainless steel, duplex stainless steel, and martensitic stainless steel. Over half of the journal articles on stainless steel are dedicated to austenitic stainless steel (61%). Other important research areas include martensitic (17%) and duplex stainless steels (13%).

Time series analysis showed a decline in the overall number of arc welding related articles in 2012, which is most likely due to non-complete Scopus records for recent years, and the decline is expected to disappear in the next couple of years, when all papers from 2012 are indexed. Articles on topics related to materials showed an interesting increase in 2008. When considering arc welding processes, the number of articles on TIG welding increased steadily from 2008 to 2012, which may be connected with intensive development of robotic welding systems, which often utilize TIG welding processes. Articles with GMAW welding as a keyword reached a maximum in 2009, since when their number has fallen. A possible reason for this trend may be that GMAW is a rather well-known procedure and therefore of decreasing research interest. The number of studies on multipass welding is increasing, mostly due to the increasing number of publications related to simulation of multipass welding and predictions of welding process outcome.

The study has some limitations, most importantly the short time period studied and the types of scientific publications considered. However, despite its limitations, the study illustrates the theoretical and practical relevance of combining bibliometric or statistical datasets, indicates how such study can be approached, and gives pointers to issues that can be addressed using analysis of a large bibliometric dataset.

Future research on similar topics can provide additional, more detailed information and can reveal more specific aspects of research trends. Future study might analyze not only keywords of journal articles but might consider bibliometrics data of other scientific publications such as references or abstracts.

Acknowledgement

Authors would like to acknowledge help of prof. Leonid Chechurin for the idea formulation and advices during the beginning stage of the current research.

References

[1] A. K. Sheela, Global Welding Products Market is Expected to Reach USD 23.78 billion by 2020, Transparency Market Research, 2015, http://www. Transparency market research. com/pressrelease/welding-products-market. htm.

[2] T. Zhang, Z. Li, F. Young, H. Jin Kim, H. Li, H. Jing, W. Tillmann, Global progress on welding consumables for HSLA steel, ISIJ International, 54 (7), 2014, pp. 1472-1484, DOI: 10.2355/isijinternational.54.1472, 2014.

[3] A. Kawamoto, Newly arc welding equipment, Keikinzoku Yosetsu/Journal of Light Metal Welding and Construction, 51 (6), 2013, pp. 5-9.

[4] A. Haelsig, M. Kusch, P. Mayer, New findings on the efficiency of gas shielded arc welding, Welding in the World, 56 (11-12), 2012, pp. 98-104, DOI: 10.1007/BF03321400.

[5] P. Kah, H. Latifi, R. Suoranta, J. Martikainen, M. Pirinen, Usability of arc types in industrial welding, International Journal of Mechanical and Materials Engineering, 9 (1), 2014, 12 p., DOI: 10.1186/s40712-014-0015-6.

[6] S. Hiroyuki, Materials and processes for arc welding, Yosetsu Gakkai Shi/Journal of the Japan Welding Society, 80 (8), 2011, pp. 21-30.

[7] R. E. Gliner, Welding of advanced high-strength sheet steels, Welding International, 25 (5), 2011, pp. 389-396, DOI: 10.1080/09507116.2011.554234.

[8] L. F. Jeffus, Welding: Principles and Applications, Cengage Learning, 2004, 904 p.

[9] R. Singh, Applied Welding Engineering: Processes, Codes, and Standards, Elsevier, 2011, 349 p.

[10] L. Bornmann and R. Mutz, Growth rates of modern science: A bibliometric analysis based on the number of publications and cited references, Journal of the Association for Information Science and Technology, DOI: 10.1002/asi.23329 [in press, available online only].

[11] X. Jin, B. W. Wah, X. Cheng, Y. Wang, Significance and Challenges of Big Data Research, Big Data Research, ISSN 2214-5796, http://dx.doi.org/10.1016/j.bdr.2015.01.006 [in press, available online only].

[12] F. W. Fosso, S. Akter, A. Edwards, G. Chopin, D. Gnanzou, How 'big data' can make big impact: Findings from a systematic review and a longitudinal case study, International Journal of Production   Economics, http://dx.doi.org/10.1016/j.ijpe.2014.12.031 [in press, available online only].

[13] A. Vera-Baquero, R. Colomo-Palacios, O. Molloy, Towards a Process to Guide Big Data Based Decision Support Systems for Business Processes, Procedia Technology, vol. 16, 2014, pp. 11-21,   ISSN   2212-0173, http://dx.doi.org/10.1016/j.protcy.2014.10.063.

[14] R. K. Perrons, J. W. Jensen, Data as an asset: What the oil and gas sector can learn from other industries about "Big Data", Energy Policy, vol. 81, June 2015, pp. 117-121, ISSN 0301-4215, http://dx.doi.org/10.1016/j.enpol.2015.02.020.

[15] A. Gandomi, M. Haider, Beyond the hype: Big data concepts, methods, and analytics, International Journal of Information Management, vol 35 (2), April 2015, pp. 137-144, ISSN 0268-4012, http://dx.doi.org/10.1016/j.ijinfomgt.2014.10.007.

[16] W. Li, Y. Zhao, Bibliometric analysis of global environmental assessment research in a 20-year period, Environmental Impact Assessment Review, vol 50, January 2015, pp. 158-166, ISSN 0195-9255, http://dx. doi. org/10. 1016/j. eiar.2014. 09. 012.

[17] M. Y. Han, X. Sui, Z. L. Huang, X. D. Wu, X. H. Xia, T. Hayat, A. Alsaedi, Biblio-metric indicators for sustainable hydropower development, Ecological Indicators, vol 47, December 2014, pp. 231-238,   ISSN   1470-160X, http://dx.doi.org/10.1016/j.ecolind.2014.01.035.

[18] D. Boanares, C. Schetini de Azevedo, The use of nucleation techniques to restore the environment: a bibliometric analysis, Natureza & Conservação, vol 12 (2), July–December 2014, pp. 93-98,   ISSN   1679-0073, http://dx.doi.org/10.1016/j.ncon.2014.09.002.

[19] Q. Wang, Z. Yang, Y. Yang, C. Long, H. Li, A bibliometric analysis of research on the risk of engineering nanomaterials during 1999–2012, Science of The Total Environment, vol. 473–474, 1 March 2014, pp. 483-489, ISSN 0048-9697, http://dx. doi. org/10.1016/j. scitotenv. 2013. 12. 066.

[20] J. Young, R. Chi, Intercultural relations: A bibliometric survey, International Journal of Intercultural Relations, vol. 37 (2), March 2013, pp. 133-145, ISSN 0147-1767, http://dx.doi.org/10.1016/j.ijintrel.2012.11.005.

[21] VOS viewer software, Centre for Science and Technology Studies, Leiden University, The Netherlands, 2015, http://www.vosviewer.com/.

[22] J. I. Granda-Orive, A. Alonso-Arroyo, F. Roig-Vázquez, Which Data Base Should we Use for our Literature Analysis? Web of Science versus SCOPUS, Archivos de Bronconeumología (English Edition), vol. 47, issue 4, 2011, p. 213, ISSN 1579-2129, http://dx.doi.org/10.1016/S1579-2129(11)70049-0.

[23] Ulrichsweb, ProQuest. Avaliable at http://www. proquest. com/products-services/Ulrichsweb. html, [accessed on 5.08. 2015].

[24] A. Paul-Hus and P. Mongeon, The journal coverage of bibliometric databases: A comparison of Scopus and Web of Science, Proceeding of METRICS 2014 workshop at ASIS & T, 5 November 2014, Seattle, USA. 1751-1577, http://dx. doi. org/10.1016/j. joi. 2015. 05. 002.

[25] L-E. Svensson, Control of Microstructures and Properties in Steel Arc Welds, CRC Press, 1993, p. 256.

[26] S. Kou, Welding Metallurgy, John Wiley and Sons, 2003, p. 480.

[27] H. Bhadeshia, R. Honeycombe, Steels – Microstructure and Properties, Elsevier, 2006, p. 354.

[28] J. C. Valderrama-Zurián, R. Aguilar-Moya, D. Melero-Fuentes, R. Aleixandre-Benavent, A systematic analysis of duplicate records in Scopus, Journal of Informetrics, vol. 9, issue 3, July 2015, pp. 570-576, ISSN.

23

Intelligent Control Mechanism for Underwater Wet Welding

Joshua Emuejevoke Omajene[1], Paul Kah[1], Huapeng Wu[1], Jukka Martikainen[1], Christopher Okechukwu Izelu[2]

[1]LUT Mechanical Engineering, Lappeenranta University of Technology, Lappeenranta, Finland
[2]Department of Mechanical Engineering, College of Technology, Federal University of Petroleum Resources, Effurun, Delta State, Nigeria

Email address:

Joshua.omajene@yahoo.com (J. E. Omajene)

Abstract: It is important to achieve high quality weld in underwater welding as it is vital to the integrity of the structures used in the offshore environment. Due to the difficulty in ensuring sound welds as it relates to the weld bead geometry, it is important to have a robust control mechanism that can meet this need. This work is aimed at designing a control mechanism for underwater wet welding which can control the welding process to ensure the desired weld bead geometry is achieved. Obtaining optimal bead width, penetration and reinforcement are essential parameters for the desired bead geometry. The method used in this study is the use of a control system that utilizes a combination of fuzzy and PID controller in controlling flux cored arc welding process. The outcome will ensure that optimal weld bead geometry is achieved as welding is being carried out at different water depth in the offshore environment. The result for the hybrid fuzzy-PID gives a satisfactory outcome of overshoot, rise time and steady error. This will lead to a robust welding system for oil and gas companies and other companies that carry out repair welding or construction welding in the offshore.

Keyword: Control System, Bead Geometry, Fuzzy Logic, Process Parameter, Underwater Welding

1. Background

Owing to the environmental conditions in which structures operate in the offshore, it is important that a high structural integrity is guaranteed. It is evident that structural failures can arise as a result of poor weld quality and other mechanical properties of the loaded structures operating in the offshore. The quality of welds achieved underwater experience a major setback because of the unique feature of the weld metal fast cooling rate and other factors such as the stability of the welding arc, loss of alloying elements and difficulty of a good visibility to weld underwater. This research paper addresses the issue of controlling the welding parameters at different water depth in achieving a desired weld bead geometry that is reasonably satisfactory of the weld quality that can operate in the offshore. A control mechanism which incorporates the design advantages of fuzzy logic control and PID control is implemented in this study. Experimental data to be analyzed in this paper is adopted from the work of Chon L. Tsai et. al. It is a well-known fact that high cooling rate and hydrogen embrittlement are characteristics of underwater wet welding (UWW). Rapid cooling mechanism and their effects have been studied by Chon L. Tsai and Koichi Masubuchi [1]. The final microstructure of the heat affected zone (HAZ) for a given material is determined by the composition, peak temperature and cooling rate. It is possible to control the weld metal composition, the peak temperature and the cooling rate to yield favorable microstructure. However, it is not possible to control the composition of the HAZ of the parent material. Fast cooling effect of the water environment in UWW results in a martensitic heat affected zone having high hardness and poor notch toughness. In UWW, the dissociation of water is a source of hydrogen and this subjects the microstructure of the HAZ to hydrogen cracking. This makes it important to control the weld metal's cooling rate [2]. The ability to reduce cooling rate during underwater welding will ensure a decrease in the content of martensite and upper bainite, increase in proeutectoid ferrite and acicular ferrite [3].

2. Underwater Welding

The application of underwater welding for the repair of

ships and offshore structures like oil drilling rigs, pipelines, and platforms is of high importance in today's welding activities. The demand for quality wet weld at a greater depth and variety of materials is continually on the increase [4]. Nowadays, shielded metal arc welding (SMAW) and flux cored arc welding (FCAW), are the most widely used underwater welding process. FCAW has a high prospect in the future because of its high production efficiency and ease to be automated [5 - 8]. The water surrounding the weld metal reduces the mechanical properties of weld done underwater due to the effect of rapid cooling of the weld. Heat loss by conduction from the surface of the base metal directly into the moving water surrounding and heat loss by radiation are the major channels in which heat is lost during underwater welding. In order to achieve higher heat input in underwater welding, it is important to apply higher current to a comparable arc voltage for welds done in the air. Underwater welding high rate of cooling of the base metal results in the creation of unfavorable microstructure such as martensite and bainite for conventional welding of steels. Martensitic and bainitic constituents are high strength and brittle which are prone to cracking in the presence of high hydrogen content [3]. Underwater wet welding bead geometry have weld bead shape that are wider and lower penetration which is the opposite situation in air welding. The welding arc in underwater welding is constricted at a higher water depth. However, shallow water depth welding is more demanding than higher depth. The welding arc instability results in porosity which affects the soundness of the weld. Weld metal carbon content increases with increase in water depth. Also, deoxidizers such as manganese and silicon are lost in higher amounts at increased water depth [9] [10].

3. Flux Cored Arc Welding Process

FCAW is a semiautomatic welding process and the operation continues until completing the weld pass, whereas SMAW process requires changing of the electrodes from time to time. The weld crack sensitivity for FCAW is reduced compared to SMAW because of the interruption of the welding pass in SMAW, especially for steels with CE less than 0.4 %. However, one major disadvantages of FCAW is the difficulty in tracking the joint precisely under the condition of poor visibility underwater. Another challenge is that the diver/welder has difficulty in hearing the arc sound or viewing the plasma during underwater welding and this poses the challenge of him having information regarding the frequent changes in the welding current and voltage. For this reason, it is very important to design a robust control system to control the welding process in the underwater environment [11].

4. Mathematical Model of FCAW

In FCAW, a constant voltage power source and constant wire feeding rate are usually used [11]. The output characteristics of the power supply at the working point is described in the equations below. The mathematical model for the design of the controller in this paper is adopted from the work of Chon L. Tsai et. al [11].

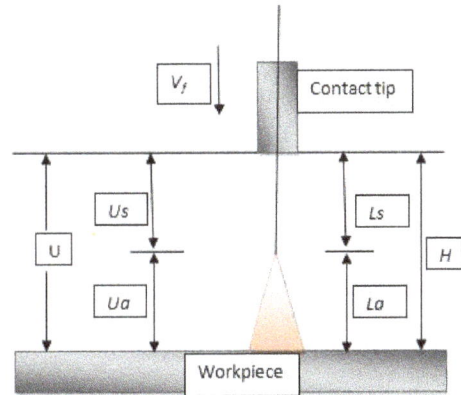

Fig. 1. *Illustration of arc parameters adapted Chon L. Tsai et. al [11].*

$$U = U_{ol} - R_s I \qquad 1$$

Where U_{ol} is the equivalent open loop voltage, and R_s is the equivalent source resistance including the cables.

The output voltage is distributed in two parts which are the arc voltage and U_{ol} and the wire stickout voltage U_s.

$$U = U_a + U_s \qquad 2$$

FCAW arc voltage-current characteristics can be described as in equation 3:

$$U_a = K_a L_a + R_a I + U_c \qquad 3$$

Where K_a is electrical field intensity which is the voltage drop per unit length of arc column. R_a is the arc resistance. U_c is a constant that is related to the anode and cathode voltage.

The voltage drop on the wire stickout is described as equation 4:

$$U_s = \rho \frac{L_s}{A} = K_s L_s I \qquad 4$$

Where ρ is the electric resistivity of the wire over the temperature ranges. A is electrode or wire cross-section area. K_s is average resistance per unit length of wire stickout.

The melting rate can be described as:

$$V_m = K_m I + K_e L_s I^2 \qquad 5$$

Where V_m is the melting rate, $K_m I$ is the melting rate from the arc heat, K_m is the constant related to the anode voltage drop $K_e L_s I^2$ is the stickout wire resistance heat contribution to the melting rate, K_e is also another constant.

$V_m = V_f$ (when in steady state, the melting rate, if represented in wire length per unit time, equals the wire feed rate).

The contact-tip-to-workpiece distance comprises of the arc length and stickout length.

$$H = L_s + L_a \qquad 6$$

In the welding process, the power source voltage, CTWD, wire feed rate, arc length and current are controllable parameters. In welding practice, it is observed that an increase in the length of CTWD will at the same time increase the arc length and will later shorten when a steady state is reached, that is at the state of fixed CTWD, wire feed rate, power source setting, and the arc is stable. The dynamic model of a welding arc power describes the transient characteristics in the change of one or more parameters. The dynamic model is based on the equation in the static model deviation [11].

The dynamic equation of power source

$$U = U_{ol} - R_sI - M_s \frac{di}{dt} \qquad 7$$

Where M_s is power source inductance.
Its frequency domain expression (Fourier Transform) is:

$$\Delta U(s) = - R_s(T_pS + 1)\Delta I(s) \qquad 8$$

The final dynamic model after setting a reasonable range of GMAW is given by the transfer function as represented in equation 9.

$$\frac{I(S)}{H(S)} = -3.46 \frac{0.0168s^2+0.457s+1}{1.28e-005\,s^3+0.01576\,s^2+0.1776s+1} \qquad 9$$

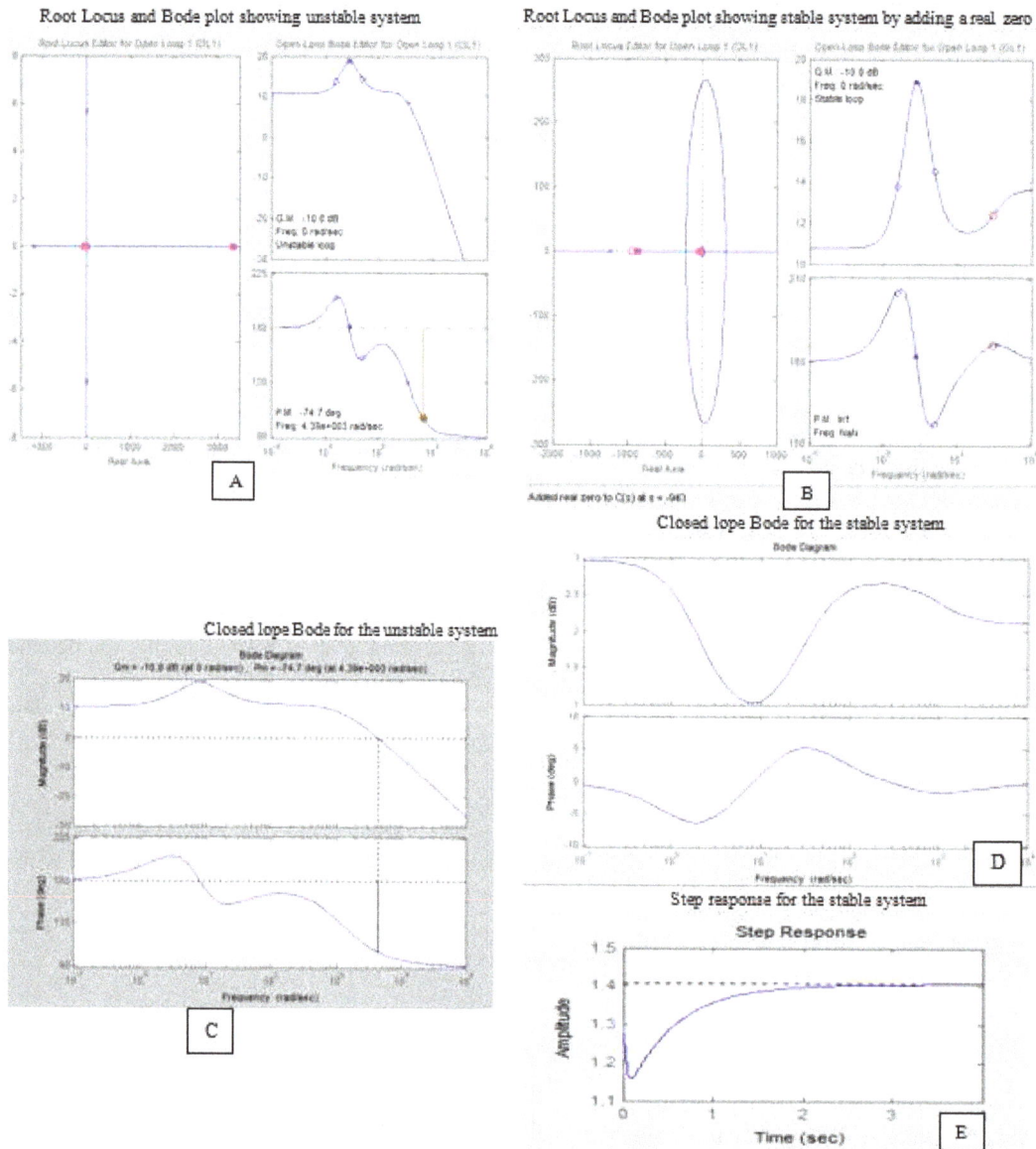

Fig. 2. Stability analysis using Bode plot, Root Locus and step response.

This transfer function is used as the plant model in the control system. The transfer function is the relationship between the arc current and CTWD. The transfer function of the system is unstable but can be adjusted by adjusting the position of the poles by adding a real zero. This is done by

the implementation Table. 1 of the following algorithm below. The SISOTOOL in MATLAB helps us to adjust the position of the poles or zeros in the Root Locus. The Bode plot Fig. 2 shows the relationship between the output signal $I(S)$ and the input signal $H(S)$ that describes the linear system. The Bode plot for the transfer function of the plant shows that the system is unstable. The closed loop Bode plot in Fig. 2A has a gain margin of -10.8 dB at 0 rad/sec and phase margin of 74.7 degrees at 4390 rad/sec shown in Fig. 2C. This means that for the system to be stable, we need to decrease the gain by 10.8 dB. However adding a real zero at s= -943 shown in Fig. 2B will make the transfer function of the system stable. The closed loop Bode Fig. 2D of this system has a peak response of 2.96 dB at a frequency of 3.71e-008 rad/sec. The step response Fig. 2E of the stable system has input amplitude of 1.4. The output response follows closely the input response without an overshoot.

Table 1. MATLAB implementation

NUM=-3.46*[0.01685 0.4575 1]
DEN=[1.28e-5 1.576e-2 1.776e-1 1]
sys=tf(NUM,DEN)
sisotool(sys)

5. Controller for Underwater Wet Welding Process

Fuzzy controller: This control system is a fuzzy logic controller that controls the plant which is the welding machine. The SMAW and FCAW mostly used in underwater welding are dependent on several process parameters that usually vary over a wide domain. Fuzzy logic technique is able to learn the relationship between the welding process input variable and output variable. Fuzzy set theory application is valuable in experimental data modeling which involves unpredictability between the relationships of the welding process input variables and the subsequent bead geometry output. The fuzzy model is used to analyze the appropriateness of the fuzzy relations in predicting the characteristics of the weld bead geometry profile. Fuzzy control is effective for systems which have dissimilarities of system dynamics. The model of the system can perform well for processes that are not precisely defined unlike PID controller. Fuzzy controllers are suitable in achieving a decreased rise time and slight overshoot [12].The structure of the two inputs (error e and error change Δe) and three output (proportional gain K_p, integral gain K_i and derivative gain K_d) are designed for the fuzzy rules used in the hybrid fuzzy controller. The structure for the fuzzy logic controller designed for this research paper is two inputs (error e and error change Δe) and single output of the error and error change. The linguistic variables defined are sentences in normal English language such as negative big (NB), negative small (NS), zero (Z), positive big (PB), postive small (NS), which are expressed by fuzzy sets. The fuzzy sets are characterized by fuzzification (assigning input variables), membership functions (mapping of the input space to a membership value), fuzzy rule (IF-THEN conditional statements), inference system (mapping inputs to outputs) and defuzzification (quantification of expressions). The outputs of the fuzzy sets are obtained in crisp form. The Fig.3

summarizes the operations that are carried out in a fuzzy logic controller. In this research, the output from the defuzzifier which is a proportional, integral and derivative gain is fed into the input signal of the transfer function of the welding machine. The aim of the fuzzy logic in the controller design is to tune the parameters of the PID controller. This will significantly improve the performance of the system as compared to the conventional PID controller.

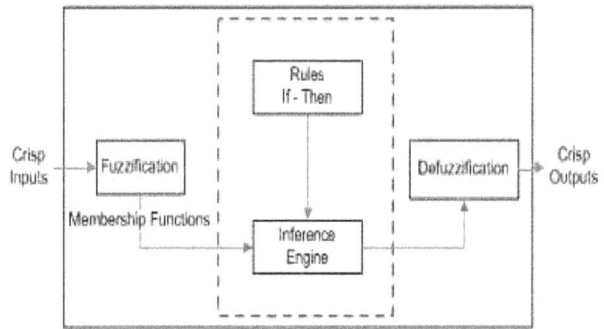

Fig. 3. Fuzzy logic controller block diagram [13].

5.1. Rules

The fuzzy control rule is a collection of fuzzy IF-THEN rules in which the preconditions are the error and error change variable and consequents are the output from the fuzzy logic variables which involves linguistic variables. The fuzzy control rule for the hybrid fuzzy PID controller have preconditions as the error and error change variable and consequence as gain parameters

5.2. Defuzzification Process

The defuzzification process converts the fuzzy output into a crisp using a defuzzification approach that relates the membership functions in Fig. 4 with the fuzzy rules [14]. In this paper, the centroid defuzzification technique is used. The defuzzification process is shown in the Fig. 5 is as a result of the assumptions reached for membership function and fuzzy rules. From the figure, it can be found that as each individual set of input parameters are changed, a subsequent change in the output parameter is effected.

PID controller: PID controller is widely utilized industrially for control applications. This controller is suitable in improving a systems transient response and steady state error simultaneously [15]. The control logic of the PID controller is implemented by finding suitable gain parameters K_p, K_i, and K_d. The transfer function of the PID controller is obtained by adding the terms of the proportional, integral and derivative controller (Equation 10). One major setback of a PID controller is that it does not effectively control a system having big lag, uncertainties and parameter variations. This makes it necessary for a fuzzy-PID hybrid control system [16].

$$PID(s) = K_p + K_i/s + sK_d \qquad 10$$

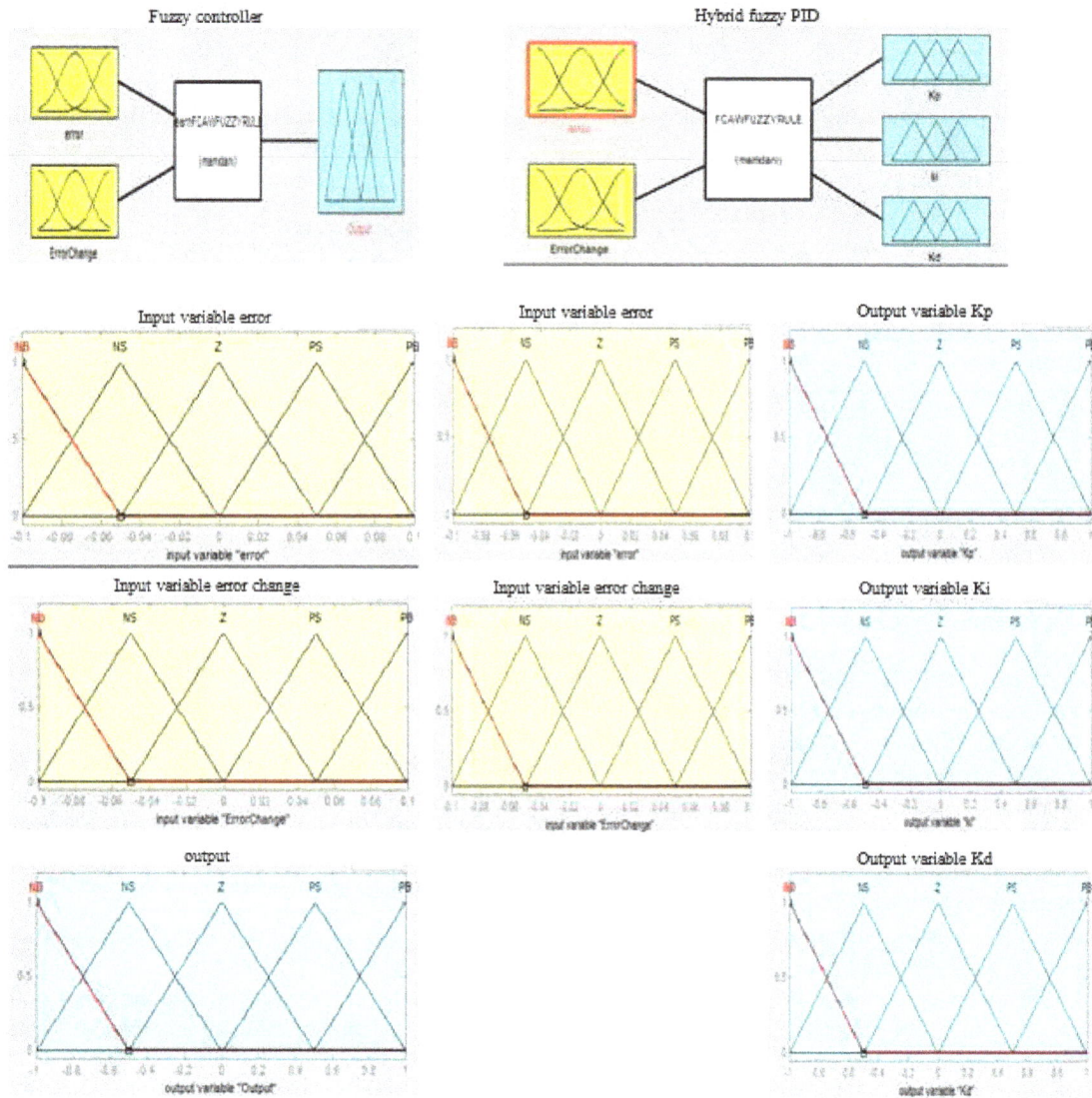

Fig. 4. *Membership function diagram*

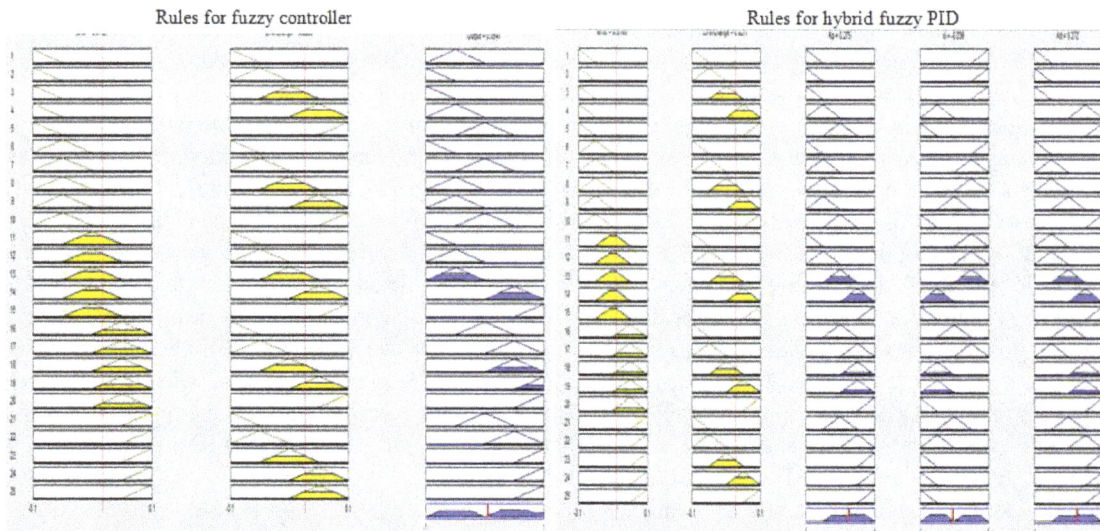

Fig. 5. *Centroid defuzzification using max-min inferencing.*

Hybrid fuzzy PID controller: The control system is a typical fuzzy-PID hybrid control system. It utilizes the advantages of a fuzzy controller and a PID controller. This controller is capable of overcoming the presence of nonlinearities and uncertainties in a system. Tuning of the different parameters of the PID controller is executed with the fuzzy logic controller. Fuzzy rules are designed for effective tuning of the parameter K_p, K_i, and K_d according to functions of the actuating error signal. The proportional term is responsible for the entire control process that is proportional to the error. The aim of the integral term is to help in reducing the steady-state error through low frequency compensation with the aid of an integrator. While the derivative term helps in improving the transient response through high frequency compensation [16].The design and implementation of the control system uses efficient technique that can achieve performance requirement in the presence of disturbances and uncertainty [17]

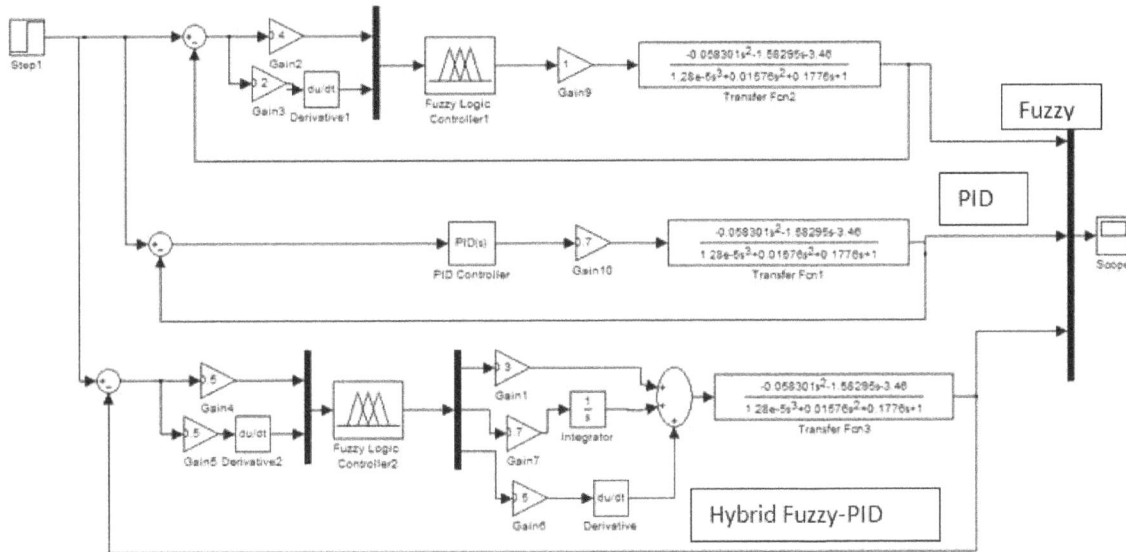

Fig. 6. Control system for underwater FCAW.

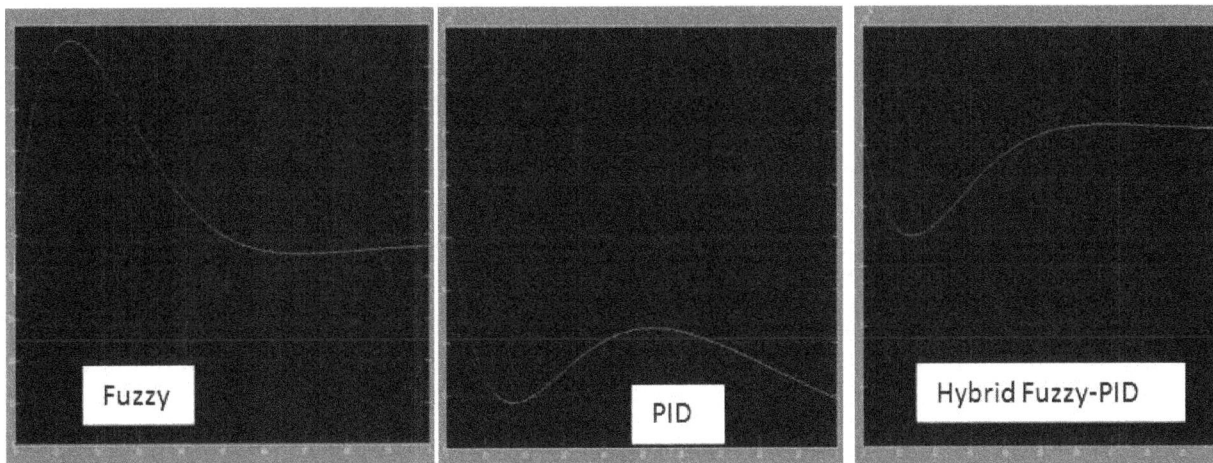

Fig. 7. Matlab simulation results of PID, Fuzzy and hybrid fuzzy PID controller.

6. Results and Discussion

The plant model executed in this paper (Figure 6) made use of the application of the control dynamics of PID, fuzzy and hybrid fuzzy PID. The performances for the different control systems were compared as can been seen in the simulation results in Fig. 7. The system response was tested using a step input signal. From the analysis of the dynamic response of the various control system indicate that fuzzy controller produces a more suitable result compared to the PID controller. From the results, it is evident that the fuzzy controller exhibits a faster response than the PID controller. Results from the PID controller gave rise to overshoot. However, results from the hybrid fuzzy PID controller gave a more satisfactory result of overshoot, rise time and steady state error. The proposed hybrid fuzzy PID controller demonstrates the advantages of fuzzy and PID controller. Application of the hybrid fuzzy PID controller is suitable for FCAW process used in underwater welding since the input current and output CTWD can be predicted and controlled.

7. Conclusion

The control of welding arc current in relationship to CTWD of FCAW for underwater welding application is effective way of ensuring the desired heat input and arc length during welding. The application of a hybrid fuzzy PID controller has the potentials of eliminating the effect of uncertainties and disturbance during underwater welding. A properly control underwater welding process will result bead geometry that is favorable for a sound and reliable weld for welding of offshore structures underwater.

References

[1] Chong-Liang Tsai, Kiochi Masubuchi, "Mechanism of rapid cooling in underwater welding," Applied ocean research, vol. 1, no. 2, pp. 99-110, 1979.

[2] Yong-hua SHI, Zei-pei ZHENG & Jin HUANG, "Sensitivity model for prediction of bead geometry in underwater wet flux cored arc welding," Transaction of nonferrous metals society of China, pp. 1977-1984, 2013.

[3] H. T. Zhang, X. Y. DAI, J. C. FENG & L. L. HU, "Preliminary investigation on real time induction heating assisted underwater wet welding," Welding research , vol. 94, pp. 8-15, 2015.

[4] J. Labanowski, "Development of Underwater Welding Techniques," Welding International, vol. 25, pp. 933-937, 2011.

[5] Yara H., Makishi Y., Kikuta Y., & Matsuda F., "Mechanical and metallurgical properties of an experimental covered electrode for wet underwater welding," Welding International, vol. 1, pp. 835-839, 1987.

[6] Liu S., Olsen D. L., & Ibarra S. J., "Designing shielded metal arc consumables for underwater wet welding in offshore applications," Journal of Offshore Mechanica and Arctic Engineering, vol. 117, pp. 212-220, 1995.

[7] Kononenko V. Y., "Mechanised welding with self-shielding flux-cored wires for repairing hydraulic installations and vessels in underwater," Welding International , vol. 10, pp. 994-997, 1996.

[8] Jia C. B. , Zhang T., Maksimov S. Y., & Yuan X., "Spectroscopic analysis of the arc plasma of underwater wet flux-cored arc welding," Journal of Materials Processing Technology, vol. 213, pp. 1370-1377, 2013.

[9] Brown R. T. &. Masubuchi K., "Effects of water environment on metallurgical structures of welds," Welding Research Supplement, pp. 178-188, 1975.

[10] Liu S. & Olsen D., "Underwater welding," ASM International, vol. 6, pp. 1010-1015, 1993.

[11] Chon L. Tsai, Baojian Liao, David A. Clukey & Josept. S. Breeding., "Development of a Microprocessor based Tracking and Operation Guidance System for Underwater Welding," Ohio State University, Ohio, 2000.

[12] Sinthipsomboon K. et al., "http://cdn.intechopen.com/pdfs/39444/A hybrid of fuzzy and fuzzy self tuning pid controller for servo electro hydraulic system," INTECH, pp. 1-16, 2012.

[13] "Tribal Engineering," [Online]. Available: http://www.tribalengineering.com/technology/fuzzy-ictl.aspx. [Accessed 9 February 2015].

[14] Ion Iancu & Mihai Gabroveanu, "Fuzzy logic controller based on association rules," Annals of the University of Craiova, Mathematics and Computer Science Series, vol. 37, no. 3, pp. 12-21, 2010.

[15] Parnichkun M. & Ngaecharoenkul C., "Hybrid of fuzzy and PID in kinematics control of a pneumatic system," in Industrial Electronics Society 26th Annual Confjerence of the IEEE, Thailand, 2000.

[16] Pornjit Pratumsuwan, Siripun Thongchai & Surapun Tansriwong "A Hybrid of Fuzzy and Proportional-Integral-Derivative Controller for Electro-Hydraulic Position Servo System," Energy Research Journal, vol. 1, no. 2, pp. 62-67, 2010.

[17] Majid Ali, Majid Ali, Saifullah Khan, Muhammad Waleed & Islamuddin,"Application of an Intelligent Self-Tuning Fuzzy PID Controller on DC-DC Buck Converter," International Journal of Advanced Science and Technology, vol. 48, pp. 139-148, 2012.

24

Experimental and CFD Investigation the Effect of Solid Particle Height in Water-Solid Flow in Fluidized Bed Column

Riyadh S. Al-Turaihi, Sarah H. Oleiwi

Mechanical Department, College of Engineering, Babylon University, Babylon, Iraq

Email address:

drriyadhalturaihi@yahoo.com (R. S. Al-Turaihi), sarah.ha369@gmail.com (S. H. Oleiwi)

Abstract: The liquid solid fluidized beds have been usually used in catalytic cracking, ion exchange, crystallization, and particle classification. Fluidized bed column has investigated experimentally and numerically. A numerical investigation has been done by using computational fluid dynamic software ANSYS Fluent 15.0. In this study, a spherical stainless steel has used as solid particle fluidized continuously by water. Three different values of water flow rate (10, 15, 20 l/min) have used with three different values of static head of the solid particles (2, 2.5, 3 cm). The pressure of the fluidized bed has been measured at four different location of the test section 20 cm apart and the flow behavior at the test section has monitored by using AOS high speed camera. The experimental data has compared with numerical solution and a good agreement has been found. It can be observed that the expansion of the solid particles and the pressure inside the bed increased as the water flow rate and static bed height increased.

Keywords: Fluidized Bed, Two Phase Flow, CFD, Ansys Fluent

1. Introduction

Circulating fluidized beds (CFBs) have been commonly operated in chemical, petrochemical, ion exchange, metallurgical, adsorption, environmental, particle classification, and energy industries for applications such as fossil fuel combustion, coal and biomass gasification, and fluid catalytic cracking (FCC). Circulating fluidized beds have several significant advantages such as increased through-put, in-bed sulfur capture, fuel flexibility, and high efficiencies in gasification and combustion with relatively low emissions of NOx [1-2]. CFB technology is more attractive in the industry of energy than several other systems. Despite its extensive applications, the complex hydrodynamics of CFBs are still not completely understood and difficult to predict, in most applications, the heat and mass transfer between the phases and the chemical reactions make it even more difficult to predict the behavior of CFB reactors [2]. The solid particles placed on porous plate called distributor subsequently the continuous phase forced through the holes of distributor causing the particles suspension in the flowing flow, the solid particles expand in the bed and the

expansion increase as the water flow rate increase. The fluidized bed subjected to research a long time ago. Tingwen et al. [2] have studied the grid refinement effect for a two circulating fluidized bed. Hashizume et al. [3] have studied experimentally the pressure drop for a liquid - solid fluidized bed with different diameters of particles. Kalaga et al. [4] have performed a liquid phase residence time distribution (RTD) study in a conventional solid liquid fluidized bed and in a solid liquid circulating multistage fluidized bed. Al-Turaihi et al. [5] have investigated experimentally the steady state gas solid fluidized bed system with various materials of solid particles. Youjun Lu et al. [6] have studied experimentally the hydrodynamics of a supercritical water fluidized bed. Loha et al. [7] have investigated the influence of specularity coefficient on hydrodynamic behavior of fluidized bed with Geldert B particles using CFD Euler-Euler model. Leckner et al. [8] have studied the pressure fluctuations in a gas - solid fluidized bed by mean of a simple model of the principal frequency of the pressure fluctuations. Ramesh et al. [9] have obtained the gas holdup from the pressure drop and investigate the effect of composite promoter presence on the pressure drop for three phase fluidized bed. In this work, the bed pressure influenced by increasing water flow rate and the solid static height has investigated, the experimental result has compared with a result obtained by ANSYS Fluent 15.0 and a good agreement has been found.

2. The Experimental Equipment and Procedure

Figure (1) shows the schematic diagram of the TWO PHASE FLOW FLUIDIZE BED. The experiments were performed using Perspex circular cross section column with 1 m height and 0.0254 m diameter. Four taps 0.2 m apart along the test section side were used for pressure sensors with ranges from 0-1 bar that used to measure the pressure along the bed, the pressure sensor connected into a personal computer through data acquisition. The spherical stainless steel solid particles placed in the test column on a net with 0.334 mm hole spacing which represents the distributor, the water pumped with maximum discharge of 500 L/min from the water tank and forced into the pipe through the holes of the distributor. 1 in valve used to control the flow rate of water Flow meter has used to measure volume flow rate of water with ranges from 5 l/min to 35 l/min .AOS high speed camera has used to record the flow behavior inside the fluidized bed, the camera has active resolution of 720x480, and 29.97 Hz (59.94 Hz interlaced) image frequency with (SDTV 480i).

Three values of static head of solid particle (H) have used and for each value the water flow rate has changed for three times. In order to make sure of the results, each test has repeated for three times. The used values are shown in table (1)

Table 1. The values used for the experiment.

H (cm)	Water flow rate (l/min)
2	10
2.5	15
3	20

3. Numerical Simulation

For the numerical simulation Software ANSYS Fluent 15.0 has used for modeling water-solid flow. Eulerian approach was employed which sometimes called granular flow model (GFM), particles in the flow has treated as magnified molecules so that on analogy of their behavior to the continuous phase molecules can be stipulated [10]. For this work, the fluidized bed flow has modeled to be turbulent according to Reynolds number (for all the cases).

$$Re = \frac{\rho u d}{\mu}$$

Consequently K-ε RNG mixture turbulent model has used.

3.1. Geometry

The test section which is 1 m height and 0.0254 m diameter pipe has modeled as two dimensional structure using Ansys Workbench 15.0. The bottom edge has been divided into 13 small edges as representation of distributor.

3.2. Mesh

Mesh has been carried out in ANSYS workbench 15.0 as well and Quad square mesh has been used to give 6500 element and 7014 node.

3.3. Initial and Boundary Condition

The bed initial volume fraction was set to be 0.7668. The bottom edge set as velocity inlet and since its already divided into 13 pieces, one piece had gave the value of inlet velocity and the next gave zero inlet velocity respectively for all of them. The sides of the bed set to be walls and the top edge which is represent outlet of the bed set as outlet pressure and the values of the outlet pressure was taken from the experimental work.

3.4. Post Processing

The value of the bed pressure was measured by setting points with the same coordinate as the pressure sensor used in experimental test and the pressure measured at this points for the bed flow behavior, image for the solid volume fraction is taken every 2.5 sec for all the experiment cases and the contour color set to be the same for all the experiment cases starting from 0 and ending with 0.2. So, the difference in packing can be seen clearly.

The other parameters have used for the numerical simulation in fluent model are shown in table (2).

Table 2. Parameters used for the simulation.

Parameter	Value
Drag model	Syamlal-O'Brien
Iteration/time step	20
Particle density	8000 kg/m³
Particle diameter	1.5 mm
Static bed height	2, 2.5, 3 cm
Time step size	0.0007 s
Under relaxation momentum	0.2
Under relaxation pressure	0.3
Under relaxation volume fraction	0.5
Water density	998.2 kg/m³
Water superficial velocity	0.33, 0.49, 0.66 m/s

3.5. The Governing Equations

The general equations used by Eulerian multiphase model are the conservation of mass and momentum. Mass conservation equation can be written in its general form as

$$\frac{\partial}{\partial t}(\alpha_q \rho_q) + \nabla.(\alpha_q \rho_q \vec{v_q}) = \sum_{p=1}^{n} m_{pq} - m_{qp}) + S_q$$

The source term S_q is zero as default value or else it can be a set constant value or define by user. Momentum conservation equation can be written in its general form as

$$\frac{\partial}{\partial t}(\alpha_q \rho_q \vec{v}_q) + \nabla.(\alpha_q \rho_q \vec{v}_q \vec{v}_q) = -\alpha_q \nabla p + \nabla.\bar{\bar{\tau}}_q +$$

$$\alpha_q \rho_q \vec{g} + \sum_{p=1}^{n}(\vec{R}_{pq} + \dot{m}_{pq} \vec{v}_{pq} - \dot{m}_{qp} \vec{v}_{qp}) + (\vec{F}_q + \vec{F}_{lift,q} + \vec{F}_{vm,q})$$

Where $\bar{\bar{\tau}}_q$ is the stress - strain tensor.

$$\bar{\bar{\tau}}_q = \alpha_q \mu_q (\nabla \vec{v}_q + \nabla \vec{v}^T_q) + \alpha_q (\lambda_q - \frac{2}{3}\mu_q)\nabla.\vec{v}_q \bar{\bar{I}}$$

q Represent the phase in all equations

4. Result and Discussion

4.1. Experimental Results

The pressure of fluidized bed is used to characterize the bed and it depends on the volume flow rate of continuous phase wither its liquid, gas or mixture of both, the physical properties of solid and continuous phases, static bed height (the quantity of solid placed in the fluidized bed) and the geometry of the test section. In this experiment the pressure has measured at four taps on the test section to find the effect of increasing volume flow rate of water and static bed height on the profile of the pressure at different location along the test pipe. Figure (2) shows the effect of increasing water flow rate on the pressure profile along the test section at four different locations where the pressure taps exist which is 0.2 m apart with respect to the static bed height. Three different values of water flow rate were shown at each value of static bed height and for three values of static bed height, as the volume flow rate increased the pressure of the bed increased due to the increase in the density of the mixture as well as the increase in the expansion of the bed as the water flow rate increase. Figure (3) shows the effect of static bed height on the pressure profile with respect to the water volume flow rate, the pressure inside the bed increases as the static bed height increases. The flow behavior in the test section as the flow rate of water increase at different values of bed height are shown in figures (4a, 5a, 6a) which is taken for the test section by using high speed camera, .The camera can monitor only 0.6 m of the test section which is 1 m. In these figures the expansion of the bed can be seen which increases as the flow rate increase. Figures (7a, 8a, 9a) shows the flow behavior in the test section as the static bed height increase with respect to the change in the water flow rate, the expansion of the bed increase as the value of static head of

solid particle(H) increased in the pipe.

4.2. Numerical Simulation

The experimental result of pressure have compared with computational fluid dynamics results obtained using ANSYS Fluent 15.0 and a good agreement has been found as can be seen in figure (2). The two phase flow is a changing phenomena due to distribution of the solid particles are different each time for the same experiment. Figure (3) shows a comparison between the pressure results as the static bed height changed with respect to the water flow rate. Figures (4, 5, 6) show the solid particle distribution of experimental test and the solid particle distribution from Fluent program for the solid volume fraction, which shows the flow behavior in the test section as the water flow rate increases for different values of the static bed height for 0.4 m for each test. Fluent program gave a good result for the expansion of the bed as the flow rate changes, however the results is not exact match between the experimental and numerical results that because the images taken from the program for the same time but in the experimental work it taken at different times during the experiment. Figures (7, 8, and 9) show compaction between the flow behavior as the value of static head of solid particle(H) increases for different values of water flow rate. The contour color for all the experiment is same which is between 0 to 0.2 it can help to see the differences in the packing and spaces between the particles as the flow rate or H increases.

5. Conclusion

In this work, the two phase fluidized bed has studied experimentally and numerically by using ANSYS Fluent. The pressure of the bed has measured at different points along the bed and the behavior of the flow has monitored (or recorded) by using high speed camera which is then compared with the solid volume fraction of bed found by ANSYS Fluent15.0. Three different values of H and water flow rate have used. It can be concluded that

- As the volume flow rate of water increase the pressure of the bed will be increase.
- As the value of H increase the pressure of the bed will be increase.
- As the volume flow rate of water increase the expansion of the bed will be increase.
- As the value of H increase the expansion of the bed will be increase.

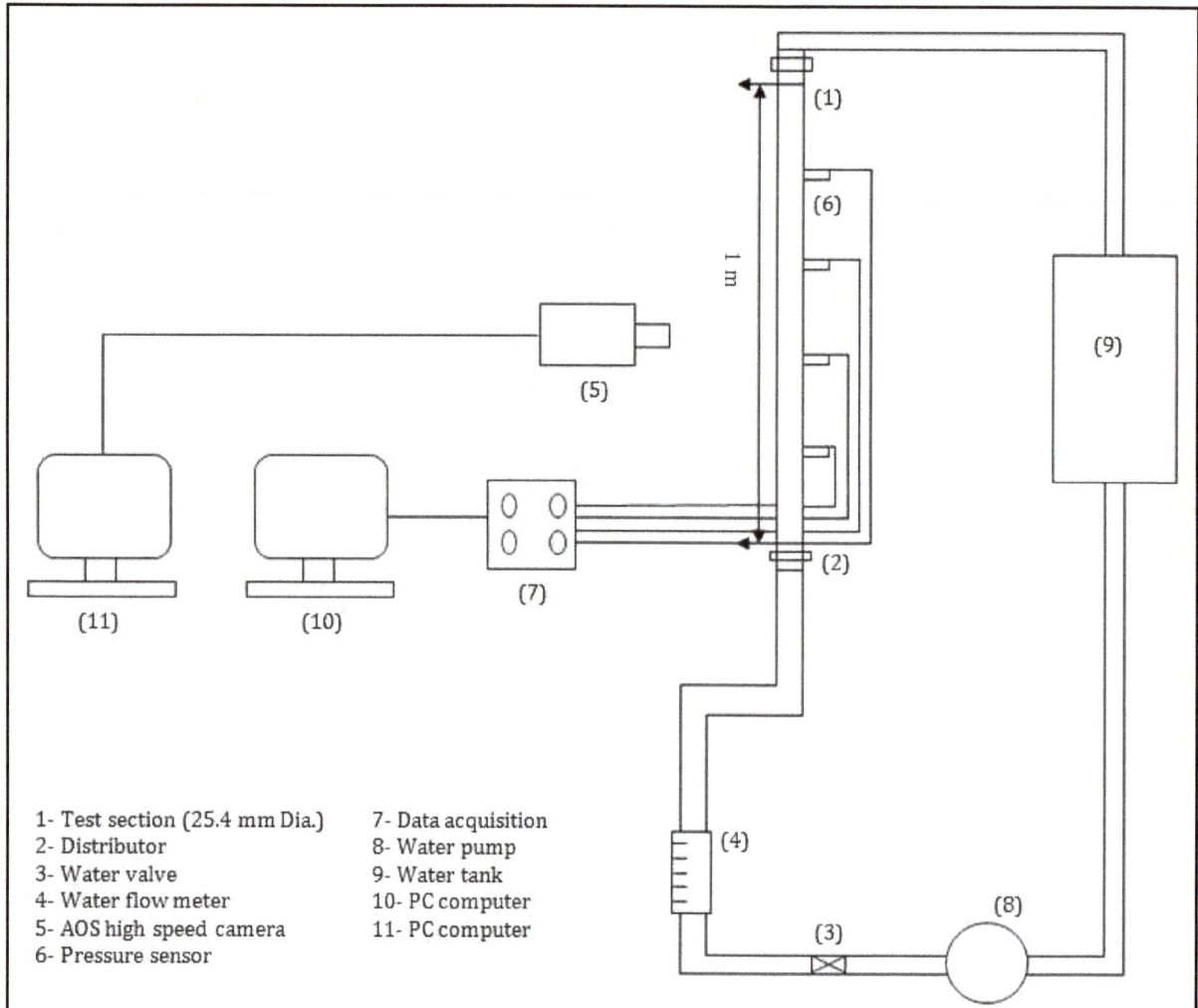

Fig. 1. *The experiment schematic diagram.*

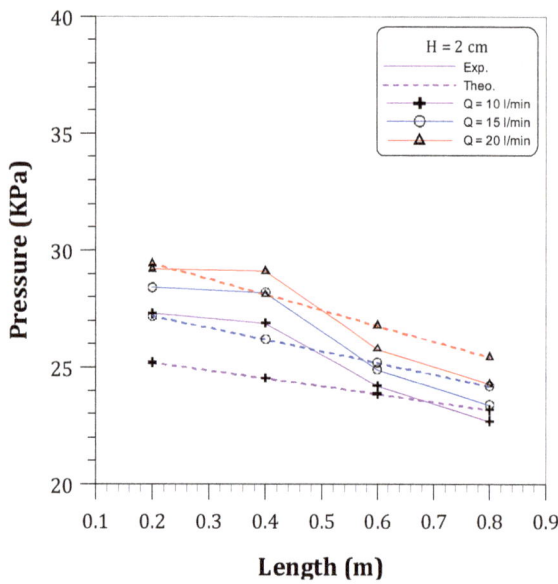

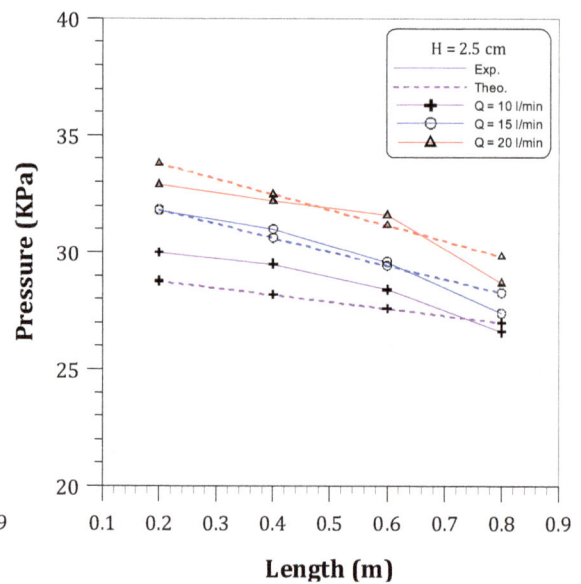

1- Test section (25.4 mm Dia.)   7- Data acquisition
2- Distributor   8- Water pump
3- Water valve   9- Water tank
4- Water flow meter   10- PC computer
5- AOS high speed camera   11- PC computer
6- Pressure sensor

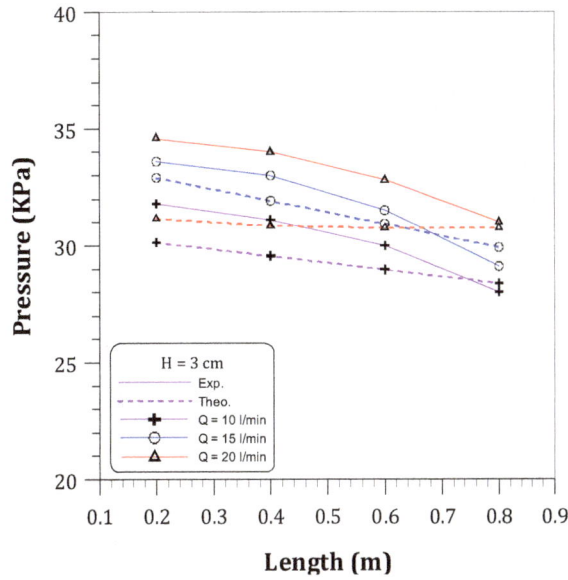

Fig. 2. *Effect of water flow rate on pressure profile.*

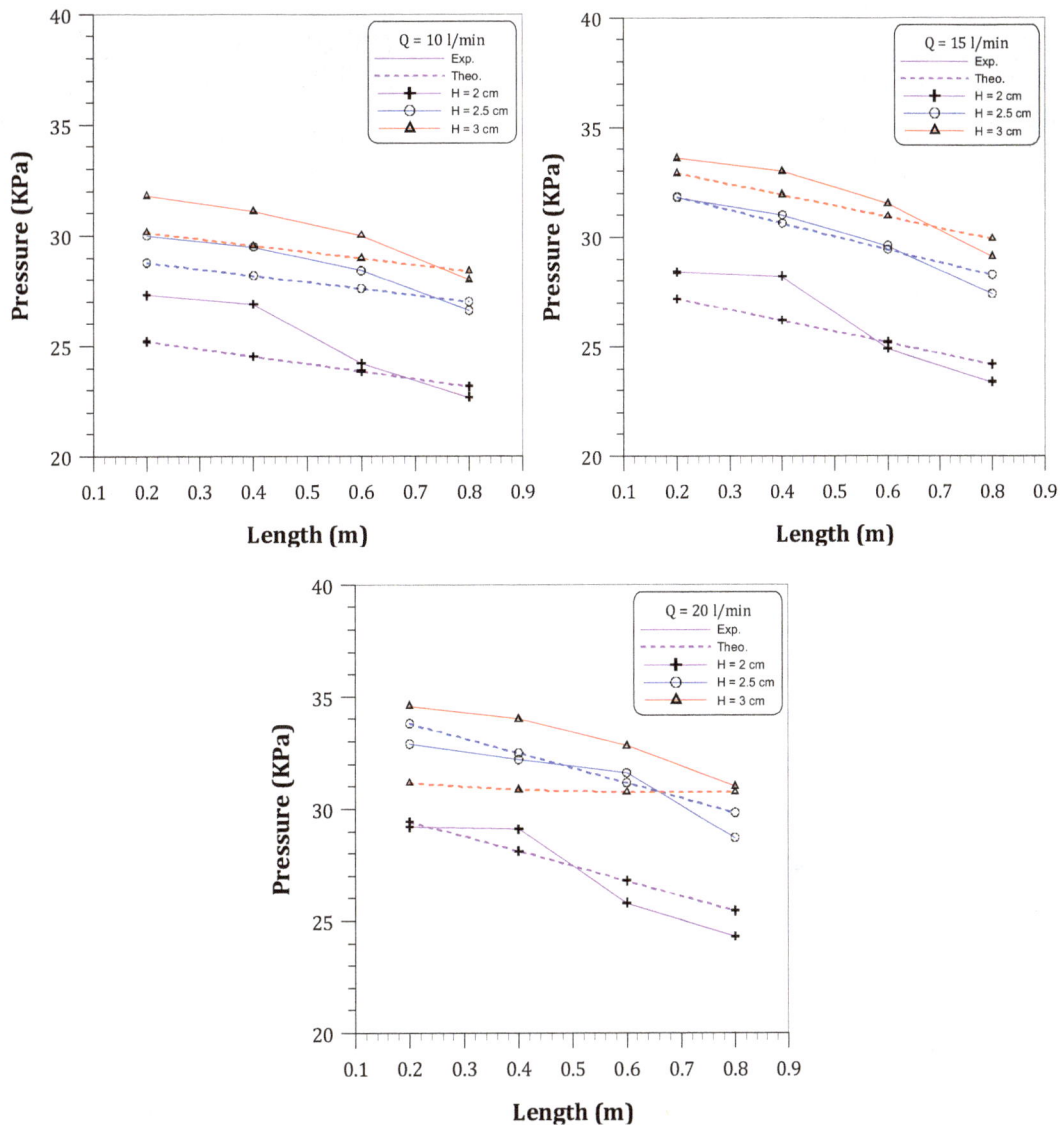

Fig. 3. *Effect of H on pressure profile.*

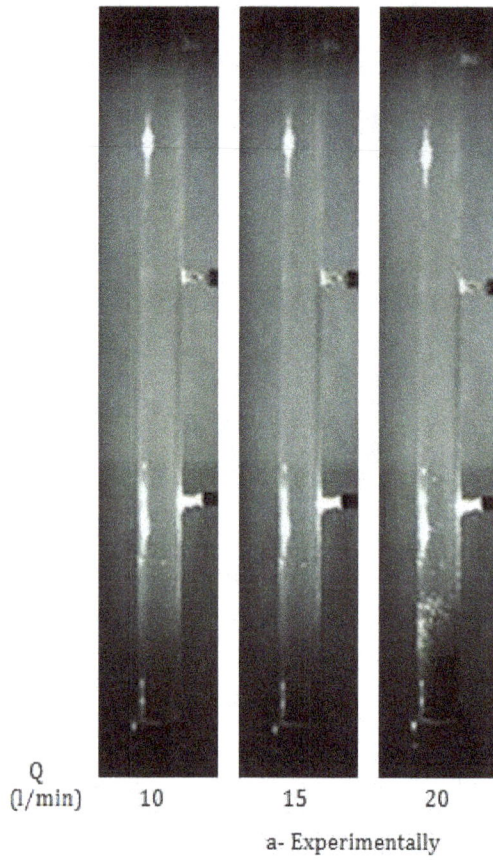

Q
(l/min)   10   15   20

a- Experimentally

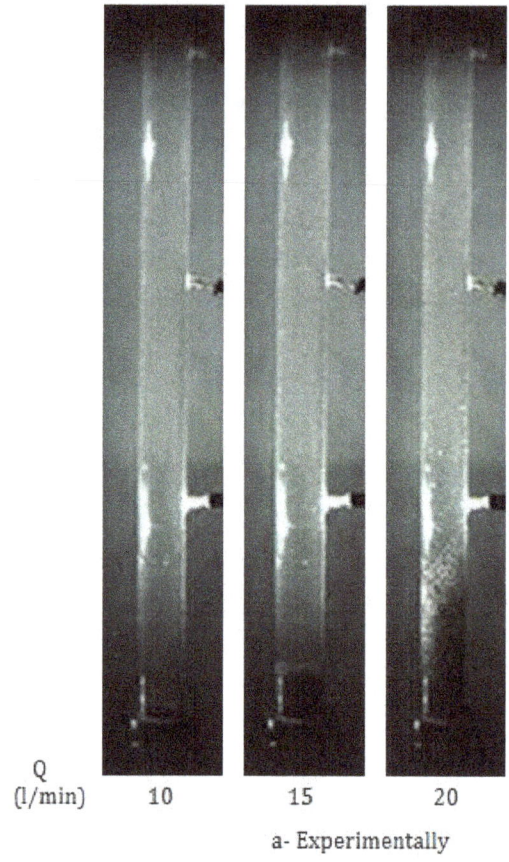

Q (l/min)   10   15   20

b- Numerically

Fig. 4. *Flow behavior at 2 cm H and different values of water flow rate.*

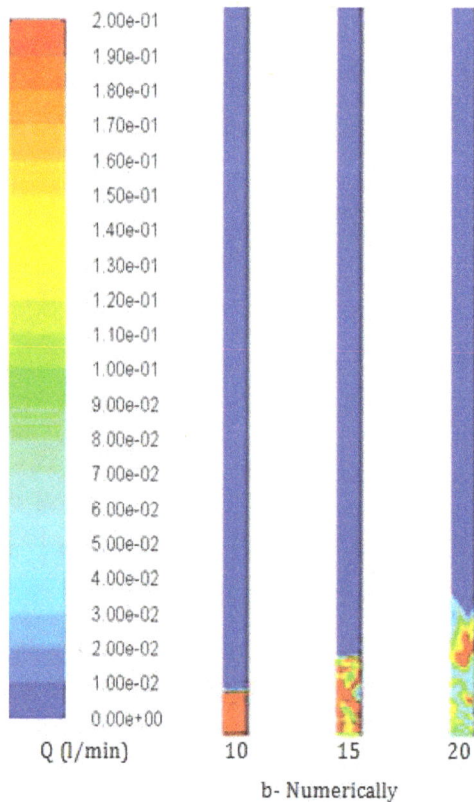

Q
(l/min)   10   15   20

a- Experimentally

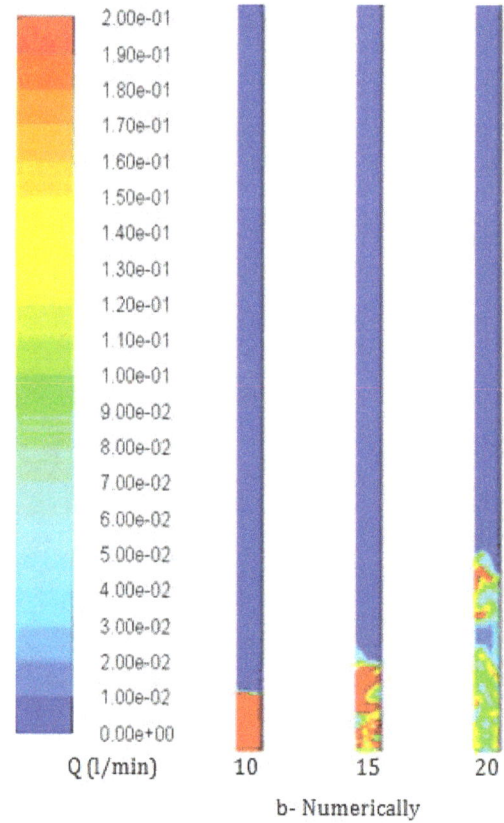

Q (l/min)   10   15   20

b- Numerically

Fig. 5. *Flow behavior at 2.5 cm H and different values of water flow rate.*

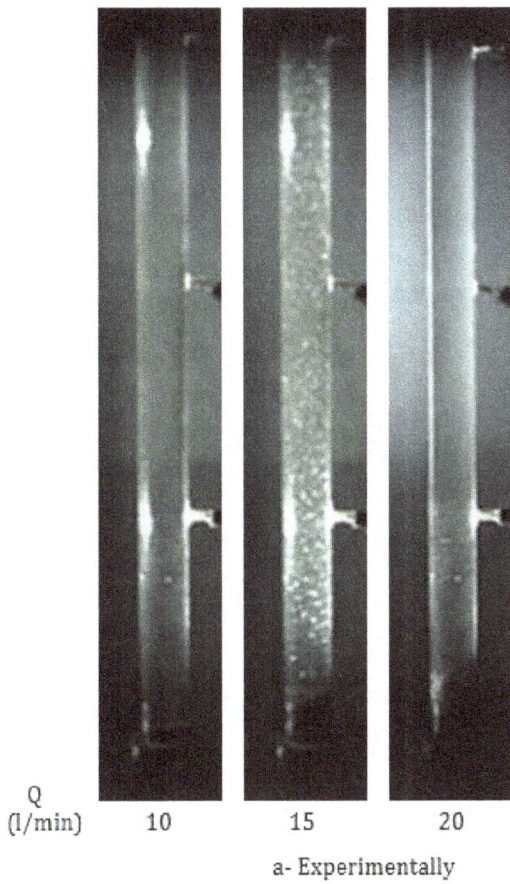

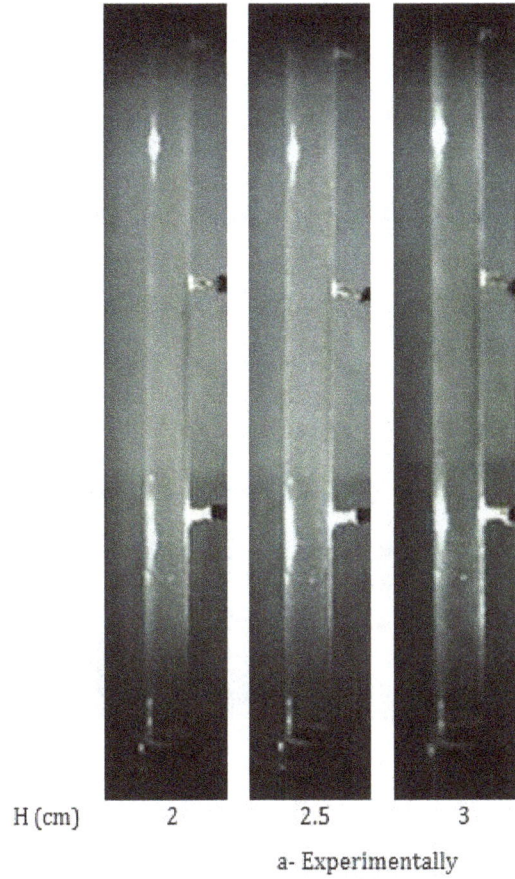

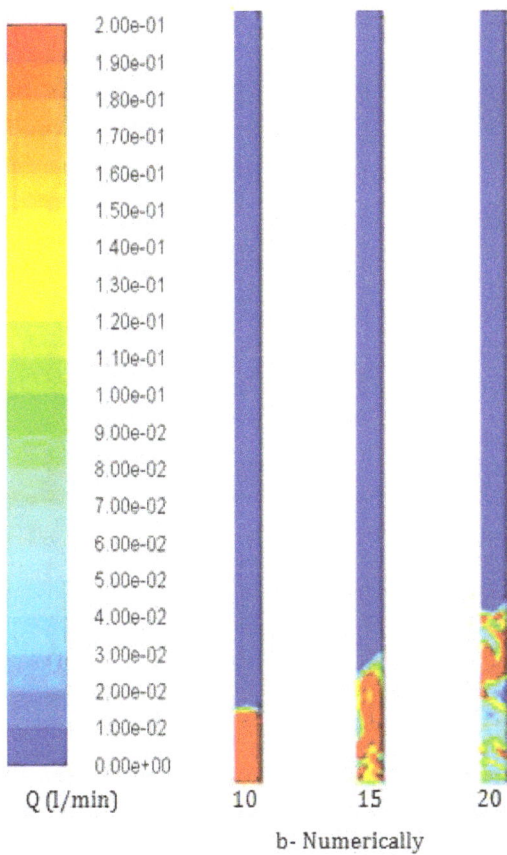

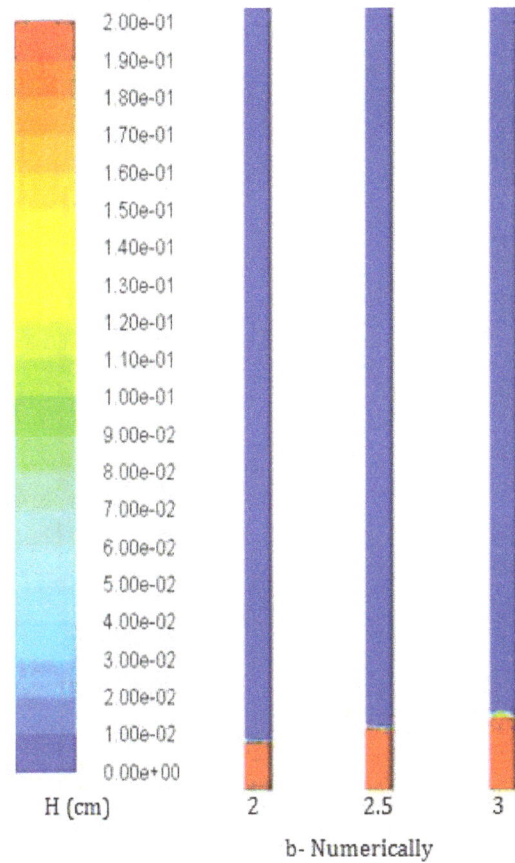

Fig. 6. *Flow behavior at 3 cm H and different values of water flow rate.*

Fig. 7. *Flow behavior at 10 l/min water flow rate and different values of H.*

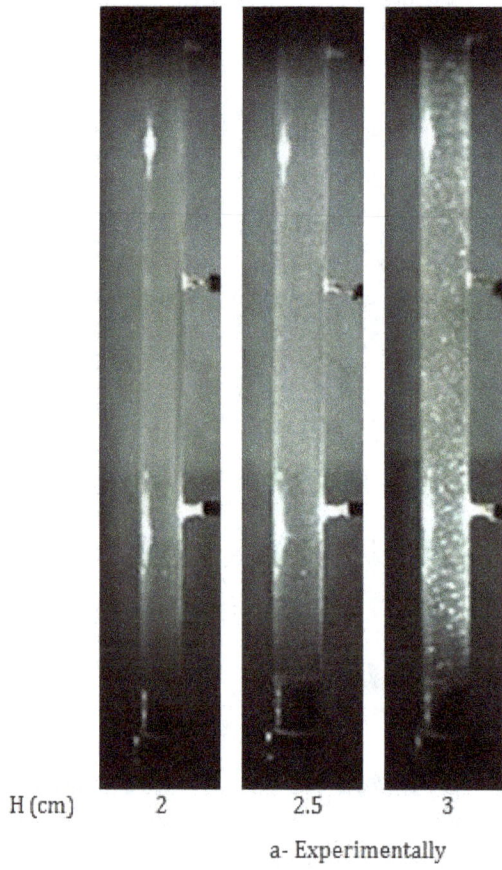

H (cm)   2   2.5   3

a- Experimentally

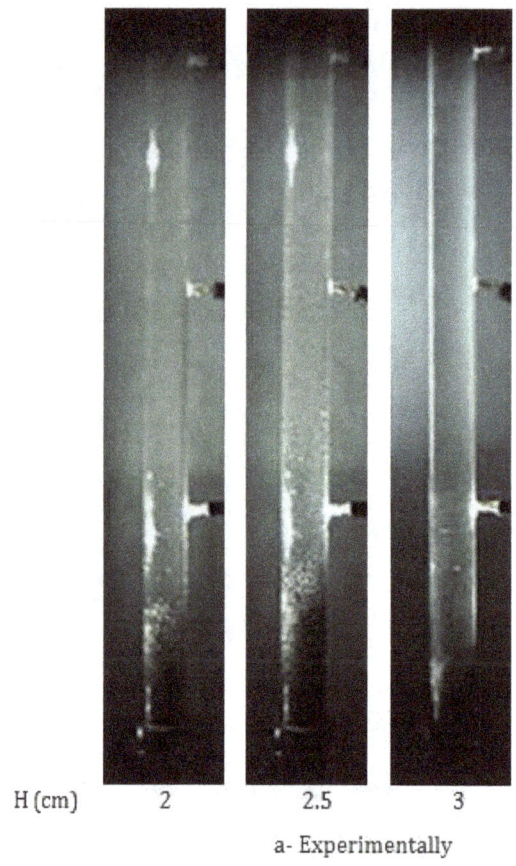

H (cm)   2   2.5   3

b- Numerically

Fig. 8. *Flow behavior at 15 l/min water flow rate and different values of H.*

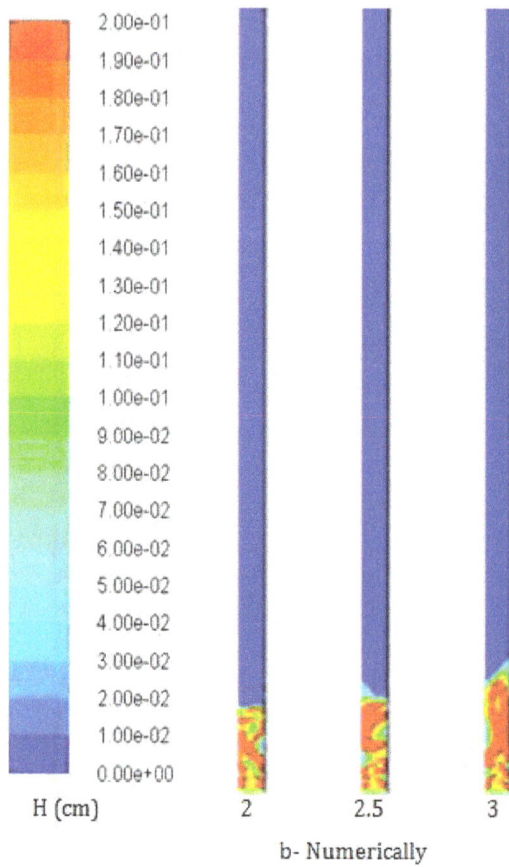

H (cm)   2   2.5   3

a- Experimentally

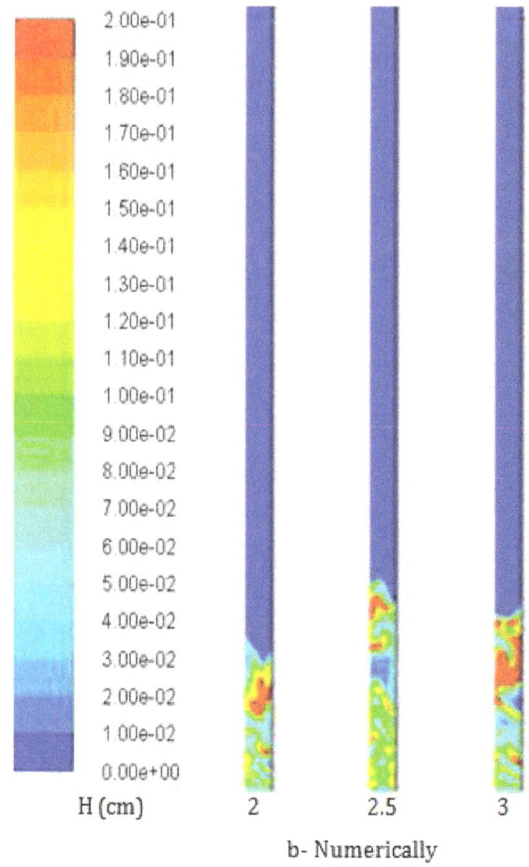

H (cm)   2   2.5   3

b- Numerically

Fig. 9. *Flow behavior at 20 l/min water flow rate and different values of H.*

Nomenclature

H: Static bed height (cm)

CFD: Computational fluid dynamics

F_q: External force (N)

G: Gravity acceleration (m/s^2)

K_{pq}: Exchange coefficient

q: Number of phases

Q: Water flow rate (l/min)

m_{pq}: Interphase mass exchange (kg/m^3.s)

Greek symbols

ε: Epsilon

$α_q$: Volume fraction for the qth phase

$ρ_q$: Mixture density for the qth phase (kg/m^3)

$τ_q$: Shear stress for the qth phase

Subscripts

A: Air

C: Continuous phase

lift: Lift force

q: Secondary phase

vm: Virtual mass force

References

[1] Reza Davarnejad, Reza Eshghipour, Jafar Abdi and Farhad Banisharif Dehkordi, "CFD Modeling of a Binary Liquid-Solid Fluidized Bed", Middle-East Journal of Scientific Research, 19 (10): 1272-1279, 2014.

[2] Tingwen Li, Aytekin Gel, Sreekanth Pannala, Mehrdad Shahnam, Madhava Syamlal, "CFD simulations of circulating fluidized bed risers, part I: Grid study", Powder Technology, 254 - 170–180, 2014.

[3] Kenichi Hashizume, Shinichi Morita, Yuki Nakamura, and Akira Wada," Pressure Drop in Liquid–Solid Circulating Fluidized Beds", Heat Transfer—Asian Research, 38 (4), 2009.

[4] Dinesh V. Kalaga, Rupesh K. Reddy, Jyeshtharaj B. Joshi, Sameer V. Dalvi, K. Nandkumar, "Liquid phase axial mixing in solid–liquid circulating multistage fluidized bed: CFD modeling and RTD measurements", Chemical Engineering Journal, 191- 475– 490, 2012.

[5] Riyadh S. Al-Turaihi, Ahmed Kadhim Hussein, Hussain H. Al-Kayiem and H. A. Mohammed, "Experimental Investigation of Various Solid Particle Materials on the Steady State Gas-Solid Fluidized Bed System", J. Basic. Appl. Sci. Res., 3(7)293-304, 2013.

[6] Youjun Lu, Liang Zhao, Qiang Han, Liping Wei, Ximin Zhang, Liejin Guo, Jinjia Wei, "Minimum fluidization velocities for supercritical water fluidized bed within the range of 633–693 K and 23–27 MPa", International Journal of Multiphase Flow, 49 - 78–82, 2013.

[7] Chanchal Loha, Himadri Chattopadhyay, Pradip K. Chatterjee, "Euler-Euler CFD modeling of fluidized bed: Influence of specularity coefficient on hydrodynamic behavior", Particuology, 11- 673– 680, 2013.

[8] Bo LECKNER, Genadij I. PALCHONOK, and Filip JOHNSSON, "PRESSURE FLUCTUATIONS IN GAS FLUIDIZED BEDS", Original scientific paper, 0354–9836, 6 , 2, 3-11, 2002.

[9] K.V. Ramesh, G.M.J. Raju, G.V.S. Sarma, C. Bhaskara Sarma, "Effect of internal on phase holdups of a three-phase fluidized bed", Chemical Engineering Journal, 145-393–398, 2009.

[10] Farshid Vejahati, Nader Mahinpey, Naoko Ellis and Mehrdokht B. Nikoo, "CFD Simulation of Gas–Solid Bubbling Fluidized Bed: A New Method for Adjusting Drag Law", Can. J. Chem. Eng., 87:19–30, 2009.

[11] Ansys 13.0 Help, Fluent Theory Guide, Mixture Multiphase Model.

[12] Ansys Fluent Tutorials, Fluidized bed.

25

Noise in Olive Mills, the Case of Jordan: Actual Measurements & Reduction Proposition

Rizeq N. Hammad, May M. Hourani*, Firas M. Sharaf

Department of Architecture, Faculty of Engineering, Jordan University, Amman, Jordan

Email address:

rizeqhammad@yahoo.com (R. N. Hammad), may_hourani@yahoo.com (M. M. Hourani), f.sharaf@ju.edu.jo (F. M. Sharaf)

Abstract: This study is concerned mainly with high noise produced by automated olive mills and its negative impact on human health. This study looks at what caused this problem such as the application of building codes concerning user exposure to noise and protective measures for workers. Long working period exposure to high noise, particularly mill workers, causes health and ear illness. Noise levels are measured in several olive mills in Jordan, recorded data was analyzed and findings indicated that noise level is higher than the maximum noise limits allowed by the Jordanian code. This study concludes design solutions and legislative procedures to improve noise control in olive mills.

Keywords: Olive Mills Design, High Noise Impact, Noise Reduction

1. Introduction

Olive oil is a basic product in the lives of Mediterranean people, both for nutrition and other related products. Olive oil is used as a raw material for many industries such as soap, cosmetics and medicines. The Mediterranean region is a main source for olive cultivation. Since the olive tree is considered a blessed tree in many religions, it is widely planted in the Middle Eastern region for hundreds of years.

In Jordan, there are approximately 20 million olive trees planted in 1.3 million acres, and almost 77% of olive trees are rain fed cultivated. The olive sector is a main contributor to the economy and national gross production in Jordan. The amount of annual income from the olive industry in Jordan is about 100 million Jordanian Dinars (JDs, about 150 million US dollars). Investment in the olive industry is increasing and the size of this investment reaches one billion JDs, this amount includes the value of planted land and industries. The olive industry in Jordan has social importance as sustenance for a large segment of farmers and other involved people [1].

A considerable change is taking place in olive mill design in Jordan because of the development in the olive industry in the last decade. More olive trees are planted and increased number of modern olive mills producing more oil and better quality. Modern automatic pressing machines have invaded the Jordanian market and replaced traditional olive oil extraction methods. The ratio of modern olive mills in Jordan is 93% while ratio of traditional olive mills is 7%. There are about 124 olive oil mills in Jordan with a production capacity of approximately 352 tons/hour. These mills include 245 advanced production lines of high quality olive oil.[1]

2. Architectural Design of Olive Oil Mills

A typical shape of a modern olive oil mill is rectangular; a fore-entrance leads to a main hall where olive oil is filled into cans under supervision of olive holders. A rear entrance is used to bring in olives inside the mill. The area used for pressing olives is not open to public, but visually connected to the entrance hall, so that everyone can observe olive oil production processes. Entrance hall is used for filling oil into metal containers and for loading and unloading bottled and metal containers. "Fig. 1" shows a typical layout of a modern olive oil mill in Jordan.

Employees of the mill are responsible for monitoring, accounting and weighing; therefore the employees' offices are located at the front of mill to allow monitoring the processes of olive pressing and filling oil in steel containers. These functions are located within one space in the mill of about 4 meters height and roofed by concrete slab, or steel roof with frames covered with aluminum panels. Olives are put in bags

1 Royal Scientific Society, Environmental Record Centre, Clear Production Unit. Amman - Jordan

in a storage area for few days awaiting pressing. The processes of producing olive oil in a modern automated mill includes washing, crushing, grinding, centrifugal oil extraction and dumping remains (peat) out of the production line "Fig. 2" The size of olive oil mill depends on the number of production lines which ranges from one to three lines. The length of the mill ranges between 15-30 m and width between 10-15 m. "Fig. 3": A, B, C show the interior of a modern olive oil mill in Jordan. Interior finishing of the mill is basic, such as a painted cement plastered surface and in some cases the concrete structure is left unfinished. Perhaps, this is because mills are used for three months only every year and close until next olive season.

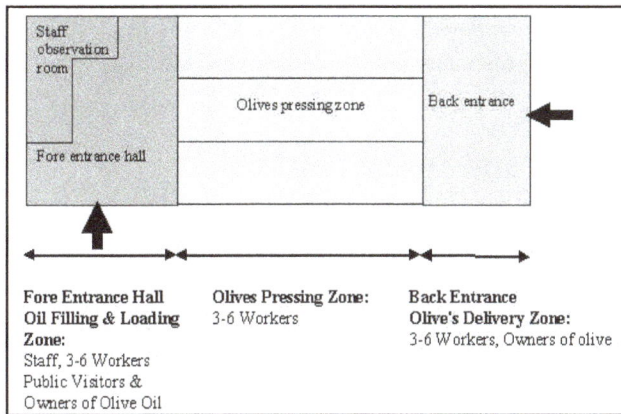

| Fore Entrance Hall Oil Filling & Loading Zone: Staff, 3-6 Workers Public Visitors & Owners of Olive Oil | Olives Pressing Zone: 3-6 Workers | Back Entrance Olive's Delivery Zone: 3-6 Workers, Owners of olive |

Figure 1. Typical layout of modern olive oil mill.

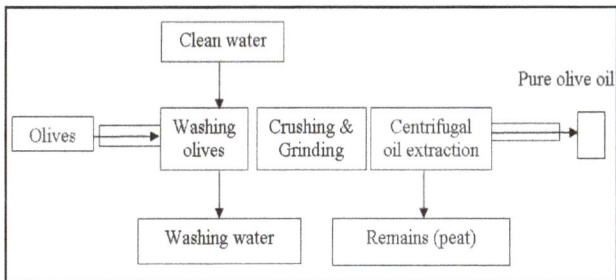

Figure 2. Olive oil production processes in a modern mill.

3. Noise Problem in Modern Olive Oil Mill

Introducing modern olive oil pressing machinery into mills has increased the production volume of olive oil while maintaining good quality and reducing human effort and time to extract olive oil.

However, modern machines have caused negative consequences for environment and user health. High noise levels rising to risk limits in olive oil mills is considered the most serious negative impact on workers, and other users such as visitors. In addition to the other problems of modern oil mills that are not the subject of this study, such as the greater consumed amount of water compared to traditional methods and producing residues which negatively affect environment and people.

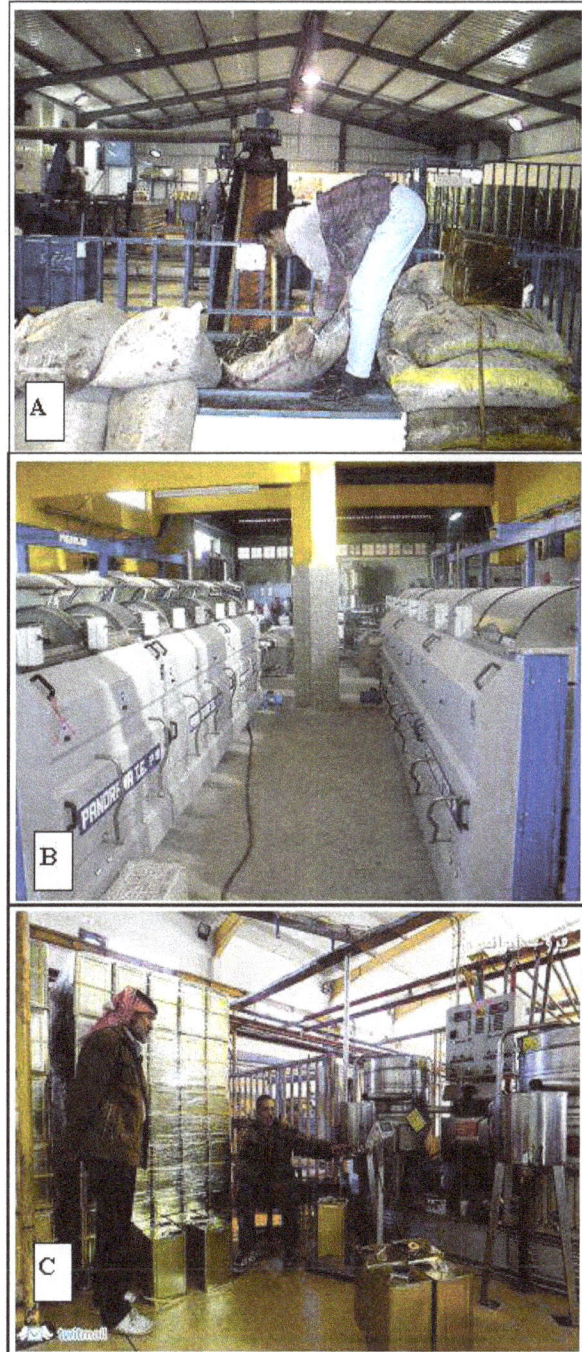

Figure 3. Olive oil mill. A: Back-house: loading olives in the mill. B: pressing machines in the mill. C: Fore-house: filling oil into cans.

The International Olive Council guidelines [2] recommends reducing harmful noise impact of oil mill by separating the grinding and crushing machines from the olive delivery area at rear entrance and also from the front area at the entrance where olive oil is collected and filled in containers ready for transportation. Workers in olive oil mills need to wear suitable ear protection at all time to minimize negative effects of high noise levels.

Although modern olive mills produce high continuous noise, little research has been conducted on this subject, which this paper seeks to address.

The duration of exposure, or daily dose (DOSE), to noise

level 95dB(A) allowed by The International standard is 3 hours a day per five days a week [3]; and according to American standards is 4 hours a day per working week [4].

4. Previous Studies

A common conclusion in previous studies [5, 6, 7, 8] is that noise levels are generally high in olive oil mills and create risk for workers, who generally work more than 8 hours a day for about two months during olive season. In a study which investigates noise levels in several olive oil mills in Italy [9], a country known for its wide spread of olive mills and large production of olive oil, noise levels were recorded in different locations inside each mill to find out the average noise level.

Conclusions of this study in Italy are:

1. Noise levels in the study mills are generally high and exceed maximum noise limits of international standards.
2. The main source of noise is from the centrifugal extractor, washing machines and steel hammer crushing machine.
3. The size of mill's building has a significant influence on noise levels because of reverberant noise.
4. The Average recorded noise level was 93.7 dB(A) at some mills which exceeded the allowed exposure limits that should not last more than 1.1 to 1.7 hours per day at maximum.

The study of olive oil mills in Italy suggested noise treatments such as sound absorption panels fixed on walls to limit reverberation. Results from this study are presented in "Fig. 4".

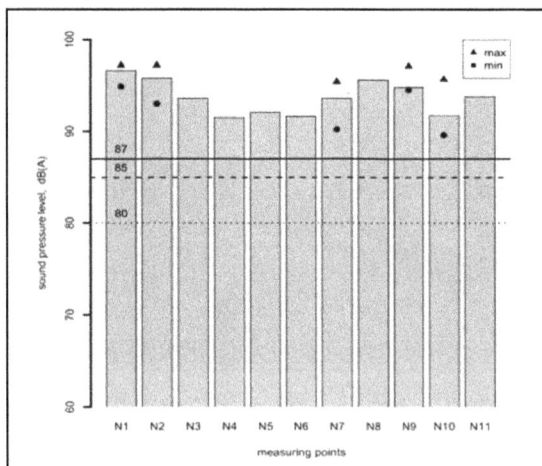

Figure 4. Results of sound pressure levels from oil mills' study in Italy [9].

5. Methodology and Procedure

Field visits to olive oil mills in Jordan were conducted to record noise levels. The surveyed mills mostly use the same equipment and olive pressing methods; the main difference between mills was the number of production lines which ranges from one to four production lines.

Noise of mill equipment is generally constant; therefore this noise could be stated by the weight A dB(A) which is certified

in Jordanian and International Standards.

A Sound Pressure Level Meter (B & K type 2215) was used to measure noise level in the surveyed mills of a frequency range 63-8000Hz. A One Third Octave Band Filter was used to assess different frequencies especially for less than 500 Hz.

Filed visits to the mills were carried out at different times in the week, as mills operate 24 hours a day in olive harvesting season. Time and date of measurement were recorded. Noise level values were recorded at the front area of the mill where many people such as workers and buyers use to collect and buy olive oil. Selection of the front area of the mill to conduct noise measurements is because of the larger number of people use this area such as workers, staff, olive owners and oil buyers. Other areas of the mill are used by limited number of people such as workers and mill owners who are responsible for supervising the maintaining of equipment and production lines.

6. Results and Discussion

Recorded sound levels are high especially at low frequencies which vary from 85 to 95 dB(A) "Fig. 5".

A comparison between measured noise level values in olive oil mills in Jordan "Fig. 5" and those in Italy "Fig. 4" show similarities in the recorded values.

Workers in the surveyed olive oil mills in Jordan work for at least 8 hours a day in high noise level conditions in the mills; some workers even work for two shifts of 16 hours a day.

Comparing the measured values in the study mills in Jordan to international standards show that workers in olive mills are exposed to high sound levels for long hours which can cause hearing problems, such as partial or total deafness and physical and psychological risks according to The World Health Organization (WHO) [10]. It was noted that there was no ears protective tools or work clothes for staff and workers in the surveyed mills. Owners of mills often show little interest to protective measures from high noise in the mill. Architectural design of surveyed olive mills in Jordan show little consideration to protect users from high noise. The pressing machinery area is completely open towards the public and workers. Quality of buildings and interior finishing in the surveyed mills is generally basic just to provide a roof and protection from sun and climate conditions.

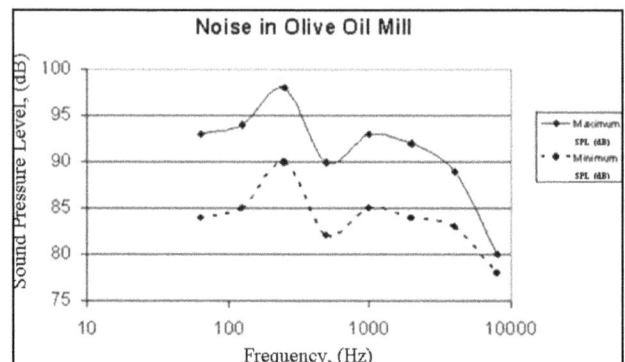

Figure 5. Noise level Measurements in olive oil mills- Jordan (85-95) dB(A).

This study would present some proposals to improve the acoustical conditions in olive oil mills. Proper architectural design of olive oil mills can help to separate users from the exposure to high noise levels produced by machines. It is also possible to organize the working hours of the staff to reduce the duration of noise exposure; aiming to meet the allowed duration of noise exposure related to a measured noise level (Dose). The International standards limit the exposure to 85dB(A) noise level for a maximum duration of 8 hours a day during five working days a week. Lack of awareness about serious health threats caused by high noise levels is noted in the surveyed mills in Jordan. Building legislation should enforce safety codes on mills to force owners to implement safety measures and subject disobedient to legal and financial penalties. Raising awareness of owners about noise risks in mills is necessary but needs time and effort.

This research suggests noise reduction in olive mills by using practical and low cost solutions to limit possible health risks of users caused by long exposure to noise in mills. "Fig. 6" shows a vertical section of a proposed wall that can separate the pressing machines area from the area used by workers and customers in mills. The wall has large glazed area which helps keeping visual connectivity between the pressing machines and olive owners to monitor and maintain quality control during olive oil extracting process. The separating wall can reduce noise level by at least 25dB(A) and lowering noise level to 70dB(A). This solution allows 8 hour exposure to noise a day without causing health implications to users.

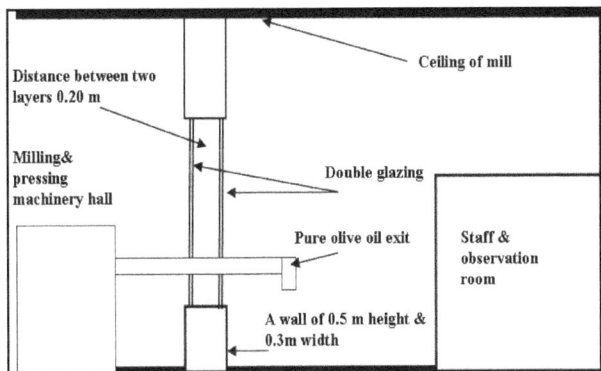

Figure 6. Proposed vertical section in olive oil mill.

7. Conclusion

1. Noise levels were measured in several olive oil mills in Jordan, the recorded noise levels and exposure duration exceeded maximum amounts as allowed by Jordanian code. International standards limit maximum exposure duration to 85dB(A) for 8 hours a day five working days a week. Workers in the study mills are found to be exposed to noise level of (85-95) dB(A) for a duration reaching up to three times than the allowed duration.

2. Noise levels recorded in the study mills in Jordan came close to noise levels recorded in the case of olive oil mills in Italy. This means that high noise levels in olive mills is a problem in many countries that needs to be addressed.

3. This study suggests reducing olive oil mills' noise through isolation of the equipment that produces noise from users, while keeping direct visual observation of olive oil production lines.

References

[1] A. Khadair, Cleaner Production in Olive Oil Industry in Jordan, In The 2nd Arab Clear Production Workshop. Amman: Jordan. August 28-30, 2007.

[2] International Olive Council, Quality Management Guide for the Olive Oil Industry: Olive Oil Mills. Madrid, Spain, 2006.

[3] International Organization for Standardization, ISO 1999: 1990 Acoustics- determination of occupational noise exposure and estimation of noise-induced hearing impairment. Geneva: Switzerland, 2006.

[4] Occupational Safety and Health Administration, Occupational noise exposure: (OSHA 1910.95 App). Department of Labor, USA

[5] U. Landstrm, E. Akerlund, A. Kjellberg, M. Tesarz, "Exposure Levels, Tonal Components, and Noise Annoyance in Working Environments," Environment International journal 1995; vol. 21 no 3: 265–275

[6] PJ. Middendorf, "Surveillance of Occupational Noise Exposures Using OSHA's Integrated Management Information System", American Journal of Industrial Medicine 2004; vol. 46, pp. 492–504.

[7] GV. Prasanna Kumar, KN. Dewangan, A. Sarkar, "Noise exposure in oil mills", Indian J. Occup. Environ. Med, 2008; 12 no1, pp. 23–28.

[8] V. Panaro, S. Pascuzzi, F. Santoro, "Analysis of exposition to noise in olive oil production sector: measurements during mechacnical harvesting and oil extraction [Apulia]", Journal of Agricultural Engineering 2004, Vol 35 issue2

[9] G. Manetto, E. Cerruto, G. Emma, Noise operator exposure in olive oil mills, In: International Conference RAGUSA SHWA 2012. Ragusa – Italy. 2012. "Safety Health and Welfare in Agriculture and in Agro- food Systems".

[10] The World Health Organization (WHO), Occupational Health, chapter 21: Selected occupational risk factors.

26

Potato Harvester for Smallholder Producers

Dagninet Amare[1], **Geta Kidanemariam**[2], **Wolelaw Endalew**[1], **Seyife Yilma**[3]

[1]Bahir Dar Agricultural Mechanization and Food Science Research Centre, Bahir Dar, Ethiopia
[2]Institute of Technology, Addis Ababa University, Addis Ababa, Ethiopia
[3]Institute of Technology, Bahir Dar University, Bahir Dar, Ethiopia

Email address:
dagnnet@gmail.com (D. Amare)

Abstract: The post harvest loss of potato in Ethiopia is more than 25% which includes harvesting loss. To minimize the harvesting loss, providing appropriate equipment (technology) is essential. As a result, a study was carried out to develop and select suitable potato digger for small holder potato producers. Comparisons were conducted on three types of potato diggers; traditional plow, AIRIC potato (ground nut) digger and third newly developed potato digger (BD digger). Bio physical and socio economic data were taken using a structured data sheet. Data was analyzed using descriptive statistics and non parametric statistical tests. The BD digger gave the highest average exposing efficiency (92.40%) and lowest tuber damage (0.81%). The average digging loss for BD digger, with 0.25 ha/hr working capacity, was the lowest (7.61%). The Kruskal Wallis analysis revealed highest positive rank sum (107) for the digger. Overall, farmers' ranked BD digger as best performing. The financial analysis indicated that BD digger has a net benefit advantage of ETB 522 in one harvesting season from a quarter hectare of land. Thus, it is important to promote the technology as a means of post harvest loss reduction.

Keywords: Potato Digger, Exposing Efficiency, Digging Loss, Kruskal Wallis, Likert Scale

1. Introduction

Potato (*Solanum tuberosum L.*) is the fourth most important food crop in the world [1]. It provides high nutrition and is an adaptive species for climate change. Potatoes use less water per nutritional output than all other major food sources and can be grown across Africa [2]. Potato provides more food per unit area than any other major staple crop. They are the perfect food and one of the few that can actually sustain life on its own. Thus, it has significant impact on providing nutrition to families, increasing household income and providing surplus to the wider market [2].

Ethiopia has possibly the highest potential for potato production than any country in Africa with 70% of the 13.5 million ha of arable land suitable for potato cultivation. Over one million highland farmers could grow potatoes in Ethiopia. Two of the three known agro-ecologies woyina dega (1500-2300 masl) and dega (above 2300 masl) exhibit the best out grower potato production [3]. However, the potato is widely regarded as a secondary non-cereal crop in part because it has never reached the potential that it has in supporting food security. It is estimated that 160,000 ha are now planted

annually by approximately one million potato farmers [2]. The Amhara region produces potato on 71325.18 ha land [4]. The total yield harvested was 339353.37 tons with average productivity of 4.8 tons per hectare. The west Amhara, where this research was conducted, accounted for 98.79% of the regional hectarage and 98% of the production volume. Potato is grown mainly on small farms. Ethiopia has a much higher potential to increase agricultural production of the crop through use of improved seeds and undertaking technological innovation that facilitate the management and reduce post harvest losses [2] [5] [1].

Post harvest loss (20 -25%) is one of the major problems in the potato production. Among this is physical damage [6], due to the digging (lifting) of the tubers by hoe or local plow [3] [1]. This entails that significant loss is incurred to the small holders that could have helped in nutrition, food security and income generation [7]. Potato yield productivity has increased far more than 24 tons per hectare due to adoption of new varieties [2]. However, post harvest loss reduction efforts have not been tailored well. Harvesting loss reduction helps increasing income, achieve food security, and subsequent storage lose reduction [1]. Thus, to reduce harvesting losses, appropriate technologies should be developed and promoted. Hence, this study was initiated with the aim of selecting, evaluating and demonstration of potato diggers that reduce current harvest loss, to smallholder producers.

2. Material and Methods

2.1. Study Site Description

This experiment was conducted in Amhara national regional state ($9^0 21'$ to$14^0 0'$ N latitude and $36^0 20'$to $40^0 20'$ E longitude) of Ethiopia. West Gojam and Awi zones are the two locations [8].

2.2. Methods

In order to achieve the development of the desirable potato lifter that could be adopted widely by the farmers, three steps were followed. They are assessment of existing skill and knowledge, modification from existing technologies, testing of the newly modified technology and ultimately on farm evaluation and demonstration.

Phase I. assessment of the traditional practice and the existing technology

During assessment of the traditional potato harvesting practice of two types of potato harvesting techniques namely using traditional plow and by using hoe were found being implemented.

Figure 1. Hand hoe.

Hand hoe: It is manual and low efficient. It is simple and can be used by children that are usually the dominant harvesters of potato. Its cost is 40ETB.

Figure 2. Potato digging in Awi.

Traditional plow: the implement is described as having poor exposing efficiency, incurs significant tuber damage and low working capacity.This implement is commonly used for hoeing maize crop and some other weeding activities. Average cost is 200ETB.

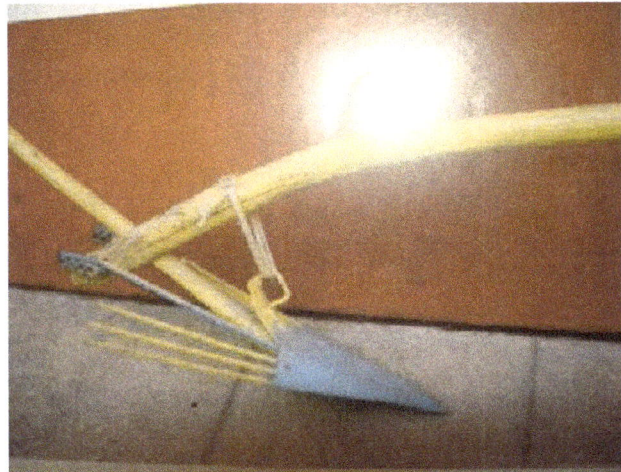

Figure 3. Melkassa ground nut digger.

Ground nut digger: it is made of a 2.5 mm thick sheet metal and deformed bar with diameters of 10 mm and 12 mm with overall dimension of length of 830 mm and 520 mm. It has a field capacity of 72 m^2/hr and exposing efficiency of 67.47%. It is animal drawn and has estimated cost of 200 ETB.

Phase II. Deciding on the design and the factors

The basic design parameters were selected based on the constraint and the capacity to mitigate. They included smooth or rounded lifter edge, increased lifter depth (by 2-5 cm), reduced damage level by 5-10% and increased exposing efficiency by 5% compared to the available technology. In general, the new potato digger was projected to have a maximum tuber damage of 5% and exposing efficiency of 85%-90%. The power source was supposed to be animal drawn, as it is the common draft force, and a price of less than 10USD which is the current price of the local plow. Working width and depth of 30-40cm and 15cm-20cm were set as target points while the raw material for the production was considered to be locally available.

The general principle for soil and implement interaction was recognized. The draft was influenced to a greater extent by the lift angle than by side angle. Based on this theory, different options were used to set angle of lifting (α) and cutting (β) angles [9] [10]. For this implement $\alpha=15°$ was selected while β has different values for different shape of the tip which spans from 45°-90°. Referring on mechanized system of potato digger or the theoretical technology options, the desired implement was designed to have either triangular tip with bent sheet, triangular tip with rod, ring wing or oval tip with rod. Ultimately, decision Matrix (Table-1) was used for the selection of the desired Potato digger. Accordingly, the oval tip with rod was ranked first. The draft force required [11] was calculated as 42.66 kgf.

Table 1. Decision matrix.

Design criteria	Exposing efficiency	Draft	Damage	Working width	Working depth	Cost	Weight	Easiness to tighten	Overall satisfaction
Weight factor	0.30	0.15	0.15	0.10	0.10	0.08	0.06	0.06	1.0
Alternative 1 (oval tip with rod)	90%*0.3= 0.27	75%*0.15= 0.112	90%*0.15=0. 135	90%*0.01=0. 90	90%*0.10=0. 90	80%*0.08=0. 064	80%*0.06=0 48	90%*0.06=0. 054	0.863
Alternative 2 (Triangular tip with rod)	85%*0.3=0.2 55	80%*0.15=0. 120	85%*0.15=0. 127	85%*0.10=0. 85	85%*0.10=0. 085	85%*0.08=0. 068	85%*0.06=0. 051	90%*0.06=0. 054	0.845
Alternative 3 (ring wing)	75%*0.30=0. 230	90%*0.15=0. 135	85%*0.15=0. 127	85%*0.10=0. 085	85%*0.10=0. 085	90%*0.08=0. 072	90%*0.06=0. 054	75%*0.06=0. 045	0.833
Alternative 4(triangular tip with curved edge)	80%*0.30= 0.240	85%*0.15=0. 127	80%*0.15=0. 120	80%*0.10=0. 080	80%*0.10=0. 080	80%*0.08=0. 064	88%*0.06=0. 052	90%*0.06=0. 054	0.817

Phase III. Development and testing of the BD digger

Based on the parameters set above, the new digger was developed. After the development, the digger was tested on the center's farm land. Ultimately, the digger was considered as another technology option and was included as a treatment. Measurements on soil moisture, row spacing, working depth, working width and digging depth were measured accordingly. Exposing efficiency was calculated using amount of exposed tubers at first and by hand digging at last. Field capacity of the implement, tuber damage, exposing efficiency, digging loss and damage were calculated accordingly.

2.3. Treatments

The treatments were Melkassa (AIRIC) potato lifter, traditional plow (Awi and Adet area plows) and the new potato lifter (BD potato digger).

3. Testing Condition

3.1. Plot Size and Preceding Crops

The first three tests were conducted Beata kebele of Kosober area in Awi Zone. The plot sizes were 32m*36m, 26m*25.6m and 24.7m*20m in the first year of the project. The weeding frequency for both plots was two times. Previous crop grown was field pea and Teff, Barely for first and second plots. The second two tests were conducted at Adet agricultural research center in West Gojam Zone. The sample plot sizes were 27m*32m and 25m*30m in the second year of the project.

3.2. Potato Varieties

In Awi, the local variety called Abalo or Ater Abeba was the variety where the implements were tested for both plots. However, at Adet agricultural research center, improved potato varieties named Gudenie for the first plot and Jalenie for the second were used. These two improved potato varieties productivity was low, compared to the research output, during testing due to irrigation water shortage. This may influence the efficiency of the implements. However, it was assumed to influence all of the implements similarly.

As a result, the output during this testing was considered as the most proximate.

3.3. Draft Force

In Awi, the implements were pulled by pair of horses, the weight of the horses were approximately 320 kg and 350 kg. In Adet, the implements were pulled by pair of Ethiopian oxen, with estimated weight of 450kg and 480 kg respectively. The traditional plow shear used in Awi area is small and its length is 26 cm. The traditional plow shear used in Adet area is large and its length is 60cm. The trail was done by the farmers themselves. During the test, the animals breathing and walking condition was considered. No signs of stress were observed.

3.4. Soil Type

At Adet, the type of soil was clay and previous crop grown was Finger Millet for both plots. Soil moisture was 13.63% and 18.65% for the first and second plots. The trail was done by the farmers. In Awi, the type of soil is sandy loam and moisture content was between 20-28%.

3.5. Data Collection and Data Analysis

Physico- mechanical data collection: measurements were performed from 3 plots and 3 rows at each treatment. A total of 27 observations in Awi and 18 observations at Adet area for each implement were taken. Blocking was done on row bases of each plot as a replication. The treatments were assigned randomly for each plot. Three rows for each replication were taken. All tubers at the first digging operation were collected and weighed. Then the measurement was repeated with the same row for the second digging operation by the implements. To evaluate the digging (exposing) capacity of each technology, hand digging using hoe was done over a depth deeper than 1st and 2nd digging operation. This was repeated at three places of each sample rows. Estimation of human labor requirement for the hoe operation was conducted using male and female adult farmers using a stopwatch on sample rows of known length and width.

Users' feedback collection: feedbacks from seven farmers

who were directly involved in the utilization and testing of the implement were collected immediately after the test. A formal data sheet was prepared and used to harness the reflection of farmers on the three diggers. A 3 level likert scale was prepared to harness the attitude of farmers. The scales used are fair (1), good (2) and very good (3). Heavy (1), medium (2) and light (3) likert scales were assigned for weight, draft requirement and tuber damage. Individual rating was made by each of the farmers for each of the three diggers on eight (8) performance criteria of the implements.

Data Analysis: AutoCAD 10 was used to analyze the cutting angle implement relationship and to design the new digger. Data were analyzed using simple descriptive statistics and non parametric tests. Stata 11 was used for data analysis.

Financial analysis: The financial analysis of a typical Ethiopian smallholder farmer is calculated. Average potato land of 0.25ha and average output of 20 tons/ha are assumed. The average yearly farm gate price is estimated to be 6ETB.

4. Results and Discussion

4.1. General Observation

The traditional plow has significant difference in area both in size and draft animal used. Around Adet the farmers use a big plow shear with length of 50 -60cm and ox as draft animal. In Awi, the farmers use a relatively smaller shear with length of 25-30cm and the draft animal used are horses. As an obvious fact, as the size of the plow shear increases the exposing efficiency increases correspondingly.

4.2. Size and Exposing Efficiency Interaction

Table 2. Exposing efficiency and tuber damage of the diggers.

| Treatment (digger) | Potato variety | | | | | |
| | Ater Abeba (Local variety) | | Gudenie (Improved variety) | | Jalenie (Improved variety) | |
	Exposing Efficiency,%	Damage tubers,%	Exposing Efficiency,%	Damage tubers,%	Exposing Efficiency,%	Damage tubers,%
Traditional plow (Awi)	85.91	1.69	-	-	-	-
Traditional plow (Adet)	-	-	95.96	0.23	94.63	0.02
AIRIC digger	87.06	1.61	95.97	0.19	93.27	0.14
BD digger	89.12	1.21	97.19	0.28	97.40	0.12

The working speed influences the exposing efficiency. In most cases, when the speed decreases the exposing efficiency increases. Thus, speed and exposing efficiency are inversely related. However, the working capacity decreases as the speed decreases. On the improved varieties (Gudenie and Jalenie, have bigger tuber size than local variety), the exposing efficiency of all technologies become effective and higher than 90% (Table-2). This shows that the size of the variety is directly related to the exposing efficiency.

The improved variety has less mechanical damage than the local variety. So, it seems the improved variety tubers have hard skin. In all cases, BD digger has better exposing efficiency than all implements. This is due to its shape and big size of the shear. It also has better working depth and width. The damage was lowest due to the shape and smoothness of the cutting edge. Our hypothesis were realized,by having appropriate angle of the digger shear that the depth of lifter was increased by 2.4cm and the damage was decreased by 3-5% compared to the traditional system and the available technology. Further, the exposing efficiency (output) was increased by 2-3%. During the demonstration, the farmers reflected similar performance evaluation results.

4.3. Draft Animals and Working Capacity

The horses have a better working capacity than those in Adet, pair of oxen (Table-3). The reasons for the higher working capacity could be the smaller shear size of the plow that requires low draft force compared to the one with larger shear size. Further, due to the mild weather conditions in Awi, the soil could be wet compared to that of Adet that was dry at the time of the test.

The performance of the BD digger on exposing efficiency, digging loss and working capacity is higher than the alternative technologies evaluated. Besides, the damage loss during harvesting is the lowest compared to others except hand hoe.

Table 3. Working depth and width of the diggers.

| Treatment | Drawn by horses Awi zone | | | Drawn by oxen Adet | | |
	Working depth (cm)	Working width (cm)	Working capacity (ha/hr)	Working depth (cm)	Working width (cm)	Working capacity (ha/hr)
Traditional plow(Awi)	7.46	25.48	0.20	--	-	-
Traditional plow (Adet)	-	-	-	9.5	25.75	0.185
AIRIC digger	7.90	35.49	0.208	10.95	28.08	0.174
BD digger	9.67	37.38	0.23	11.65	32.16	0.187

Table 4. Overall performance.

Parameter	Damage (%)	Exposing efficiency (%)	Digging loss (%)	Working capacity (ha/hr)
Hand hoe	0.78	100	-	0.0025
Local Maresha	1.06	89.164	10.836	0.219
AIRIC	1.03	90.069	9.923	0.217
BD digger	0.81	92.391	7.609	0.247
Average	0.92	92.906	9.456	0.171

4.4. Farmers' Feedback

The result indicated that exposing efficiency, working width and easiness of tightening were rated as having a significant difference ($x^2 = 14.74, p = 0.005$) among the local, BD and AIRIC diggers signifying that the AIRIC digger was found superior than the others according to the rating done by farmers. Similarly, the chi-square analysis of the working depth of the different diggers indicated the presence of significant ($x^2 = 10.25, p = 0.036$) difference according to their rating. In contrast, the three scale likert measurement for draft requirements of the three diggers showed absence of such significant ($x^2 = 6.25, p = 0.181$) difference in their draft requirements. On the other hand the rating for tuber damage showed presence of significant ($x^2 = 21, p = 0.000$) difference and making the local plow as having the highest tuber damage. An overall rating on the weight of the plows for transportation indicated the absence of statistically significant ($x^2 = 4.77, p = 0.311$) difference in weight among the three plows. However, the appearance rating indicates that there is significant difference ($x^2 = 17.67, p = 0.001$) in the appearance of the plows which showed that the BD digger is somehow better than others.

Table 5. Kruskal-Wallis equality-of-populations rank test.

Plow type	Observation	Rank Sum	x^2	x^2 with ties	df
Local plow	7	34.00	10.827***	12.738***	2
BD digger	7	107.00			
AIRIC digger	7	90.00			

*** Significant at 1% level of confidence

The results of the Kruskal Wallis analysis indicated that the BD digger has the highest positive rank sum and it is significantly higher than the other two diggers. This implies that according to the farmers ranking, the BD digger is the most efficient potato digger among the diggers incorporated in this test. This was also confirmed during the overall ranking where the farmers put the BD digger, the AIRIC digger and the local plow in order of highest to lowest preference respectively.

4.5. Financial Analysis

The net benefit from BD digger is higher than the most common harvester, traditional maresha.

Table 6. Comparative financial profitability.

Parameter	Hoe	Traditional maresha	AIRIC	BD digger
Output	5 tons	5 tons	5 tons	5 tons
Income (ETB)	30000	30000	30000	30000
Material cost(ETB)	40	200	250	250
Labor cost at working capacity(ETB/ha)	2000	68.50	69.12	60.73
Tuber damage cost (ETB)	234	2120	2060	1620
Digging loss(ETB)	-	216.72	198.46	152.18
Total variable cost	2474	2605.22	2577.58	2082.91
Net benefit	27526	27394.8	27422.4	27917.1

5. Conclusion and Recommendation

As to the assumption, the exposing efficiency was averagely increased by around 4% compared to the traditional lifter. Further, the working depth was increased by more than 3cm compared to the traditional maresha lifter. The modified BD potato digger has the qualities desired at the start of the project. Farmers ranked it the best among the available technologies that fit their production system. The financial analysis showed an advantage net benefit of more than 500 ETB from a hectare of potato by using the BD digger. Given the issue of food security in the country, a small increase in total output contributes a lot to the overall food availability in the country. Even in the production areas where there is the highest degradation and food productivity is low, a small increase in quality and quantity of potato produced helps to tackle food insecurity. Hence, reduction of this post harvest loss becomes very indispensable. Therefore, demonstration of the technology with better performance is essential. According to the test results it is better to demonstrate the best technology in regard of exposing efficiency (BD Digger) to the farmers. Thus, demonstration at large scale to all potato producing areas of the region as well as to the country should be undertaken. Besides, the preference of the small scale farmers along with the durability of the implement should be studied for further improvement.

References

[1] Hakan Kibar, 2012. Design and Management of Postharvest Potato (*Solanum Tuberosum* L.) Storage Structures. *Ordu Univ. J. Sci. Tech.*, 2(1):23-48.

[2] Vita and IPF (Irish Potato Federation), 2014. Potatoes in Development: A Model of Collaboration for Farmers in Africa. Pdf doc.

[3] Tesfay, A., 2008. Potato Production Manual. Amharic Version printed in 1999 Ethiopian Calendar.

[4] CSA, 2003. Ethiopian Agricultural Sample Enumeration 2001/02. Results for Amhara Region. Statistical Report on Area and Production of Crops. Part II. A.

[5] Tewari V. K., A. Ashok Kumar, Satya Prakash Kumar, Brajesh Nare, 2012. Farm mechanization status of West Bengal in India. *Journal of Agricultural Science and Review*, 1(6):139-146.

[6] Akeson W. R., Fox S. D., Stout E. L., 1974. *Journal of the American Society Sugar Beet Technologists,* 18:125–135. Dawit, A., 2004. Agricultural Technology evaluation, adoption and marketing. Part 2. Proceedings of the workshop held to discuss the socioeconomic research results of 1998-2002. August 6-8, 2002, Addis Ababa, Ethiopia, ERO, 2004.

[7] BoFED (Bureau of Finance and Economic Development), 2007. Annual statistics for Amhara National Regional State. Bahir Dar, Ethiopia.

[8] Kaburaki, H., and Kisu, M., 1959. Studies on Cutting Characteristics of Plows. J. Kanto-tosan Agric. Exp. Stn. (Translation 79, Scientific Information Dep., NIAE, Silsoe, UK).

[9] Sommer, M., 1999. Animal Traction in Rain Fed Agriculture: In Africa and South America Eschborn.

[10] Newbauer, K., 1989. Agricultural Machinery, Prague.

27

Prediction of cycle life of flexible pipe bellows

Abdulrahman Th. Mohammad[1], Jasim Abdulateef[2], Zaid Hammoudi[2]

[1]Baqubah Technical Institute, Middle Technical University, Baghdad, Iraq
[2]Mechanical Engineering Department, Diyala University, Diyala, Iraq

Email address:
abd20091976@gmail.com (A. Th. Mohammad), jmabdulateef@gmail.com (J. Abdulateef), zshaaa@yahoo.com (Z. Hammoudi)

Abstract: This paper aims to investigate the relationship between maximum stresses produced and cycle life of different shaped bellows expansion joints. Flexible pipe bellows which have been selected in the present study are made of stainless (STS 304) and (STS 316) material based on two types of (axial, and axial with control ring) with three shapes of bellow section (U-shape, Ω-shape and disc-shape). Our calculation is done using simulation model written by MATLAB. The simulation results show that, U-shaped bellow has smaller internal pressure-induced stress, longer fatigue life, and is more suited for higher internal pressure situations. The most essential bellow design factor is the correct specification of the bellow movement requirements.

Keywords: Cycle Life of Bellows, Flexible Pipe, Bellows Movement

1. Introduction

Bellows are widely used as the element of expansion joint in various piping systems. Piping systems for industrial plants often suffer excessive deformations or displacements caused by heat expansion, vibration, non uniform subsidence of ground, etc. [1]. Bellows function is to absorb regular or irregular expansion and contraction in such piping systems.

A metal bellows consist of individual convolutions which have a uniform inside and outside diameters, constant pitch or spacing, and the parallel planes of each convolution. The flexibility of the convolutions gives the bellows the ability to absorb axial, angular and lateral displacement individually or in combination [2]. Axial motion is extension or compression of the bellows along the longitudinal centerline (x-axis). The motion is absorbed equally by the convolutions.

The bellows cycle life or bellows fatigue life is defined as the total number of complete cycles which can be expected from the expansion joint based on data tabulated from tests performed at room temperature. A bellows cycle can also be defined as one complete movement of an expansion joint from initial position to the extreme position and return to initial position. The bellows cycle life is affected by the following design factors [3].

- Operating pressure
- Operating temperature
- Bellows Material
- Thickness of bellows and number of plies/layers
- The movement per convolution
- Pitch, depth and shape of convolutions
- Heat treatment of bellows

Besides, the physical limitations of deflecting a bellows without damaging it, the design is normally based on specified cycles movements for a given fatigue life at the operating pressure and temperature.

The correct specification of bellows movement requirements is one of the most essential factors in the successful application of this product. The axial, lateral and angular movements must be realistically stated along with the corresponding cycle life. One of the most common mistakes made is to overstate these values in an attempt to obtain a conservative design [2].

EJMA [4] defined fatigue life analysis in terms of meridional stresses and have mentioned that the other type of stresses, if signification, can also be caused a fatigue failure.

A number of studies have been done to predict the fatigue failure in bellows expansion joints [5-7].

Tingxin et al. [8], studied experimentally the toroid-shaped bellows behavior. The results showed that, compared with U-shaped bellows, toroidal bellows have longer fatigue life, smaller internal pressure-induced stress, stronger ability to resist internal pressure instability.

Zhu et al. [9], studied experimentally the effect of environmental medium on corrosion fatigue life for U-shaped bellows expansion joints. The results showed that the presence of corrosive medium accelerates both crack initiation and propagation rates and reduces the failure life for the expansion joints.

Pierce and Evans [10], analyzed the failure of a metal bellows flexible hose which subject to multiple pressure cycles. It was found that by cycling the pressure on the hose and applying a pressure much greater than the yield capabilities of the hose, this combination of yielding and cycling, caused a burst failure.

The metal bellows presents an ongoing challenge: the prediction of an accurate cycle life. Bellows manufacturers require a method for detection of bellows cycle life to failure. The objective of the present work is to study the relationship between maximum stresses produced in the bellows and their cycle life. The flexible pipe bellows used are made of (STS 304) and (STS 316) material based on two types of (axial, and axial with control ring). The solution process involves three shapes of bellow section (U-shape, Ω-shape and disc-shape).

2. Cycle Life of Flexible Pipe Bellows

When bellows deflect, the motion is absorbed by bending of the sidewalls of each convolution. The associated stress caused by this motion is the deflection stress or EJMA stress [11]. This stress runs longitudinal to the bellows centerline. The maximum value of deflection stress is located in the sidewall of each convolution near the crest or root.

Expansion joints are designed to operate with a value for deflection stress that far exceeds the yield strength of the bellows material. This means that most expansion joints will take a permanent set at the rated axial, angular or lateral motion. Expansion joint bellows are rarely designed to operate in the elastic stress range [11]. Therefore the bellows will eventually fatigue after a finite number of movement cycles. It is important to specify a realistic cycle life as a design consideration when ordering an expansion joint. An overly conservative cycle life requirement can result in a bellows design that is so long and soft that it is subject to squirm failure

3. Mathematical Model

A bellows is a flexible seal and convoluted portion of an expansion joint which is designed to flex when thermal movements occur in the piping system. The number of convolutions depends upon the amount of movement the bellows must accommodate or the force that must be used to accomplish this deflection. The convoluted element must be strong enough circumferentially to withstand the internal pressure of the system. The longitudinal load (pressure thrust) must then be absorbed by some other type of device [11].

The internal pressure exerts an axial force equal to the pressure times the internal cross-section area of the pipe and also induces stresses in the circumferential direction. It can be assumed that the displacement (e) mm per one bellow corrugation gives a stress due to deflection produced in the

bellows, (σ_e) kg/mm^2, and the internal pressure or the external pressure can also be added to these stresses is (σ_p) kg/mm^2. The maximum stress (σ_m) produced for three shapes of bellow section (U-shape, Ω-shape and disc-shape) are approximately expressed as follows [12,13]:

$$\sigma_m = \sigma_e + \sigma_p \tag{1}$$

For U-shape corrugation

$$\sigma_{m_1} = \frac{1.5 * E * t * e}{2 * \sqrt{W} * H^{1.5}} + \frac{P * H^2}{2 * t^2 * m} \tag{2}$$

For Ω-shape corrugation

$$\sigma_{m_2} = \frac{1.5 * E * t * e}{2 * H^2} + \frac{P * H^2}{t * m} \tag{3}$$

For disc-type corrugation

$$\sigma_{m_3} = \frac{3 * E * t * e}{2 * H^2} + \frac{P * H^2}{2 * t^2} \tag{4}$$

Bellows meridional bending stress due to internal pressure (kg/mm^2) is written as[13]:

$$\sigma_P = \frac{P * H^2}{2 * t^2 * m} \tag{5}$$

where $\sigma_P \leq \sigma_y$

Also, the bellows circumferential membrane stress due to internal pressure (kg/mm^2) is written as [13]:

$$\sigma_e = \frac{P * D_p * W}{t * m * (1.142 * W + 2 * H)} \tag{6}$$

where $\sigma_e \leq \sigma_a$

Finally, the relationship between maximum stresses (σ_m) produced in the bellows and its life cycle (N) is expressed as [12]:

$$N_A = \left(\frac{563}{\sigma_m}\right)^{3.5} \tag{7}$$

$$N_B = \left(\frac{1125}{\sigma_m}\right)^{3.5} \tag{8}$$

4. Calculation Procedure

Figure (1) shows the proposed bellow design that used in our calculation. The materials of bellows that used are both of (STS 304) and (STS 316). The solution procedure is done by solving the above equations using simulation model written by MATLAB [14]. The design parameters of bellows are presented in Table (1).

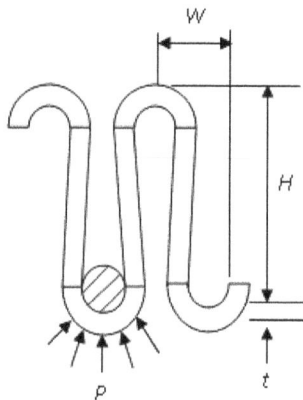

Figure 1. *Bellow design*

5. Results and Discussion

It is important to predict a realistic cycle life as a design consideration when ordering bellows expansion joints. Thus, a life cycle prediction program by MATLAB is done to investigate the relationship between maximum stresses produced in the bellows and the predicted cycle life. The model depends on the data that mentioned in Table (1) as input parameters under different operating temperature [13].

Tables (2 and 3) show that the maximum total stresses acts on the flexible pipe bellows is limited by the internal and external pressure or circumference tensile stress. When the design temperature increases, the value of (σ_y) decreases. As a result, the pressure and σ_m decrease while both of allowable and broken cycle lives increase.

Table 1. *Design parameters of bellow*

PARAMETERS	VALUE	UNIT
Height of bellow corrugation, (H)	60	mm
Half pitch of bellow corrugation, (W)	30	mm
Thickness of bellow,(t)	1.5	mm
Effective diameter of bellow, (D_P)	346	mm
Allowable tensile stress, (σ_a)	10.1	Kg/mm^2
Yield strength, (σ_y)	16.1	Kg/mm^2
Number of plies, (m)	1	-----

Table 2. *The maximum total stresses applied on the flexible pipe bellows of both (axial, and axial with control ring) type, (STS 304)*

Axial type		100	200	300	400	500	600
	$T(C^0)$	100	200	300	400	500	600
U-shape	σ_m (kg/mm^2)	35.71	33.35	31.65	24.91	29.81	23
	N_A	15560	19767	23739	54887	29276	72235
	N_B	175496	222945	267747	619050	330192	814713
Ω – shape	σ_m (kg/mm^2)	64.49	59.293	56	38.62	53.25	36.15
	N_A	1965	2637	3209	11823	3842	14898
	N_B	22172	29752	36203	33350	43342	167987
Disc- shape	σ_m (kg/mm^2)	42.78	40	38	31	36	29
	N_A	8268	10329	12380	25527	15229	32006
	N_B	93251	116496	139628	287908	171761	360982
Axial type with control ring							
U-shape	σ_m (kg/mm^2)	99.47	90.79	86.25	84.5	81.73	83.73
	N_A	431	593	710	763	857	803
	N_B	4865	6697	8014	8607	9676	9064
Ω – shape	σ_m (kg/mm^2)	382.87	346.39	331.46	323.74	312.68	322.86
	N_A	4	5	6	7	7	7
	N_B	43	61	72	78	88	79
Disc- shape	σ_m (kg/mm^2)	106.55	97.55	93.33	90.68	87.85	88.93
	N_A	339	461	539	596	666	638
	N_B	3824	5208	6080	6725	7515	7200

Table 3. *The maximum total stresses applied on the flexible pipe bellows of both (axial, and axial with control ring) type, (STS 316)*

Axial type		100	200	300	400	500	600
	$T(C^0)$	100	200	300	400	500	600
U-shape	σ_m (kg/mm^2)	36.27	34.79	33.65	32.5	31	26.47
	N_A	14735	17048	19459	21613	25499	44375
	N_B	166194	192284	216065	243766	287592	500489
Ω – shape	σ_m (kg/mm^2)	66	63.34	61.66	60	55.28	45.8
	N_A	1806	2093	2300	2530	3371	6512
	N_B	20371	23609	25943	28543	38022	73452
Disc-shape	σ_m (kg/mm^2)	43.34	41.58	40.1	38.6	37.13	32.5
	N_A	7900	9130	10351	11850	13575	21636
	N_B	89100	102973	116744	133651	153106	244023

Table 3. *(continued)*

Axial type with control ring							
U-shape	σ_m (kg/mm²)	104.27	98	94.85	93.3	87.3	70.4
	N_A	365	454	509	539	680	1441
	N_B	4125	5127	5746	6085	7672	16256
Ω – shape	σ_m (kg/mm²)	404.47	378.79	367.46	363.34	337.88	265.2
	N_A	3	4	4	5	6	14
	N_B	35	45	50	52	67	157
Disc-shape	σ_m (kg/mm²)	11.35	104.78	101.33	99.48	93.36	76.1
	N_A	296	359	404	431	538	1099
	N_B	3278	4055	4559	4863	6073	12404

Figures (2 and 3) show the effect of total stress on the allowable and broken cycle life for the three shapes of bellows (STS 304) with both of axial and axial with control ring types respectively. As shown in Figures, both the N_A and N_B in axial type are increased with decreasing the total stress. It can be seen inFig.2, when the total stress changes from 30 to 60 Kg/mm², the N_A and N_B reach to the maximum value of about 90×10^3 and 800×10^3 respectively in case of U-shape corrugation. In axial type with control ring, the N_A and N_B reach to the maximum value of about 1×10^3 and 10×10^3 respectively in case of U-shape bellows as shown in Fig.3.

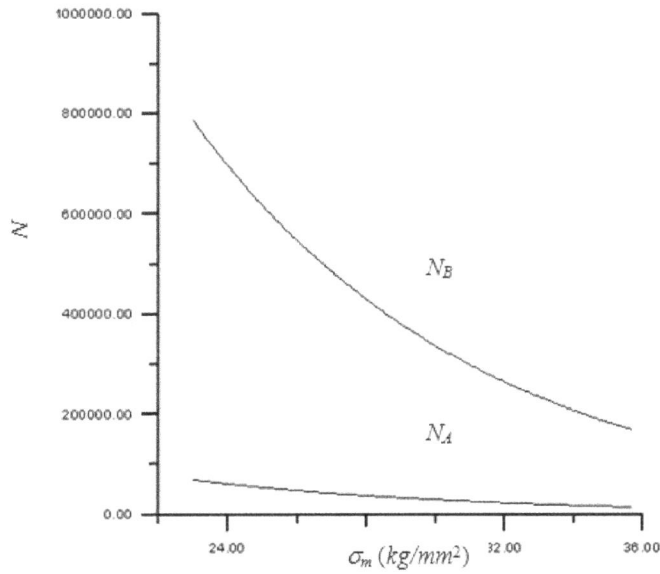

(a) U-shape bellow

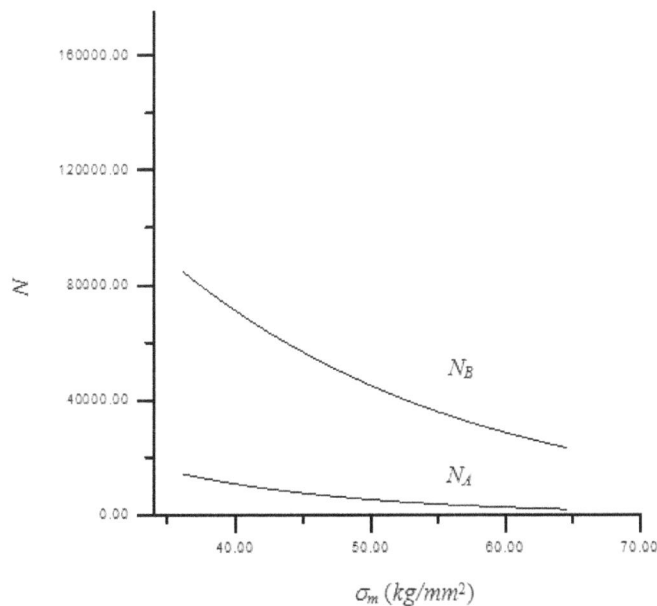

(b) Ω-shape bellow

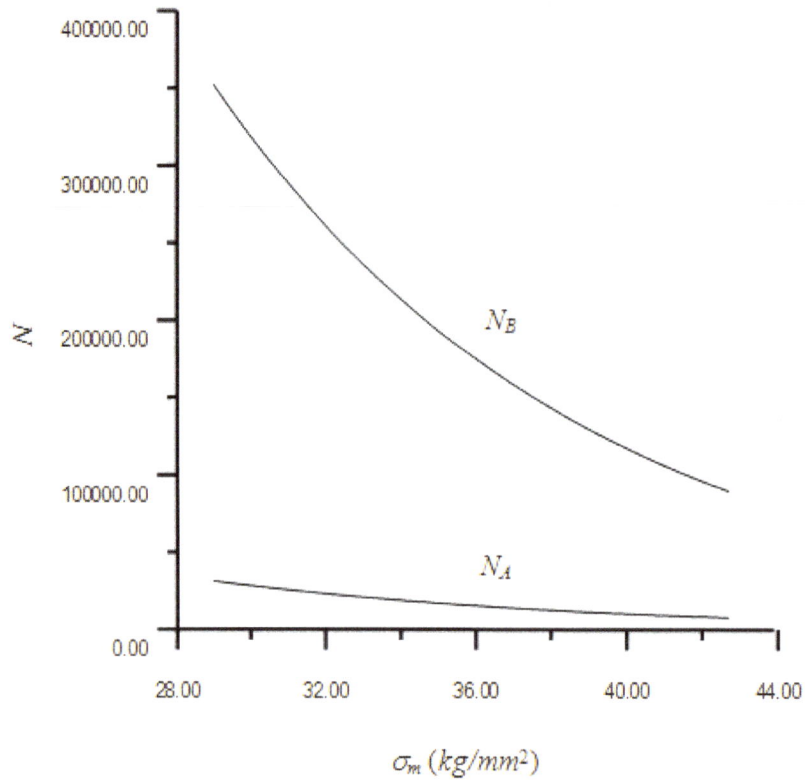

(c) disc-shape bellow

Figure 2. *The relationship between (σ_m and N) for axial type, (STS 304)*

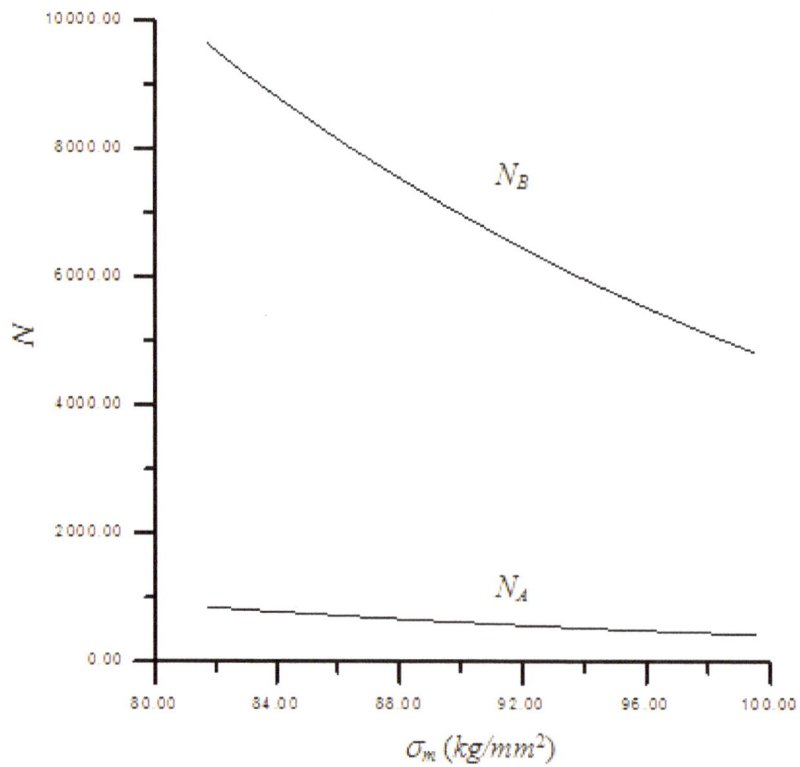

(a) U-shape bellow

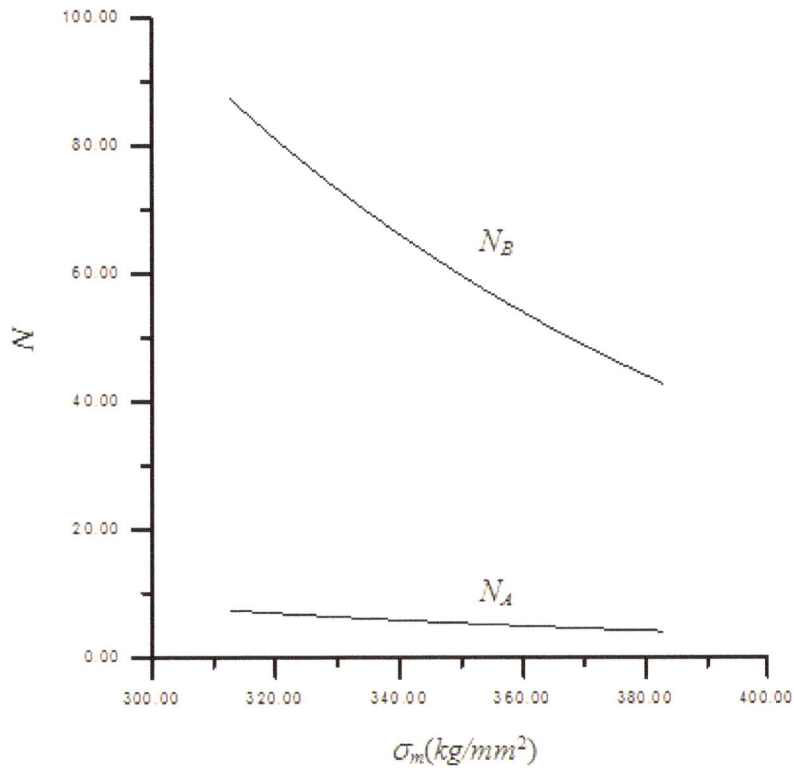

(b) Ω-shape bellow

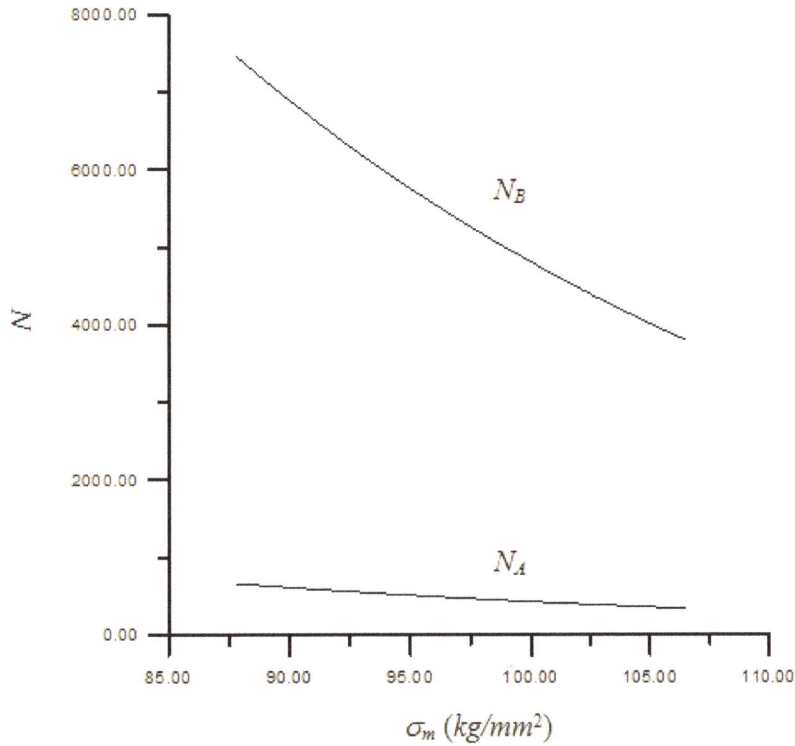

(c) disc-shape bellow

Figure 3. *The relationship between (σ_m and N) for axial type with control ring, (STS 304)*

For the bellow with (STS 316) material, Fig. (4) shows that the N_A and N_B reach to the maximum value of about 50×10^3 and 500×10^3 respectively in case of U-shape corrugation with axial type when the total stress changes from 30 to60kg/mm².

In case of the axial type with control ring, the N_A and N_B reach to the maximum value of about 1.7×10^3 and 16×10^3 respectively in U-shape corrugation as shown in Fig. (5).

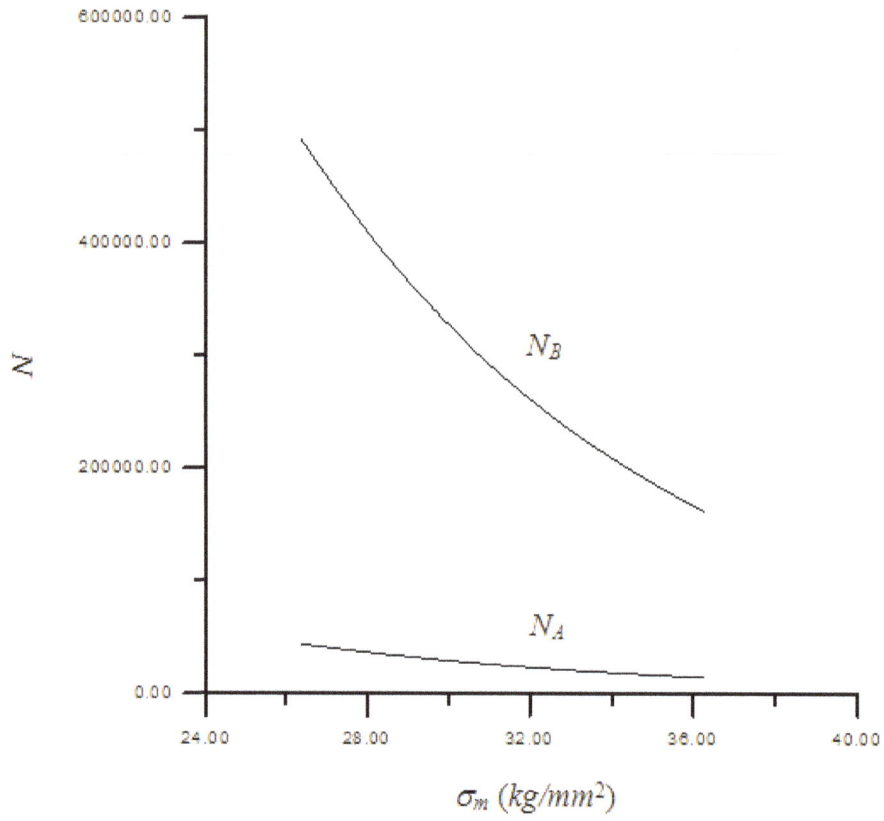

(a) U-shape bellow

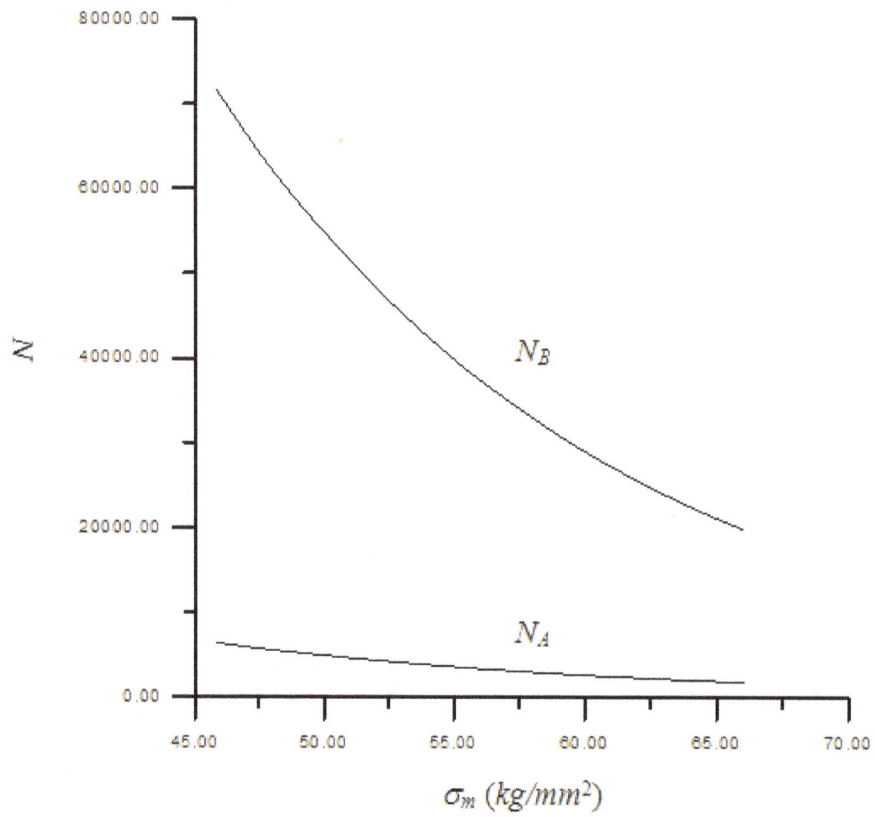

(b) Ω-shape bellow

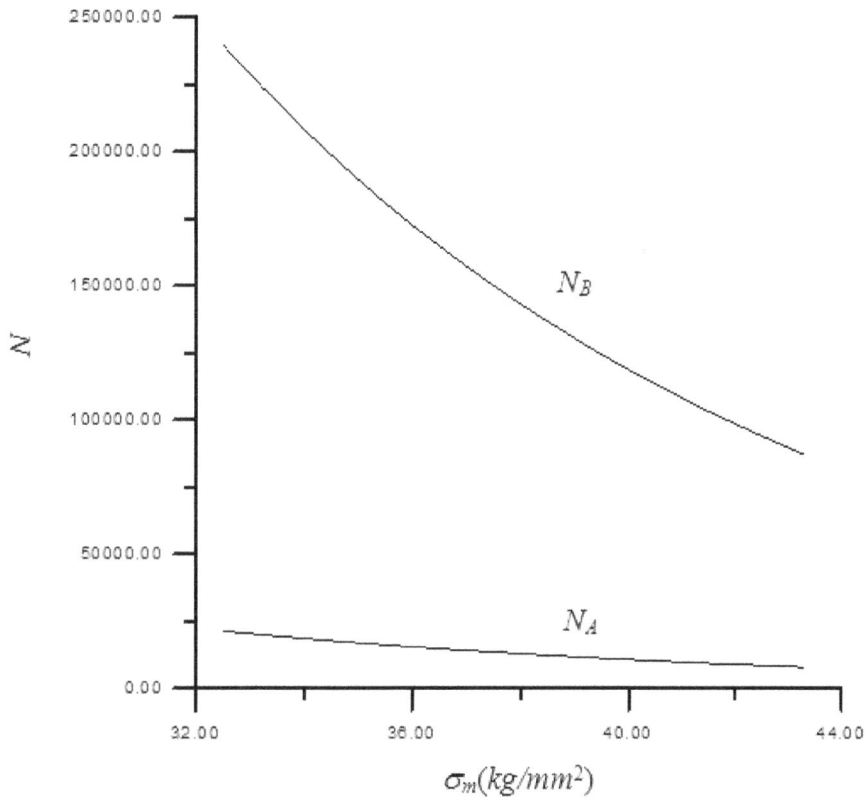

(c) disc-shape bellow

Figure 4. *The relationship between (σ_m and N) for axial type, (STS 316)*

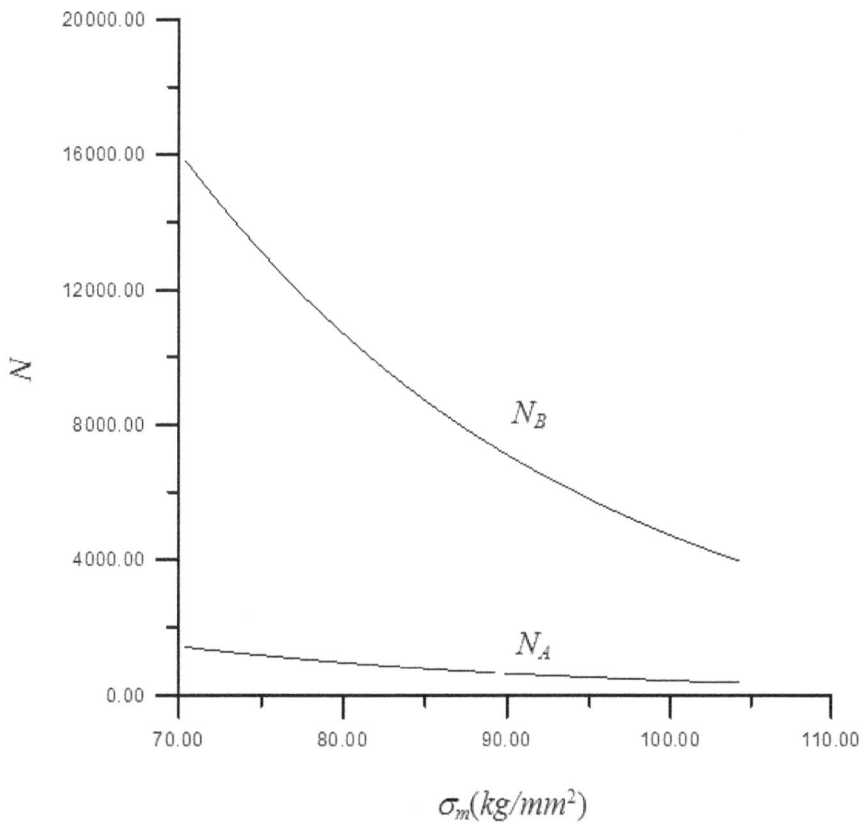

(a) U-shape bellow

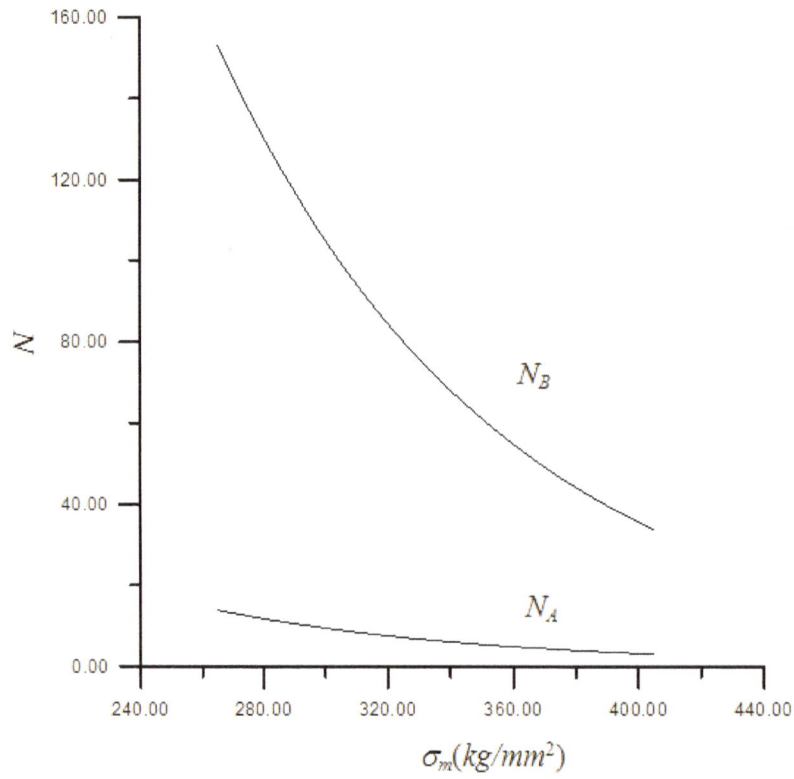

(b) Ω-shape bellow

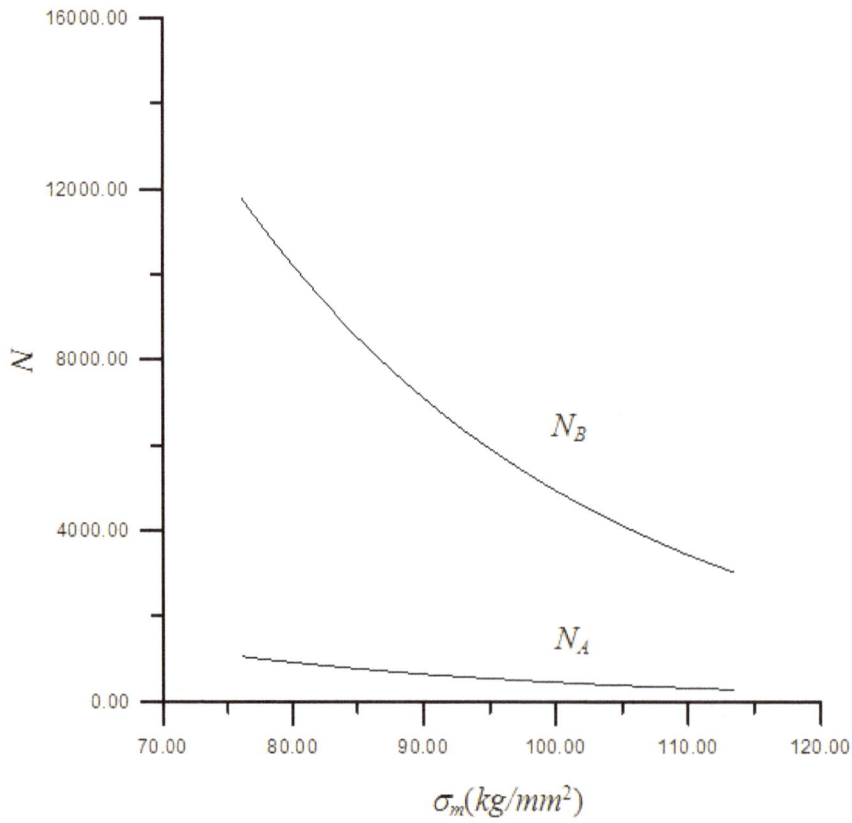

(c) disc-shape bellow

Figure 5. *The relationship between (σ_m and N) for axial type with control ring, (STS 316)*

6. Conclusions

The relationship between maximum stresses produced and the predicted cycle life of bellows which are made of (STS) material has been presented. Simulation model is written by MATLAB based on specific design parameters of bellows. The solution procedure involves three shapes of bellow section (U-shape, Ω-shape and disc-shape).

The following conclusions can be summarized:

- It is important to predict a realistic cycle life as a design consideration when ordering an expansion joint.
- Both the N_A and N_B are increased with decreasing of the maximum stress (σ_m).
- In the case of bellows of (STS 304) material, the N_A and N_B reach to the maximum value of about 90×10^3 and 800×10^3 in case of U-shape corrugation with axial type, while in case of axial type with control ring, reach to the maximum value of about 1×10^3 and 10×10^3 in U-shape corrugation.
- In the case of bellows of (STS 316), the N_A and N_B reach to the maximum value of about 50×10^3 and 500×10^3 in case of U-shape corrugation with axial type and reach to the maximum value of about 1.7×10^3 and 16×10^3 in U-shape corrugation with axial type with control ring.
- U-shaped bellows have smaller internal pressure-induced stress, longer fatigue life, and is more suited for higher internal pressure situations.
- The correct specification of bellows movement requirements can be considered as one of the most essential factors in the design.

Nomenclature

D_P	Effective diameter of bellow (mm)
e	Axial movement per one bellow corrugation (mm)
E	Young's modulus of elasticity (kg/mm^2)
H	Height of bellow corrugation (mm)
m	Number of plies
N_A	Allowable cycle life
N_B	Broken cycle life
P	Internal pressure (kg/mm^2)
t	Thickness of bellow(mm)
T	Temperature ($^\circ$C)
W	Half pitch of bellow corrugation (mm)

Greek symbols

σ_a	Allowable tensile stress (kg/mm^2)
σ_e	Stress due to deflection (kg/mm^2)
σ_m	Total stress (kg/mm^2)
σ_P	Stress due to internal pressure (kg/mm^2)
σ_y	Yield stress (kg/mm^2)

Abbreviates

EJMA   Expansion Joint Manufacturing Associate

References

[1] Igi, S., Katayama, H., and Kawahara, M., "Evaluation of mechanical behavior of new type bellows with two-directional convolutions," Nuclear Engineering and Design, vol. 197(1), pp. 107-114, 2000.

[2] Hyspan Technical Notes, "Bellows Movements", http://www.hyspan.com/pdfs/BellowsMovemStTechNote.pdf, 2005.

[3] http://www.flexpertbellows.com/products/technical-informatio n/cyclic-or-fatigue-life-of-metal-bellows-expansion-joints/

[4] EJMA, "The Expansion Joint Manufacturing Association Standard", http://www.ejma.org/, 8th Edition, 2003.

[5] Li, Y., and Sheng, S., "Strength analysis and structural optimization of U-shaped bellows," International Journal of Pressure Vessels and Piping, vol. 42(1), pp.33-46, 1990.

[6] Subramanian, G., and Raghunandan, C., "On improving the fatigue life of U-form bellows," Journal of Materials Processing Technology, vol. 41(1), pp. 105-114, 1994.

[7] Becht IV, C., "Fatigue of bellows, a new design approach," International Journal of Pressure Vessels and Piping, vol.77 (13), pp. 843-850, 2000.

[8] Tingxin, L., Xiaoping, L., Tianxiang, L., Xigang, H., and Xinfeng, L., "Experimental research of toroid-shaped bellows behavior," International Journal of Pressure Vessels and Piping, vol. 63(2), pp. 141-146, 1995.

[9] Zhu, Y. Z., Wang, H. F., & Sang, Z. F., "The effect of environmental medium on fatigue life for u-shaped bellows expansion joints," International Journal of Fatigue, vol. 28(1), pp. 28-32, 2006.

[10] Pierce, S. O., & Evans, J. L., "Failure analysis of a metal bellows flexible hose subjected to multiple pressure cycles," Engineering Failure Analysis, vol. 22, pp. 11-20, 2012.

[11] http://www.expansionjointsindia.com/Questionnaire.pdf

[12] Hammond M. G., "Quite Hydraulic Actuation Bellows Design Considerations", www: http://www.ligo.caltech.edu, 2002.

[13] Expansion Joint & Flexible Product, Fabrication procedure, http://www.megaflexon.com/design/default/images/Introducti on_Metallic%20Expansion%20Joint.pdf

[14] Edward B.M., Shapour A., Balakumar B., James D. Keith H. Gregory W., An engineer's guide to MATLAB : with applications from mechanical, aerospace, electrical, and civil engineering, Prentice Hall ©2011, 3rd edition, 2011.

28

Connection Between the Number of Complaints About Welding Suppliers and End Product Quality: The Case of Customized Welding Production

Jenni Toivanen, Harri Eskelinen, Paul Kah, Jukka Martikainen

Mechanical Engineering, Lappeenranta University of Technology, Lappeenranta, Finland

Email address:
jenni.toivanen@lut.fi (J. Toivanen)

Abstract: Although undesirable, manufacturing defects leading to complaints are almost inevitable in the production of manufactured products, and consideration of manufacturing quality is therefore an essential aspect of management of supply chains with multiple suppliers. This study evaluates the relationship between complaints about the end product and welding production in a multiple supplier chain. In the studied case, it was noticed that there is potential for improved welding production management by suppliers to increase profitability by decreasing the number of welding defects that cause complaints. This study shows one approach to analysis of the relation between complaints in the supply chain and their effect on the end product.

Keywords: Welding Production, Welding Suppliers, Complaints, Welding Quality, Productivity Improvement, Empirical Study

1. Introduction

Supply chain quality and manufacturing have been the subject of active research interest and the topics have been examined from many perspectives, for instance, from the viewpoint of management [1, 2, 3], partnership [4], products [5, 6] and costs [7, 8]. From the point of view of manufacturing quality, the quality of supplier relationships [9] is an important aspect in supply chain management. Quality management is usually separated into segments such as internal actions of operations managers within organisations and external actions with purchasing and logistic functions [1].

Researchers have studied issues related to the supply chain such as production with defects [10], customer complaints about service [11, 12] and data modelling for organizational learning from complaints [13]. Researchers' interest has also been drawn to complaints related to latent defects with no discovered reason [14], false failure returns [15] and the impact of supplier production rate on defects and production costs [16]. When studying human factors, Mateo et al. [17], using production data, found that there is no significant relation between complaints and absenteeism, and Gajdzik and Sitko [18] found a relation between complaints and human errors in steel sheets manufacturing.

Generally, existing research on complaint management and complaints in production are based on modelling [19] and interviews or questionnaires [3, 20]. Furthermore, the studies tend to be general in nature, giving generic recommendations. Despite the importance of general results, which give valuable information about issues of concern and suggestions for management of relevant aspects of complaints handling, there is a need for more substantive data-based results and more profound observation of complaints and their effect on production and costs. Of particular interest is mirroring complaints data to complete product data as a means of discovering prospective processes to increase profitability.

This paper evaluates the effect of complaints on the end product in a welding production supply chain with multiple suppliers. The approach provides new insights into the effect of complaints about a specific manufactured product and manufacturing process and indicates actions that may lead to increased profitability. The study considers the following research questions: firstly, how the characteristics of complaints about welding reflect the quality and costs of the end product; and secondly, whether there is potential to influence the quality and profitability of manufacturing of the end product by improved control of the welding supply chain. Results of the studied case show clearly the role of welding production in end product quality in multisectoral manufacturing and thus its importance in manufacturing quality. The paper consists of two parts: a theoretical part considering the theoretical background and an empirical research part utilizing complaints data of the case focal company in the welding supplier chain.

The novelty value of this research is based on the new viewpoint of complaint analysis, which is supported by statistical data and reliability calculations to which data is collected from both the PDM system and complaint system. The results of this analysis are verified with more than 14 000 real case examples.

2. Complaints Run Out from Profitability

Quality of manufacturing is important thus complex with multiple manufacturer in supply chain. Coordination of supply chain [21], strategic supply chain management [22] and thus, strategic supplier selection is acting more significant role in manufacturing [23]. In welding manufacturing the one typical structure of manufacturing is structure where the focal company dominates the manufacturing with multiple suppliers [24]. This kind of structure leads to especially dyadic contacts but the increase of cooperation between network members also enables multi-tiered relationships and augmented manufacturing. However, its complexity introduces multiple functions and linkages that can have an effect on defect incidence and end product quality [25].

The focal manufacturer in such a network defines the product quality and price [26], which encompass, among other costs, the costs of development and manufacturing. Inadmissible parts in production inevitably cause disturbances in the manufacturing chain, and defects and failures generate unnecessary expense in the form of rework and waste. The costs arising from defects consist of internal failure costs from scrap, rework and delay, and external failure costs from repairs, warranty claims and lost custom [27]. These additional costs reduce the profitability of the end product and therefore complaints do not promote profitable outcomes. Complaints leading to additional waste also complicate efforts to reduce the negative environmental impacts of manufacturing [28].

However, defects leading to complaints are almost inevitable in production with multiple member manufacturing chains and the effect of inadmissible results can magnify across the multiple actors involved. Complaints can originate for many different reasons, for example, product defect, damage during transportation, or even as a result of misunderstandings. It should be noted that even false failures or returns for no reason are detrimental to suppliers [15], and they cause unnecessary expenses for both the supplier and the focal company. Complaint behaviour and complaint management are essential to ensure effective relationship in business [20] and effective relationships enable essential in profitable manufacturing. Competitive advantage in manufacturing need continuous improvement of product quality [29] and thus cost of quality seems to have more strategic and economic importance compared to previous time [7].

Quality design plays an important role in business decisions concerning quality level and actions, and, thus, the costs of quality. In a dynamic manufacturing chain, far-sighted (i.e., economically long-term focused) quality behaviour results in a more price-sensitive demand than a myopic (i.e., economically short-term focused) approach. The myopic approach provides consumers with a higher quality–price ratio and more quality sensitive but less price-sensitive market [26]. The balance between quality and product costs is difficult to define when coordinating a complex manufacturing chain. Product quality data can be valuable for managing suppliers and product quality. Based on information from monitoring the quantity and type of complaints generated, the focal company can take remedial and optimization actions to gain improved profitability and enhanced production quality at the network level of the dynamic manufacturing chain. Quality data thus form an important basis for decision making regarding activities related to quality control, management and improvement [29]. Effective utilization of complaints data can assist the focal company in control of quality costs.

3. Research Methods

In this study, numerical data of complaints relating to welding suppliers and two welded case products were gathered and analysed. The case products, which are two different mobile machines, are end products with a large number of different parts that are sourced from many suppliers. Analysis of the complaints data focused on items manufactured by welding suppliers and their effect on quality and profitability in the welding network.

3.1. Data Collection

Case end products, Machine A and Machine B, are mobile machines designed for work in demanding environments, and the machines consist of multiple welded parts and structures with multiple items. The products consist of 3 891 different items and the total number of parts is 14 907. Data about the parts used in this study are from the company's PDM (Product Data Management) system and data about complaints are from the company's complaints system. The gathered data were tabulated and percentage portions were calculated and scaled to find links between complaints and potential for increased production profitability. Identification of the links was based on observations of curve shape variations together with peak values found from tabulated data. Finally, a tentative curve fit to describe the changes in the number of complaints was tested based on learning curve. The data also contain summarized numerical information about items and complaints but the focus is on welding supplier information.

The studied data comprise information about the number of individual items, where the total number of items in the end product is not given. The category total number of items includes also the number of the same items (Table 1). Complaints are observed similarly at a general level and also focusing on items manufactured by welding suppliers. In this analysis, welding supplier items are outsourced parts manufactured by a supplier who does welding manufacturing of parts or subassemblies for the focal company. Complaints values show the number of individual complaints and do not take into account the number of items within a single complaint. Either observing costs are only indicated items, not manufacturing activities. The welding supplier data were divided into different categories for analysis with a mixed method approach [30] with numerical data and clarification results according to the root cause of the complaints.

Table 1. Data calculation example of number of individual items and total number of items. The number of individual items describes every new item number in the end product.

Item number	Item description	Quantity
34966	Fastener	12
42815	Cover sheet	6
43467	Shaft	2
21201	Sleeve	4
Number of individual items		4
Total number of items		24

* The data are collected from an item group of 3 891 individual items where the total number of items is 14 907.

The machines studied illustrate the number and cause of complaints in the end product. The complaints data used are general information about complaints regarding items in production and, in this work, are not assigned to particular machines. However, the data show the connection between welded items and the volume of complaints and thus indicate prospects of improving production profitability. Complaints can be observed with general categorization of all complaints in a welding network whereas this study focuses on studying complaints from the viewpoint of an example end product.

3.2. Finding Connections Between the Content of Complaints and the Items of the End Product

This research concentrates on three main areas. One focus is summarization of numerical information about items and complaints concerning two example end products (Machine A and Machine B) over an eight year period. This part also shows information about complaints in the launching year of the product. The second focus is study of costs related to items that have been the subject of complaints and analysis of the correlation of these costs to the end product. The third focal point is categorization of complaints on the basis of root cause.

The number of complaints relative to the number of items for Machine A and Machine B is presented in Table 2. The table includes the percentage share of items manufactured by welding suppliers and the total number of items, and also the number of complaints relative to the number of items manufactured by the welding suppliers. The share of items related to welding suppliers relative to the number of items about which complaints are received is considerable, thus the impact of welding suppliers on the end product is evident. As explained earlier, these results are the share of complaints linked to an individual item and as the end products may include several of the same items, the value does not take into consideration the total number of items or complaints. Therefore, Table 3 shows the share of the total number of items manufactured by the welding suppliers. The total number of items in Machine A is 4.06 times and in Machine B 3.74 times the number of individual items. The total number of complaints, including complaints not related to the welding suppliers, was for Machine A 1.38 times and for Machine B 0.65 times the total number of complaints about individual items. From Table 2 and Table 3 it can be seen that the share of individual items related to welding suppliers has reduced by 13% - 15% but the complaints to welding suppliers is still prominent when the difference in composition of the items, e.g. multipart welded structures and bulk items, is taken into consideration.

Table 2. The share of number of complaints, items related to welding supplier and complaints related to welding supplier for Machine A and Machine B.

Description	Machine A	Machine B	Average
Number of complaints / Number of individual items	24.18%	14.65%	19.42%
Items related to welding suppliers / Number of individual items	22.55%	27.73%	25.14%
Complaints of items related to welding suppliers / Number of complaints	29.70%	37.65%	33.68%

* The data are collected from an item group of 3 891 individual items where the total number of items is 14 907.

Table 3. Total number of items manufactured by welding suppliers and complaints related to welding suppliers for the case end product.

Description	Machine A	Machine B	Average
Total number of items related to welding suppliers / Total number of items	8.74%	10.74%	9.74%
Total number of complaints related to welding supplier / Total number of complaints	21.62%	19.29%	20.45%

* The data are collected from an item group of 3 891 individual items where the total number of items is 14 907.

Total complaints in the year of launch of the machines, 2008, the number of complaints related to welding suppliers was 24% and in 2014 25%. The total number of complaints was 4.68-fold (for Machine A) and 2.05-fold (for Machine B) the number of complaints about individual items. However, the number of complaints in the year of launch related to case products is presented in Table 4. The difference in the results for Machine A and Machine B can be explained by the totally new design and assembly of Machine B and therefore the sensitivity of the welded items to complaints.

Cost of items of particular end products are counted using information about the total number of related items. The complete end products contain some very expensive items as welded items, e.g. motor and power transmission. Excluding the three most expensive items, the total cost of items manufactured by welding suppliers raises, and by excluding the five most expensive items, the amount continues to grow. Table 5 shows the relative cost of items about which complaints have been received by welding suppliers for all parts and when the most expensive items are excluded.

Table 4. Launching year information of complaints of items related to welding supplier and total quantity of welding supplier complaints.

Description	Machine A	Machine B	Average
Number of individual complaints related to welding suppliers in launching year	14.29%	37.02%	25.65%
Total number of complaints related to welding suppliers in launching year	5.34%	37.74%	20.04%

* The data are collected from an item group of 3 891 individual items where the total number of items is 14 907.

Table 5. Relative cost of items in the case end products for all items and without the most expensive items.

Description	Machine A	Machine B	Average
Cost of items related to welding suppliers	9.93%	13.28%	11.61%
Without 3 the most expensive items	29.31%	26.91%	28.11%
Without 5 the most expensive items	33.38%	28.21%	30.80%
Cost of complained items related to welding suppliers	6,35%	5,08%	5,72%
Without 3 the most expensive items	18,74%	10,30%	14,52%
Without 5 the most expensive items	21,35%	10,79%	16,07%

* The data are collected from an item group of 3 891 individual items where the total number of items is 14 907.

The ratio of the number of complaints related to the number of items in the end products manufactured by welding suppliers to all complaints to the welding suppliers divided in years is given in Fig. 1. General market conditions and investment in product development is reflected in the number of complaints in a particular year. Machine A has been under development for several years with revision of many items, which affects the rate of complaints. Machine B is a totally new design structure and the number of complaints first rises continuously before settling down. One approach to interpretation of the curves in Fig. 1 is utilization of learning curve theories. The learning process is complex [31] and learning rate is not constant [32]. It can be expected that quality is lower and costs are higher when there is a lack

of learning [32]. Consequently, after initial launch of a product, the trend of complaints is a rising curve that peaks before starting to decrease. The continuous line in Fig. 1 describes production of Machine A, where the number of welding errors increases to a peak in 2012 before falling due to the effect of organizational learning. Assuming that organizational learning for Machine B follows a similar pattern, the dashed line can be expected to peak after a few years and then decline. The welding error lines resemble each other but learning is at different stages of the learning life cycle with different intensity of development. Curve behaviour depends on how much development and how many revisions are made and therefore the number of complaints will probably never drop to zero.

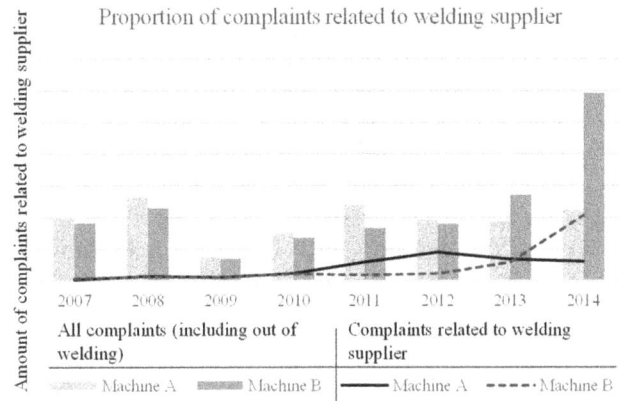

Figure 1. Complaints related to welding suppliers in the studied machines for 2007–2014. From the two curves it can be concluded that the trend of complaints is rising and presumably after reaching a peak it will start to decrease.

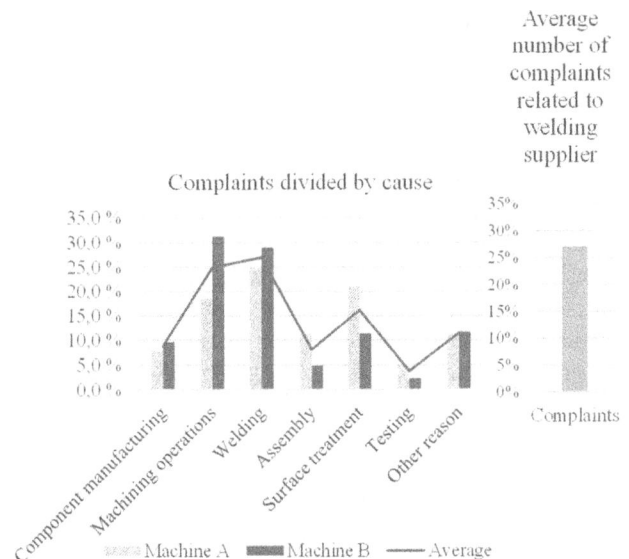

Figure 2. Proportion of complaints related to welding supplier in particular machines divided by root cause phases in welding manufacturing in years 2007–2014. From the diagram it can be concluded that welding-related functions are important factors in welding manufacturing.

The data in Fig. 1 do not relate to a specific reason or cause of complaint. However, to improve manufacturing productivity and decrease the number of defects, it is important to establish the underlying cause of complaints. Complaints for 2007–2014 related to welding suppliers are categorized by root cause and phase in welding manufacturing in Fig. 2. Welding includes supporting activities and actual welding is only one part of the welding functions needed to reach the quality requirements of the manufactured product. In Fig. 2, welding activities are divided by actions related to welding activity. The figure also shows the average proportional share of the average number of complaints related to welding suppliers when the number of the items and total quantity of items are included.

4. Discussion

Defects are common in production [10] and despite existing models of quality costs [7] companies are still required to deal with defective manufacturing. Liu et al. [26] states that the manufacturer is responsible for product quality and price, and this responsibility drives companies to make efforts to ensure conformity of manufacturing throughout the supply chain. Complaints are non-compliant actions in manufacturing and inevitably cause extra costs. Reducing such non-lucrative activity might increase production profitability. The achievable gain depends on the efforts required to reduce the number of complaints by investing in quality actions and improvements in different functions of the production chain.

In the study, the end products are two different machines constructed of multiple items and designed for equivalent end use. Approximately 25% of the individual items are manufactured by welding suppliers, which forms approximately 10% of the total number of items utilized in production. This correlates with the notable role of welding in manufacturing because the number of welded parts relative to all items is disproportionate. Welded items usually entail bigger workload and the required number of pieces is smaller than ancillary items like bulk items, which have all individual item number. The coefficient of the number of items relative to the individual number of items confirms this finding.

Observing the complaints more closely, individual items manufactured by the welding supplier (Machine A and Machine B) caused 34% of the total complaints about individual items. Over the period studied, complaints made to welding suppliers show that the mean amount of complaints dealing with Machine A and Machine B is about the same as in the whole welding production (Fig. 3). The number of complaints varies depending on production volume but the rate is visible. In Fig. 3, the launching year position of welding supplier complaints can be seen. Even though the complaints are not related to particular machines and show the number of marked complaints and do not take into account the number of items inside a single complaint, the results gives an overview of the significance of complaints in welding production.

Cost of items of end products is not contained in manufacturing costs and therefore is not relevant for observing the total cost of the end product. However, the cost

of items indicate the material investment and therefore have a big impact on total costs. As noted earlier, the end products contain multiple items and different items will involve different workload. The cost of items manufactured by welding suppliers is approx. 12% of total cost of items of product. Excluding the five most expensive items, the figure is 2.7-fold compared previous and excluding only the three most expensive items, the figure is still 2.4-fold. This shows that welding suppliers are responsible for a third of the manufacturing potential of the end product, and therefore managing supplier quality occupies a very important position in the manufacturing operations. The cost of items manufactured by welding suppliers about which complaints are received follows the same rate as total items manufactured by welding supplier with 2.8-fold (excluding five items) and 2.5-fold (excluding three items) results compared to without excluding any of items. Excluding the five most expensive items gives the result of 16% of total items that welding suppliers are responsible for.

The total number of complaints to the welding suppliers varies over the eight years period. It is difficult to analyse any regular trend by items of particular end product of these years. Although, these case end products seem to be close to general trend of all complaints. The launching year of the end product and the general market situation also have an impact on production. Categorizing the complaints of case machines by root cause of manufacturing phase indicates the importance of welding-related functions. Actual welding is near third of reason for complaints related to welding suppliers.

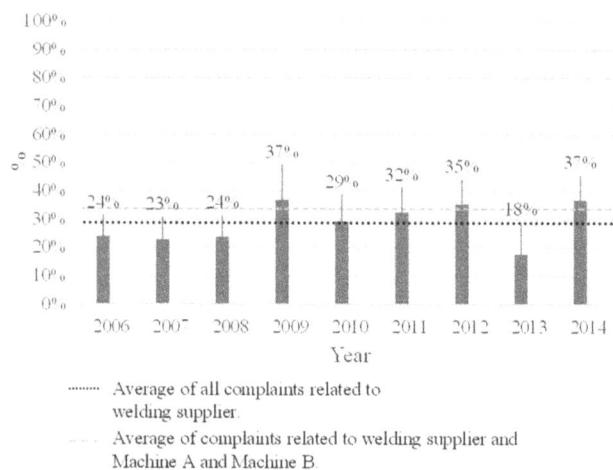

.... Average of all complaints related to
         welding supplier.
 - - -   Average of complaints related to welding supplier and
         Machine A and Machine B.

Figure 3. The number of complaints related to welding suppliers seems to remain relatively stable year-on-year.

5. Conclusion

Pettersson and Segerstedt [33] state that manufacturing costs are only one part of supply chain costs. This study shows the potential of welding production functions to increase production profitability by reducing manufacturing defects that cause complaints. In the light of the results presented, the effect of complaints on quality and costs is evident. The results also indicate the potential to influence the quality and profitability of the end product by control of complaints in the supplier chain. This shows the prospect to impact disadvantages on welding production.

This research observed complaints and costs from the viewpoint of items, in future research it would be interesting to focus on costs of manufacturing through complained items and therefore their impact on total costs of the end product. Such information is essential when making efforts to improve manufacturing quality and profitability from whole welding network viewpoint. This research shows one approach to study of the effect of manufacturing in the supply chain on end product. Other studies could concentrate on categorization of complaints in a welding supplier network by root cause and finding the link between complaints, knowledge transfer and competencies. Additionally, the effect of complaints on revisions of welded items in a welding supplier network would be interesting to define. Wider observation is needed in the state of network pictures of welding network and influence on the amount of complaints in welding production.

References

[1] S. T. Foster Jr. and J. Ogden. On differences in how operations and supply chain managers approach quality management, International Journal of Production Research, Vol 46 (24), 2008, pp. 6945–6961. doi: 10.1080/00207540802010815.

[2] S. T. Foster Jr., S. T., C. Wallin and J. Ogden. Towards a better understanding of supply chain quality management practices, International Journal of Production Research, Vol 49 (8), 2011, pp. 2285–2300. doi: 10.1080/00207541003733791.

[3] R. Schmitt and A. Linder. Technical complaint management as a lever for product and process improvement, CIRP Annals - Manufacturing Technology, Vol. 62 (1), 2013, pp. 435–438. doi: 10.1016/j.cirp.2013.03.040.

[4] M. Srinivasan, D. Mukherjee and A. S. Gaur. Buyer–supplier partnership quality and supply chain performance: Moderating role of risks, and environmental uncertainty, European Management Journal, Vol. 29 (4), 2011, pp. 260–271. doi: 10.1016/j.emj.2011.02.004.

[5] L. J. Zeballos, M. I. Gomes, A. P. Barbosa-Povoa and A. Q. Novais. Addressing the uncertain quality and quantity of returns in closed-loop supply chains, Computers and Chemical Engineering, Vol. 47, 2012, pp. 237–247. doi: 10.1016/j.compchemeng.2012.06.034.

[6] G. Xie, W. Yue, S. Wang and K. K. Lai, Quality investment and price decision in a risk-averse supply chain, European Journal of Operational Research, vol. 214, 2011, pp. 403–410. doi: 10.1016/j.ejor.2011.04.036.

[7] K. K. Castillo-Villar, N. R. Smith and J. L. Simonton, The impact of the cost of quality on serial supply - chain network design, International Journal of Production Research, vol. 50, 2012, pp. 5544–5566. doi: 10.1080/00207543.2011.649802.

[8] W. Wang, R. D. Plante and J. Tang, Minimum cost allocation of quality improvement targets under sup-plier process disruption, European Journal of Operational Research, vol 228, 2013, pp. 388–396. doi: 10.1016/j.ejor.2013.01.048.

[9] B. Ivens and C. Pardo, Are key account relationships different? Empirical results on supplier strategies and customer reactions, Industrial Marketing Management, vol. 36 (4), 2007, pp. 470-482. doi: 10.1016/j.indmarman.2005.12.007.

[10] R. Xiao, Z. Cai and X. Zhang, A Production Optimization Model of Supply-driven Chain with Quality Uncertainty, Journal of Systems Science and Systems Engineering, vol. 21 (2), 2012, pp. 144–160. doi: 10.1007/s11518-011-5184-8.

[11] T. Garín-Muñoz, T. Pérez-Amaral, C. Gijón and R. López, Consumer complaint behavior in telecommunications: The case of mobile phone users in Spain, Telecommunications Policy, 2015. doi: 10.1016/j.telpol.2015.05.002i.

[12] P-W Dong and Y-Q Huang, Research of Customer Complaints and Service Recovery Effects, Management Science and Engineering, 2006, pp. 958–962, October 2006 [ICMSE '06 International Conference, Lille]. doi: 10.1109/ICMSE.2006.314008.

[13] M. A. Lapré, Reducing Customer Dissatisfaction: How Important is Learning to Reduce Service Failure?, Production and Operations Management, vol. 20 (4), 2011, pp. 491–507. doi: 10.1111/J.1937-5956.2010.01149.

[14] A. Yim, Failure Risk and Quality Cost Management in Single versus Multiple Sourcing Decision, Decision Sciences, vol. 45 (2), 2014, pp. 341–354. doi: 10.1111/deci.12070.

[15] X. Huang, S-M. Choi, W-K. Ching, T-K. Siu and M. Huang, On supply chain coordination for false failure returns: A quantity discount contract approach, International Journal of Production Economics, vol. 133 (2), 2011, pp. 634–644. doi: 10.1016/j.ijpe.2011.04.031.

[16] S. Sharma, Effects concerning quality level with the increase in production rate, International Journal of Advanced Manufacturing Technology, vol. 53 (5–8), 2011, pp. 629–634. doi: 10.1007/s00170-010-2851-8.

[17] R. Mateo, M. Tanco and J. Santos, Less Expert Workers and Customer Complaints: Automotive Case Study, Human Factors and Ergonomics in Manufacturing & Service Industries, vol. 24 (4), 2014, pp. 444–453. doi: 10.1002/hfm.20396.

[18] B. Gajdzik, B. and J. Sitko, An analysis of the causes of complaints about steel sheets in metallurgical product quality management systems, Metallurgia, vol. 53 (1), 2014, pp. 135–138.

[19] B. C. Giri and S. Sharma. Optimizing a closed-loop supply chain with manufacturing defects and quality dependent return rate, Journal of Manufacturing Systems, vol. 35, 2015, pp. 92–111. doi: 10.1016/j.jmsy.2014.11.014.

[20] T. Gruber, S. C. Henneberg, B. Ashnai and P. Naudé, Complaint resolution management expectations in an asymmetric business-to-business context, Journal of Business & Industrial Marketing, vol. 25 (5), 2010, pp. 360–371. doi: 10.1108/08858621011058124.

[21] G. A. Akyuz and T. E. Erkan, Supply chain performance measurement: a literature review, International Journal of Production Research, vol. 48 (17), 2010, pp. 5137–5155. doi: 10.1080/00207540903089536.

[22] N. Panayiotou and K. G, Aravosis, Supply Chain Management, in Theory and Practice of Corporate Social Responsibility, S. O. Idowu and C. Louche, Eds. Berlin Heidelberg: Springer-Verlag, 2011, pp. 55–70. doi: 10.1007/978-3-642-16461-3_4.

[23] B. Nepal and O. P. Yadav, Bayesian belief network-based framework for sourcing risk analysis during supplier selection, International Journal of Production Research, vol. 53 (20), 2015, pp. 6114–6135. doi: 10.1080/00207543.2015.1027011.

[24] J. Toivanen, J. Martikainen and P. Heilmann, From supply chain to welding network: A framework of the prospects of networks in welding, Mechanika, vol. 21 (2), 2015, pp. 154–160. doi: 10.5755/j01.mech.21.2.8463.

[25] J. Toivanen, P. Kah and J. Martikainen, Quality Requirements and Conformity of Welded Products in the Manufacturing Chain in Welding Network, International Journal of Mechanical Engineering and Applications, vol. 3 (6), 2015, pp. 109–119. doi: 10.11648/j.ijmea.20150306.12.

[26] G. Liu, S. P. Sethi and J. Zhang, Myopic vs. far-sighted behaviours in a revenue-sharing supply chain with reference quality effects, International Journal of Production Research, 2015, doi: 10.1080/00207543.2015.1068962.

[27] C. Hicks, O. Heidrich, T. McGovern and T. Donnelly, A functional model of supply chains and waste, International Journal of Production Economics, vol. 89 (2), 2004, pp. 165–174. doi: 10.1016/S0925-5273(03)00045-8.

[28] M. Firoozi, A. Siadat, N. Salehi and S. M. Mousavi, A Novel Multi-Objective Fuzzy Mathematical Model for Designing a Sustainable Supply Chain Network Considering Outsourcing Risk under Uncertainty, Industrial Engineering and Engineering Management, 2013, pp. 88–92, December 2013 [IEEE International Conference, Bangkok] doi: 10.1109/IEEM.2013.6962380.

[29] X. Tang, X. and H. Yun, Data model for quality in product lifecycle, Computers in industry, vol. 59 (2–3), 2008, pp. 167–179. doi: 10.1016/j.compind.2007.06.011.

[30] G. Guest, E. E. Namey and M. L. Mitchell, Collecting Qualitative Data: A Field Manual for Applied Research. Thousand Oaks: Sage Publications Inc., 2013, pp. 16–17.

[31] P. S. Adler and K. B. Clark, Behind the learning curve: A sketch of the learning process, Management Science, vol. 37 (3), 1991, pp. 267–281.

[32] M. A. Lapré, A. S. Mukherjee and L. N. Van Wassenhove, Behind the Learning Curve: Linking Learning Activities to Waste Reduction, Management Science, vol. 46 (5), 2000, pp. 597–611. doi: 10.1287/mnsc.46.5.597.12049.

[33] A. Pettersson and A. Segerstedt, Measuring supply chain cost, International Journal of Production Economics, vol. 143 (2), 2013, pp. 357–363. doi: 10.1016/j.ijpe.2012.03.012.

Permissions

The contributors of this book come from diverse backgrounds, making this book a truly international effort. This book will bring forth new frontiers with its revolutionizing research information and detailed analysis of the nascent developments around the world.

We would like to thank all the contributing authors for lending their expertise to make the book truly unique. They have played a crucial role in the development of this book. Without their invaluable contributions this book wouldn't have been possible. They have made vital efforts to compile up to date information on the varied aspects of this subject to make this book a valuable addition to the collection of many professionals and students.

This book was conceptualized with the vision of imparting up-to-date information and advanced data in this field. To ensure the same, a matchless editorial board was set up. Every individual on the board went through rigorous rounds of assessment to prove their worth. After which they invested a large part of their time researching and compiling the most relevant data for our readers.

The editorial board has been involved in producing this book since its inception. They have spent rigorous hours researching and exploring the diverse topics which have resulted in the successful publishing of this book. They have passed on their knowledge of decades through this book. To expedite this challenging task, the publisher supported the team at every step. A small team of assistant editors was also appointed to further simplify the editing procedure and attain best results for the readers.

Apart from the editorial board, the designing team has also invested a significant amount of their time in understanding the subject and creating the most relevant covers. They scrutinized every image to scout for the most suitable representation of the subject and create an appropriate cover for the book.

The publishing team has been an ardent support to the editorial, designing and production team. Their endless efforts to recruit the best for this project, has resulted in the accomplishment of this book. They are a veteran in the field of academics and their pool of knowledge is as vast as their experience in printing. Their expertise and guidance has proved useful at every step. Their uncompromising quality standards have made this book an exceptional effort. Their encouragement from time to time has been an inspiration for everyone.

The publisher and the editorial board hope that this book will prove to be a valuable piece of knowledge for researchers, students, practitioners and scholars across the globe.

List of Contributors

Pham Son Minh and Tran Minh The Uyen
Department of Mechanical Engineering, HCMC University of Technology and Education, Ho Chi Minh City, Vietnam

Bui Manh Tuan
School of Mechanical Engineering, Southeast University, Nanjing city, Jiangsu Province, China
Faculty of Mechanical Engineering, Tuy Hoa Industrial College, Tuy Hoa City, Phu Yen Province, Vietnam

Chen Yun Fei
School of Mechanical Engineering, Southeast University, Nanjing city, Jiangsu Province, China

Esam Mejbel Abed and Zahra'a Aamir Auda
Babylon University-College of Engineering, Mechanical Department, Babylon, Iraq

Abderrahmane Abene
Universite de Valenciennes, Laboratoire d'Aerodynamique, d'Energetique et de l'Environnement, I. S. T. V, Valenciennes, 59300 Aulnoy lez valenciennes France

Boda Hadya and A. M. K. Prasad
Mechanical Engineering Department, U. C. E., Osmania University, Hyderabad, Telangana State, India

Suresh Akella
Sreyas Institute of Engineering and Technology, Affiliated to J. N. T. U., Hyderabad, Telangana State, India

Bo Anders Nordell
Department of Architecture and Water, Luleå University of Technology, SE-97187 Luleå, Sweden

Ragnar Oskar Gawelin
Enskilda Gymnasiet, SE-11161 Stockholm, Sweden

Liang Jian-kai, Song Wan-qing and Li Qing
College of Electronic and Electrical Engineering, Shanghai University of Engineering Science, Shanghai, P.R. China

Joshua Emuejevoke Omajene, Jukka Martikainen and Paul Kah
LUT Mechanical Engineering, Lappeenranta University of Technology, Lappeenranta, Finland

Garimella Sridhar and Ramesh Babu Poosa
Department of Mechanical Engineering, University College of Engineering, Osmania University, Hyderabad, India

H. Babaei, M. Malakzadeh and H. Asgari
Department of Mechanical Engineering, Engineering Faculty, University of Guilan, Rasht, Iran

Arturo Mendoza Castrejón and Damasio Morales Cruz
School of Mechanical and Electrical Engineering -IPN, 07738, U. P. Adolfo López Mateos, D. F., México

Herlinda Montiel Sánchez
Technosciences Department, Center for Applied Science and Technology Development -UNAM, 04510, C.U., D. F., México

Guillermo Alvarez Lucio
School of Physics and Mathematics -IPN, 07738, U. P. Adolfo López Mateos, D. F., México

Xiaochen Yang, Paul Kah and Jukka Martikainen
Laboratory of Welding Technology, Lappeenranta University of Technology, Lappeenranta, Finland

Jenni Toivanen, Paul Kah and Jukka Martikainen
Laboratory of Welding Technology, Lappeenranta University of Technology, Lappeenranta, Finland

Remi Bouttier
Ecole Nationale Superieure de Mechanique et D'Aerotechnique (ISAE-ENSMA), Département d'Energétique, France

Gabriel Lopes
Federal University of Uberlandia, Engenharia Mecânica, Santa Mônica, Uberlândia - MG, Brazil

Luke Clarke and Rocco Lupoi
Trinity College Dublin, the University of Dublin, Department of Mechanical and Manufacturing Engineering, Parsons Building, Dublin 2, Ireland

Xiaochen Yang, Paul Kah and Jukka Martikainen
Laboratory of Welding Technology, Lappeenranta University of Technology, Lappeenranta, Finland

Kishori Yadav and Jeevan Jyoti Nakarmi
Central Department of Physics, T. U., Kirtipur, Nepal

Nodar Davitashvili and Otar Gelashvili
Department of Transport and Mechanical Engineering of Georgian Technical University, Tbilisi, Georgia

Muhamad Mohd Ridha, Zou Yanhua and Sugiyama Hitoshi
Graduate School of Engineering, Utsunomiya University, Utsunomiya-shi, Tochigi-ken, Japan

Ramesh Chandra Mohapatra
Mechanical Engineering Department, Government College of Engineering, Keonjhar, India

Antaryami Mishra and Bibhuti Bhushan Choudhury
Mechanical Engineering Departmet, Indira Gandhi Institute of Technology, Sarang, India

Yi-Lung Hsu, Ming-Shyan Huang and Rong-Fong Fung
Department of Mechanical & Automation Engineering, National Kaohsiung First University of Science and Technology, Kaohsiung, Taiwan

Pavel Layus and Paul Kah
Lappeenranta University of Technology, Skinnarilankatu, Lappeenranta, Finland

Joshua Emuejevoke Omajene, Paul Kah, Huapeng Wu and Jukka Martikainen
LUT Mechanical Engineering, Lappeenranta University of Technology, Lappeenranta, Finland

Christopher Okechukwu Izelu
Department of Mechanical Engineering, College of Technology, Federal University of Petroleum Resources, Effurun, Delta State, Nigeria

Riyadh S. Al-Turaihi and Sarah H. Oleiwi
Mechanical Department, College of Engineering, Babylon University, Babylon, Iraq

Rizeq N. Hammad, May M. Hourani and Firas M. Sharaf
Department of Architecture, Faculty of Engineering, Jordan University, Amman, Jordan

Dagninet Amare and Wolelaw Endalew
Bahir Dar Agricultural Mechanization and Food Science Research Centre, Bahir Dar, Ethiopia

Geta Kidanemariam
Institute of Technology, Addis Ababa University, Addis Ababa, Ethiopia

Seyife Yilma
Institute of Technology, Bahir Dar University, Bahir Dar, Ethiopia

Abdulrahman Th. Mohammad
Baqubah Technical Institute, Middle Technical University, Baghdad, Iraq

Jasim Abdulateef and Zaid Hammoudi
Mechanical Engineering Department, Diyala University, Diyala, Iraq

Jenni Toivanen, Harri Eskelinen, Paul Kah and Jukka Martikainen
Mechanical Engineering, Lappeenranta University of Technology, Lappeenranta, Finland

Luu Hong Quan and Vu Ngoc Long
Department of Aeronautical and Space Engineering, School of Transportation Engineering, Hanoi University of Science and Technology, Hanoi, Vietnam

Nguyen Phu Hung
The Ministry of Science and Technology – Hanoi University of Science and Technology, Hanoi, Vietnam

Le Doan Quang
Faculty of Aviation Technologies, Vietnam Aviation Academy, Ho Chi Minh city, Vietnam

Index

www.ingramcontent.com/pod-product-compliance
Lightning Source LLC
Chambersburg PA
CBHW080627200326
41458CB00013B/4536